W9-BKO-449

To Cherish the Life of the World

To Cherish the Life of the World

SELECTED LETTERS OF

MARGARET MEAD

Edited by

Margaret M. Caffrey and Patricia A. Francis

With a Foreword by Mary Catherine Bateson

BASIC
BOOKS

A Member of the Perseus Books Group
New York

Published by Basic Books
A Member of the Perseus Books Group

Books published by Basic Books are available at special discounts for bulk
purchases in the United States by corporations, institutions, and
other organizations. For more information, please contact the
Special Markets Department at the Perseus Books Group, 11 Cambridge Center,
Cambridge MA 02142, or call (617) 252-5298 or (800) 255-1514, or
e-mail special.markets@perseusbooks.com.

Designed by Trish Wilkinson
Set in 11.5-point Adobe Garamond by the Perseus Books Group

Library of Congress Cataloging-in-Publication Data

Mead, Margaret, 1901–1978.
    To cherish the life of the world : selected letters of Margaret Mead /
edited by Margaret M. Caffrey and Patricia A. Francis ; with a foreword by
Mary Catherine Bateson.
        p.   cm.
    Includes bibliographical references and index.
    ISBN-13: 978-0-465-00815-5 (hardcover : alk. paper)
    ISBN-10: 0-465-00815-1 (hardcover : alk. paper) 1. Mead, Margaret,
1901–1978—Correspondence. 2. Anthropologists—United States—
Correspondence. I. Caffrey, Margaret M. (Margaret Mary), 1947–   II.
Francis, Patricia A. III. Title.
GN21.M36M4 2006
301.092—dc22   2006009544

06 07 08 09 / 10 9 8 7 6 5 4 3 2 1

In memory of our friend
Mary Margaret Wolfskill (1946–2005)
without whom this book could not have been written

Contents

Brief Chronology
of Margaret Mead's Life

December 16, 1901	Born in Philadelphia, Pennsylvania
1917	Meets her future first husband, Luther Cressman
Fall 1919–Spring 1920	Studies as a freshman at DePauw University, Greencastle, Indiana
Fall 1920–Spring 1923	Studies at Barnard College, New York City, from which she graduates
Fall 1922	Meets Ruth Benedict
September 1923	Marries Luther Cressman
Fall 1925–Spring 1926	Does field work in Samoa
Summer 1926	Meets her future second husband, Reo Fortune, on a ship to Europe after leaving Samoa
1926	Begins her career as assistant curator of ethnography at the American Museum of Natural History (AMNH) in New York City
1928	*Coming of Age in Samoa* published
Fall 1928	Marries Reo Fortune in New Zealand; they do field work among the Manus of the Admiralty Islands in the Pacific, home by Fall 1929
Summer 1930	Does field work with Fortune among the Omaha in Nebraska
Fall 1931–Fall 1933	Does field work with Fortune in New Guinea; meets her future third husband, Gregory Bateson, at Christmas, 1932.

Summer 1934	Attends the Hanover Seminar on Human Relations; afterwards, meets Bateson for a month-long visit in Ireland
1935	*Sex and Temperament* published
Spring 1936	Marries Gregory Bateson in Singapore; they do field work in Bali until 1939
December 8, 1939	Daughter Mary Catherine Bateson is born
1940	Becomes a founding member and first president of the Society for Applied Anthropology
1940–1944	Cares for two English children, Philomena and Claudia Guillebaud, with Bateson and her parents
1942	Promoted to associate curator of ethnology at AMNH
1943–Spring 1945	Takes a leave of absence to work in Washington, D.C., as executive director of the Committee on Food Habits, National Research Council; hires Rhoda Metraux as an assistant
1950	Divorces Gregory Bateson; her mother dies
1953	Restudies the Manus in the Admiralty Islands, with Theodore and Lenora Schwartz
1957–1958	Awarded the Viking Medal in General Anthropology by the Wenner-Gren Foundation for Anthropological Research
1960	Serves as president of the American Anthropological Association; her daughter marries
1962	Begins to write a column for *Redbook Magazine* with Rhoda Metraux; she continues the column until her death
1968–1970	Serves as chair of the social sciences division and professor of anthropology at Fordham University, New York City
1969	Retires from AMNH with the rank of full curator emeritus, keeping her office

1969–1971	Serves as president, World Society of Ekistics (study of human settlements)
May, 1971	Mead's Hall of Pacific Peoples opens in AMNH
1975	Serves as president of the American Association for the Advancement of Science
1975	Elected to the National Academy of Sciences
November 15, 1978	Dies of pancreatic cancer in New York City

Acknowledgments

We owe thanks to many people and institutions who assisted us in the completion of this book. We would especially like to thank the current and former members of the staff of the Manuscript Reading Room at the Library of Congress for their unfailing courtesy, helpfulness, and good humor in completing this project—in particular Fred Bauman, Ernest Emrich, Jeffrey Flannery, Joseph Jackson, Ahmed Johnson, Bruce Kirby, and Patrick Kerwin—as well as Manuscript Division Editor Janice Ruth.

We owe an enormous debt to the late Mary Wolfskill, former head of the reference and reader service section of the Manuscript Division, who archived the Margaret Mead Papers for the Library of Congress and knew the collection better than anyone else in the world. Without Mary's help and encouragement this book could not have been written, and so we have dedicated it to her memory.

We would also like to thank Daniel Metraux, who shared with us letters from Margaret Mead to his mother that the Library of Congress did not have. In addition, we are grateful to Martha Ullman West and others who offered us some of the correspondence in their possession.

We value greatly Mary Catherine Bateson's cooperation in this project, without which we could not have brought such a volume to fruition. We acknowledge gratefully permission to publish her mother's letters at such length. In addition, we extend many thanks to our agent, Katinka Matson, and to our helpful editors at Basic Books, Jo Ann Miller and Assistant Editor Ellen Garrison.

On Margaret Caffrey's part, she would like to thank the University of Memphis, which gave her a Faculty Summer Research Grant in Summer

2004 and also a Professional Development Assignment in Fall 2004 to work on this project. She would also like especially to thank her brother and sister-in-law, Jim and Susan Caffrey, and her brother and sister-in-law, Tom and Lucy Caffrey, because without them, this book could not have been written, as they took her in for extended periods of time and helped in numerous ways that can never be repaid. A special extra thanks goes to Sue for her quick aid in a medical emergency. Finally, she would like to thank Ronda Sue Campbell for her patience and helpful suggestions in reading parts of this book in progress.

Patricia Francis would like to thank Gerald Sullivan for his invaluable comments on the manuscript in progress, as well as to extend thanks to him and other anthropologists who answered technical questions beyond our knowledge of the field. She appreciates the forbearance and support of all who have followed her work on this and related projects over the years, particularly her husband, Liam Healy, who has learned more about Margaret Mead over the past two decades than he would have ever thought possible. Special thanks go to Katherine Lee for her careful reading of the manuscript at a critical time. Finally, Francis would like to express her gratitude once again to the wonderful staff at the Manuscript Division of the Library of Congress.

"I wonder, if one never did a single thing to please another person which didn't spring from one's own desire—not one's desire to please them, to fit in with their different view of what is lovely or right or necessary—would all relationships be honest?"

—*Margaret Mead to Gregory Bateson, November 7, 1934*

Foreword

by Mary Catherine Bateson

A lifetime of letters. My father and I were terrible at answering letters, and worse at saving them, but my mother was an archivist's dream. The majority of her letters were typed and their carbons saved, always carefully dated. The Margaret Mead Collection in the Library of Congress is the largest personal collection of papers in that vast institution, and much of the collection consists of letters, boxes and boxes of letters regularly used by researchers of many kinds: biographers of Mead and of her correspondents, many of whom were less careful; ethnographers of the peoples Mead studied; historians and sociologists, tracking the growth of ideas and of institutions.

When I sat down, more than twenty years ago, to write a memoir of my parents, it struck me that "in my family, we never simply live, we are always reflecting on our lives. . . ." For me, the letters evoke my mother's personality more vividly than any other body of material. They also show how her work and life were intermeshed and reveal that the stance of participant observer was not one she picked up and set aside depending on context, but a way of being in the world that extended to relationships of all kinds, from the most formal to the most intimate. Because this sense of her personality was shared by so many, I adopted early on the convention of referring to her in writing primarily by her first name, Margaret. Whatever Margaret was doing or feeling, she was noticing patterns in her own behavior and the behavior of others, and reflecting on the nature of relationships. When she spoke or wrote, her effort to understand the people in her life and how they made their choices pervaded what she said, creating an extraordinary diversity of tone.

Because of this quality, when Margaret Caffrey and Patricia Francis, who now know the Mead archive more intimately than any living person but neither of whom met my mother in life, proposed a volume of letters, I urged them not to limit the collection to chronicling particular correspondences but to look at the letters as together expressing her evolving beliefs about human relationships, including her conviction that out of intense conversations and exchanges new and important ideas could emerge.

As I have read over the letters they have selected from these many boxes, I have been struck anew by how much Margaret treasured the people in her life. She was always reluctant to let friendships fade and drift apart, often writing out of the blue or telephoning colleagues or relatives at odd hours and taking up a conversation in midstream that might have begun several years before. Although she spoke of "best friends" in her earlier letters, she was always urging me not to put friendships in a hierarchy but to value each one for what was unique and distinctive. She believed it was possible to be in love with more than one person at a time in different ways, and she believed that when marriages broke down it was important that friendship continue. She manipulated manipulators (like her father), let managers (like my godmother, Marie Eichelberger) manage her life to their hearts' content, and often knew what people were going to say before the words were spoken. She loved joining in when visionaries were weaving the future and helping them transform their imaginings into reality, and she unearthed and encouraged unrecognized potentials. She could also be pretty bossy—friends who dream about her since her death often dream that she is telling them to get on with their work or to stop talking nonsense. She sometimes showed annoyance and impatience, but generally set it aside in later letters.

The discipline of reading these letters is to avoid worrying too much about the references to events and about the order in which they unfolded, and rather to attend to the way Margaret frames her words to communicate with a particular person. To her sisters, she speaks as a sister and yet differently to each. She reaches out to different lovers as her own understanding of love developed, and yet tactfully skirts their preconceptions. Sometimes the reader will recognize wishful thinking or self-deception—and then be surprised by a stunning insight into the other or herself. She acknowledges the strength of emotional responses and tries to understand them, but she makes no sharp separation between her emotional and her intellectual life. Love and thought, affection and observation are intertwined.

It is because these letters reflect a relationship between the professional and the personal and between emotion and analysis that was fundamental to my mother's lifework that I feel it is appropriate to share them. I have tried with the editors to consider the privacy of recipients who are still living in making this selection and indeed to protect their descendants from uncomfortable revelations. Margaret had a quick temper and she does comment from time to time on the continuing hostility of individuals who expressed personal resentments in professional attacks, but it is striking to me that these letters are not haunted by continuing animus and that she always maintained in her comments her respect for the work of individuals who opposed her. Edward Sapir, with whom she had a stormy relationship that is reflected here, was one of the first anthropologists she suggested I read when I began to be drawn to linguistics.

A collection of letters tends to highlight one-on-one relationships, but this collection still reveals how much my mother shared her friends with each other, constantly elaborating them into networks, providing introductions, and bringing people into contact. As I read through these pages, I was struck by how often she was thinking about how these networks could function in the world—how the limited number of anthropologists available, for instance, could train the needed number of field workers, or how emerging disciplines could strengthen each other.

She was, however, selective in her referrals, as shown by her correspondence in the early 1960s, where she showed reluctance to introduce Derek Freeman to psychoanalytic circles in the United States, for by that time his erratic behavior was already well-known among the Australian anthropologists with whom Margaret kept in touch. Although he did not publish his attacks on Mead's work in Samoa until after her death, Mead and Freeman met in a seminar in Australia where Mead did her best to respond to Freeman as a professional colleague and to reassure the quaking students, as reflected in her letter to him included here. By the end of the 2001 centennial of Mead's birth, every substantive aspect of the debate about Samoa had been put into ethnographic perspective, but a full description of the psychological context of Freeman's behavior has only recently become available.

The collection of my mother's letters in the Library of Congress documents her life in a way that we are unlikely to see again. She lived at the end of the great era of letter writing. She also lived in the era of salvage anthropology, when field sites were remote and anthropologists rushed to document cultures being flooded with Western influence. During years spent at remote field work sites, mail often arrived only once or twice a month as had been true in earlier centuries, with all the possibilities for anxiety and misunderstanding that this created, and her earliest correspondence preceded the widespread availability of the telephone. Letters were a lifeline and the very stuff of intimacy. Even as telephones became available, long distance use was limited and costly, and letters remained the preferred form until mid-century. Thus, she tried to teach me to be the kind of letter writer she was, starting from taking letters by dictation during my childhood, and she compensated for my father's inadequacy as a correspondent by having him sign a batch of postcards that I could receive month by month while he was overseas during the war, when I was too young to worry about stamps or postmarks.

The global pattern of Margaret's networks has a very modern feel to it, but her medium reflects another age. By the time I moved away from home, first to college and then to marriage, my mother was depending more and more on telephone conversations, even when one of us was overseas. She was always interested in new technologies and adapted fluently from radio to television, and if she were living today, she would surely be using e-mail—the lifelong discipline of documentation might make her print out and save correspondence, but the pattern would inevitably have shifted toward more rapid and fragmented exchanges rather than long thoughtful letters. She became in her later years an extremely public and prolific person, full of energy and traveling constantly. Yet as these letters show, she was also a thoughtful and reflective person, for throughout her life, the writing of letters remained an occasion for the kind of reflection that is ever harder to maintain in contemporary life.

Introduction:
The Logic of Human Relationships

"The logic of human relationships is not as simple as you think."
—*Mead to Gregory Bateson, November 26, 1948*

This book explores the many dimensions of relationship in anthropologist Margaret Mead's life through annotated versions of correspondence she wrote to an inextricably interwoven network of family (both immediate and extended), friends, lovers, spouses, and colleagues. The letters show how Mead felt about such subjects as family, friendship, sexuality, marriage, children, and career. Mead was self-aware about relationships and commented self-consciously on the nature of various relationships, both generally and as they played out in her own life and in the lives of those important to her. The letters in this book show not only how Mead experienced being a daughter or a wife but lend insight into her thinking on various relational roles: how she experienced kinship—which she studied professionally in her work with other cultures—within her own life.

THE CALCULUS: SYMMETRY, BALANCE, AND RECIPROCITY

Mead lived according to a complex calculus of human relationships of her own devising. She did not feel bound—or did not like to feel bound—by the terms of "status relationships." She refused to be compelled to behave a certain way toward an individual because she was a daughter, a sister, or a wife. Although she was loyal, she took offense at the suggestion that loyalty would trump ethical or moral behavior on her part. From a young age, Mead felt a sense of obligation to do things for others and at times felt these perceived demands a burden. She sought in her adult life to maximize what she called "freedom from constraint," or release from duty.[1] This was true both in the case of purely personal relations and of professional ones as well. If she received funds to do work, she felt it her moral duty to complete that work, lest the money and the opportunity go to waste.

Mead was very aware of symmetry in relationships but did not feel obligated to match another person's affection with the same measure of words or actions. Similarly, she felt that she gave what she had to give, freely. She balanced relationships, at one point taking a concept of compensatory behavior in relationships from Papua New Guinea culture to express her understanding of reciprocity in relationships extending through her own social networks. She was very aware of and taken by "the chain effect as one life touches another,"[2] yet also worried about what scandals or troubles in her life or in others' lives might be visited upon others in the network.

Mead preferred to be the intellectual inferior in her relationships, with much to learn. Ironically, considering her public role as a teacher, she did not generally like to occupy the teacher role in her personal relationships. When she mentored, she was happiest when working with people who had the ability to equal or better her. Intellectual exchange excited her. She liked the back-and-forth of a heated discussion, yet was puzzled when others interpreted her fervent engagement in such discussions as anger.

Mead was fascinated by patterns, and many of the patterns in her relationships were rooted in the earliest relationships in her family, where she learned to observe, inquire, and debate. Her father, an economist, used money as a means of controlling his family's behavior. Mead accepted his right to determine how his money was spent but learned to manipulate him to get the money she needed. As an adult, she worked to build her own financial base so that she would not be dependent on others for money. She channeled her own money back to help finance her anthropological field trips and the work of her collaborators. This arrangement was formalized in the Institute for Intercultural Studies, which Mead established in 1944.

As a child, she was close to her grandmother, who taught and confided in her. During the periods when Mead attended school and was not taught at home by her grandmother, she found the experience odd and unchallenging. She also found that formal schooling had a theatrical quality about it; it was a pattern that could be learned and performed. In her autobiography, *Blackberry Winter,* she wrote, "I wanted to live out every experience that went with schooling, and so I made a best friend out of the most likely candidate, fell sentimentally in love with one of the boys, attached myself to a teacher and organized, as far as it was possible to do so, every kind of game, play, performance. . . ."[3] Variations on this pattern continued throughout Mead's life, often in the professional interdisciplinary working groups she favored.

Mead's sister Katherine, whom she had been allowed to name, died at the age of nine months when Mead was five years old. The event traumatized Mead and drove a wedge between her parents. Throughout her life, Mead dreamed and daydreamed of dead babies and kidnapped or missing children. After having numerous miscarriages, Mead had a daughter in 1939 with her third husband, Gregory Bateson. Mead then fantasized that the child would be stolen by her second husband, Reo Fortune, with whom she had had a tumultuous relationship and several miscarriages. Mead had a profound ambivalence about having a child with Fortune, saying at different times that she married him because she thought she could not have children and, when she discovered that she could get pregnant, saying that she was going to have children regardless of how many miscarriages she had.

Another focus of Mead's missing child fantasies was a twin sister who disappeared. She also identified as a precursor to this, "the other twin," a male twin.[4] She sought these lost siblings out at times in her intimate relationships, searching for the other part of her who shared her mind.

Although she had a brother, Richard, who was a little more than two years younger than she, Mead longed for an older brother and sought in her friendships a male who could play that role. She often took on a male confederate in her work on various projects, and she recalled in *Blackberry Winter* that her father did this in pursuing his projects. As was the case with her father's relationships, that confederate was often not in her own academic specialty or of the same professional standing, but someone from a different field or a former student. The closest readers and editors of her work, however, were women: Ruth Benedict, who, until her death in 1948, read everything Mead had ever written; Mead's godmother Isabel Ely Lord, who edited numerous manuscripts; and finally, Rhoda Metraux, who not only edited but co-wrote some of Mead's later writings, most notably her *Redbook* magazine columns of the 1960s and 1970s.

Mead's mother, Emily, had five children (four who lived to adulthood) and never finished her Ph.D. in sociology, which she worked on during Mead's childhood. When Mead herself was in college and graduate school, she encouraged her mother to complete her degree, but Emily ultimately did not. In *Blackberry Winter,* Mead paints a portrait of her relationship with her mother that is somewhat at odds with the picture that emerges from their correspondence. Although some letters are contentious, the correspondence shows a warmer, more affectionate relationship than the one portrayed in *Blackberry Winter,* and one in which Mead confides in her mother about her personal affairs.

Emily and Mead were also professional colleagues in a sense, as Mead's work for her master's thesis in psychology drew on the sociological research her mother had done on immigrants, and Mead sought to gather information through her research that her mother could use as well. Additionally, Mead's anthropology professor, Franz Boas, arranged for Emily to collect information for him from those immigrants.

Mead often deputized her mother to do such things as collect data for her, arrange speaking engagements, and assist in the preparation of her books. Mead also brought her younger sisters into the realm of anthropology from time to time. In Mead's later years, her daughter became a professional colleague, first specializing in linguistics and Middle Eastern studies, then moving into the realm of cultural anthropology.

While in her twenties and thirties, Mead spent nearly a cumulative decade in the field, far from home and any immediate connection to her family and friends. She kept in contact through letters and the occasional cable in an emergency. She liked to feel as though, through long-distance correspondence, she was carrying on a conversation with those important to her, both in the areas of personal relationships and of intellectual and academic matters. Due to the great distances involved, however, the time lag for the delivery of letters could be months. Air mail was expensive, making it a luxury both for Mead and her correspondents, but the savings in time was critical to coherent communication. Mail was sometimes lost; transport vehicles carrying mail went down. Letters arrived in unpredictable order. Some never arrived at all.

Repeated unanswered letters upset and frustrated Mead. Lacking control of the situation and powerless to force people to write, Mead would lash out, sometimes informing her wayward correspondents that she would not send them any more letters until she heard from them.

Mead did not like to be isolated from events or intellectual developments. She sought world news and personal gossip from family and friends and requested reading material. She wrote colleagues intricately detailed letters on points of theory and provided extensive analysis of materials sent to her for review. She wanted to apply the latest developments in social science theory to her work while in the field and saw her field sites as opportunities to gather information for others to use in their work. Despite personal difficulties that might interfere, Mead always accorded "the work" paramount importance.

Mead wanted those close to her to be able to follow the events of her life as she lived them while in the field, though different correspondents got different versions of events. Her correspondence to Ruth Benedict on her early trips reads almost as a diary, multi-dated letters bundled together and put into one envelope when the boat arrived to carry them to their destination. As far as receiving mail, however, she liked to get as many envelopes as possible, and she devoted rituals to organizing them for reading and rereading.

On January 7, 1926, Mead wrote Ruth Benedict from Samoa about these rituals:

This is a bad habit I've got into of always heaping the accumulated turmoil of boat day[5] on your shoulders. It's a state of mind which is quite unique and quite awful. I feel as if everyone were talking to me at once at the top of their lungs. Your letters are put to most utilitarian uses. If I'm feeling frightful and frightened of tackling the mail, then I read them first. Otherwise I try to save them for the benediction. And this time they made such a glorious pile—all in separate envelopes. I *like* lots of envelopes.

The physical form letters took mattered to Mead as well. While in the Sepik River region of Papua New Guinea early in 1933, she and Gregory Bateson, working at different field sites, sent bundles of small note slips back and forth to each other by messenger. Designed for recording field notes, these specially-sized pieces of paper facilitated, piece by piece, an ongoing intimate conversation on personal, logistical, and anthropological issues. The intimacy was not always a function of the subject matter but also of the manner of conveying it, through personal messenger and generally not meant for the eyes of Mead's husband at the time, Reo Fortune. These note slips—which Mead preferred typed for conveying factual information and handwritten for more personal matters—symbolized their private conversation, and she continued to use cut-up slips of paper even after their return from the field. She noted the eventual change to a more conventional note paper as marking the beginning of a new era in their relationship.

Mead also recorded an overview of events on each trip designed to inform and entertain a general audience of family and friends. She sent multiple typed carbon copies of these accounts out as "field bulletins." In 1977, a collection of these bulletins was published as the book *Letters from the Field: 1925–1975*. We have not included any of those letters here.

THE CHAIN

Beginning in her childhood, Mead organized people and events, putting together networks to get things done. She introduced her friends and colleagues to each other and sometimes to her family. She identified people with "special gifts" and brought them into the field of anthropology. (Dancers, she found, were especially attuned to understanding the choreography of cultural behavior.) She made new friends throughout her life while continuing to nurture older relationships. Mead sometimes diagrammed on paper the networks of people she had brought together and also diagrammed her intellectual influences. While she remained at the center of it all, she wanted those whom she brought into contact to have their own flourishing relations with each other.

Mead shared those she loved with each other, yet some who did not share her views sometimes became jealous of her nonexclusiveness in relationships.

Mead believed in what she called "polygamy," a word from the avant-garde of her young adulthood, connoting a free love philosophy whose essence was an ability to love more than one person—male or female—at a time. She found monogamy difficult and against her essential orientation. Her involvement with those who believed in monogamy imposed obstacles to relationships and sometimes caused her to question and rethink her beliefs on polygamy.

Mead adapted metaphors from the Amy Lowell poem "A Decade" to describe her intimate relationships. There were people who were exciting, sparkling wine and those who were nourishing bread. Some people who felt she was wine to them were bread to her. In her life, she needed both bread and wine.

She saw each relationship and each moment as an experience unto itself and did not care for re-evaluating the meaning of a particular moment retroactively. It was this world view that allowed her to manage multiple relationships without seeing them as being in conflict.

Feeling responsibility for those people who had been dropped as friends or lovers of her loved ones, Mead brought them into her own network, where they often stayed. She sought not only to stay in contact with her ex-husbands but also with other members of their families, passing along news and information. She stayed in contact with her own distant relatives, advised them on their problems, and sometimes gave them money. She took in people who were having personal problems and advised them on psychiatric treatment when she felt it was necessary.

Mead also gathered together people to work cooperatively at child-rearing, living in a joint household with her friend and colleague Lawrence Frank that included both of their spouses, as well as visiting friends, domestic workers, children, and foster children in various combinations, from 1942 until 1955. In addition to the Franks, Mead's college friend Marie Eichelberger and other friends from the 1930s, Sara and Allen Ullman, cooperated with Mead in helping to raise her daughter, Catherine, born in 1939.

THE COMPASS

Mead's personal and professional lives were closely intertwined, not just in terms of the people involved but in her application of psychological and anthropological concepts and theories to her own life and relationships. One of the most prominent examples of this is the theory of "the squares" (sometimes also referred to as "the points of the compass"), which emerged from intense discussions of temperament and culture that Mead had with her then second husband Reo Fortune and her future third husband Gregory Bateson along the Sepik River in early 1933. In this theory, temperament was an individual's innate disposition, and personality was the result of the interaction between individual temperament and culture.

The theory was never formally published by Mead in her lifetime, beyond a summation of it in her 1972 autobiography, but it was critical to the way she conceptualized and discussed her relations with others from 1933 into the 1940s. By the time of Mead's divorce from Bateson in 1950, the squares was no longer as prominent a feature of her personal vocabulary as it had been, but in the 1930s it was a special language she used and taught to those close to her; it was an intimate shorthand.

Some elements of the squares remained stable over time while others changed. It was a fourfold classification system that mapped individuals and cultures by temperament (individuals) and ethos (cultures): Northern and southern ran along a central vertical axis and western and eastern along a central horizontal axis, with intermediate points denoting NE, SE, NW, SW. She was fascinated by the interactions of people, whether within the same category or across the major axes and the diagonals from each other. Structured in contrast, opposing pairs also had a dialectical relationship to each other and each had characteristics that fed off the other. As well as their obvious differences, they had underlying similarities. While north and south were antithetical, they had features in common, especially in being concerned with interpersonal relationships. East and west were more concerned with the individual's connection to the world rather than to other people.[6] Mead mapped her friends, colleagues, family members, and spouses along the squares diagram and explained their troubles and felicities in the language of this theory.

Mead's friendships were interwoven with the other close relationships in her life, as she made friends with her husbands' friends, relatives, and other wives—remaining in contact even after marriages dissolved—and with the spouses and in-laws of her siblings. She remained friends with those who were lovers for a time and with her ex-husbands; brought friends and siblings into the professional realm of anthropology, making friends out of colleagues and colleagues out of friends; and wove together a large and flexible network of kith and kin composed of all these groups, a network that sustained her through the highs and lows of her life and career.

This book's six chapters focus on distinct types of relationship in her life. Many people fall into multiple categories, and letters have been positioned by their subject matter rather than exclusively by recipient. The first chapter includes letters to or about Mead's family, those people in the household in which she grew up. The second chapter includes letters to or about her three husbands, the first of whom she met while in her mid-teens. The third chapter deals with her long-term relationships with two female partners: Ruth Benedict and Rhoda Metraux. The fourth chapter includes letters to her friends, many of them from college. The fifth chapter includes letters to colleagues, both those in anthropology and outside of it. In the final chapter, Mead's extended family network comes into focus: her daughter and granddaughter; her godchildren and godmothers; her aunts, uncles, and distant relatives; and those who became "adopted" family members, such as Claudia and Philomena Guillebaud and Daniel Metraux.

This book does not represent all of Mead's existing correspondence. There were many friends and colleagues who were left out, and close friends who are barely mentioned. But the sample presented here gives a sense of the complex, interrelated life Mead created, the intensity and excitement generated by her relationships with other people, and how she perceived life as incomplete when isolated from her interrelating circles of friends and family.

NOTE ON THE SELECTION
AND ORGANIZATION OF MATERIALS

In her foreword to the 1966 edition of Mead's *An Anthropologist at Work: Writings of Ruth Benedict,* Rhoda Metraux observed that Mead "drew on the records Ruth Benedict herself had chosen to keep." [7] To a great extent, that is true of this book as well. With a few exceptions, letters included here are found—in manuscript, typed carbon, or photocopy form—in Margaret Mead's papers at the Library of Congress.[8] That leads necessarily to some idiosyncrasies in the materials utilized for this volume.

Mead believed very strongly in the importance of preserving documents for their historical value and usefulness to subsequent scholars. She retrieved a substantial number of her letters from her second and third husbands, Reo Fortune and Gregory Bateson. She did not retrieve manuscript letters from her first husband, Luther Cressman, and he had a different philosophy toward saving old correspondence than did Mead. Mead was Ruth Benedict's literary executor and she—and those who looked after Mead's affairs after her death—preserved their correspondence as well. Mead also retained original letters she had written to members of her family and to other friends whose papers came into her possession.

When Mead typed letters, she kept carbons, sometimes multiple carbons. Those are in her papers at the Library of Congress. Letters she wrote by hand and did not retrieve were generally not available for use in this volume. In later years (in some cases), her correspondents and their family members, such as Marjorie Bull, made photocopies for Mead of her letters. These are housed in the Library of Congress as well.

Like other Americans, Mead relied increasingly on the telephone for personal communications over time, so the range of personal letters available from later years is more limited than from earlier years. She still relied on letters, however, while in the field or traveling long distances. And regardless of time period, sometimes Mead wrote to people simply because she enjoyed the act of writing and the intimacy it represented.

In this book, her letters have been arranged largely in chronological order according to the topic of each chapter. They are organized thematically and not necessarily by recipient, though there is congruence between the two in many cases. Many of the recipients of these letters are now deceased, as are numerous people mentioned in the correspondence. In assembling this volume, we have, however, tried to be sensitive to the privacy of those people still living without compromising the integrity of the materials.

Note on Editorial Conventions

Mead was home-schooled for much of her childhood, and her grandmother, who taught her, emphasized experiential over rote learning. As a result, Mead's spelling and grammatical skills were irregular. For the sake of clarity, we have functioned as benevolent editors in systematizing and regularizing spelling and punctuation in this collection of letters and in adding paragraphing in places. We have, however, retained some of her idiosyncratic use of language—such as her use of British spellings through much of her life, and her making "ofcourse" one word—in order to give a sense of her distinctive style.

Mead's handwriting could be quite difficult to decipher, much to her correspondents' exasperation. Even during times Mead had a typewriter available, she sometimes wrote by hand anyway. There were a number of reasons for this: she liked the intimacy of writing by hand to those especially close to her; sitting at a typewriter for prolonged periods of time, which she had to do in the field to type up her notes, led to physical discomfort; and, when she wanted her communication kept private—particularly when it passed through many hands or might be seen by others—her poor penmanship, with its tiny cramped letters, was impervious to casual reading.

Working in tandem, we have been able to decode Mead's handwritten letters with a fairly high degree of certainty. Some of the translations are open to interpretation, however, and we have noted in the book places where we were particularly unsure of a word or phrase. We have supplied words in brackets where they appear to be missing in the text. In the case of sentences that cannot be made clear with the addition of a word or two, we have left them intact and given our interpretation in a footnote of what we believe Mead was trying to say. We have also standardized titles of books, magazines, and newspapers in italics.

While we have worked from originals when possible, a significant number of the letters we have worked from are carbons. Mead sometimes made corrections or changes on original letters before sending, changes which are not recorded on the carbons. For this reason, original copies of the same letters, if they do exist, may differ slightly from our versions.

We have edited some letters for length and to highlight the focus of the volume on relationships. Some of the letters we worked from were very long, often written journal-style over multiple days, and we have made judicious selections from these letters to best fit the theme of the book. If a letter bears multiple dates, that is noted in the introduction to the letter. Letters to different people often contained the same or similar information, and we have cut some material that was repetitious from letter to letter or within individual letters. In order to enhance the readability of the volume, we have also cut from letters passages of an excessively technical nature. As Mead's interests in anthropology, psychology, and other related subjects were critical to her intellectual engagement not only with her colleagues but also with her family and friends, we have retained some technical material in order to remain true to the tone of the correspondence. In order to distinguish our elisions in letters from places where Mead used ellipses for effect (e.g., . . .) in the original, we have marked her punctuation with [orig]. When a postscript has been omitted after the closing of a letter, it is designated with a triple asterisk (***).

We have used an [*h*] in the introduction to a letter to indicate that it was handwritten. We have included Mead's name as she wrote it where a signature appears in the original. This is generally the case with handwritten letters. Where we have worked from carbons, letters generally were not signed but ended with a typed closing. In some letters (generally of a more formal variety) Mead's name has been typed beneath the closing. To minimize confusion, letters which have Mead's typed name but no signature will carry a [*t*] beside the name. Letters which end abruptly with no closing or signature will be marked as such.

ADDITIONAL READING

A selected bibliography of Mead's publications follows the text of this book. The reader may also find the following books by other authors useful in providing additional context and background for the letters included in this volume:

Lois W. Banner, *Intertwined Lives, Margaret Mead, Ruth Benedict and Their Circle.* New York: Alfred A. Knopf, 2003.
Mary Catherine Bateson, *With a Daughter's Eye: A Memoir of Margaret Mead and Gregory Bateson.* New York: William Morrow, 1984.
Margaret M. Caffrey, *Ruth Benedict: Stranger in This Land.* Austin: University of Texas Press, 1989.

Jane Howard, *Margaret Mead: A Life.* New York: Simon and Schuster, 1984.

Hilary Lapsley, *Margaret Mead and Ruth Benedict: The Kinship of Women.* Amherst: University of Massachusetts Press, 1999.

David Lipset, *Gregory Bateson: The Legacy of a Scientist.* Englewood Cliffs, N.J.: Prentice-Hall, 1980.

Judith Schacter Modell, *Ruth Benedict: Patterns of a Life.* Philadelphia: University of Pennsylvania Press, 1983.

1

Family: Bread and
Butter and Applesauce

❖

Although there are some letters that exist from Mead's childhood, the bulk of her correspondence began when she went away to college, first as a freshman at De-Pauw University in Indiana, her father's alma mater, then at Barnard College in New York City, from her sophomore to senior years. Mead began her letter-writing career with correspondence intended for various members of her immediate family: her parents; her paternal grandmother, Martha Ramsay Mead; her brother Richard, just over two years younger; and her two sisters, Elizabeth and Priscilla, seven and nine years younger, respectively.

Her letters give us a sense of both family dynamics and the relationships she had with the other members of her family.

A growing theme as she moved through her life was her sense of responsibility for the family, for its smooth running, and for straightening out various family problems. The letters also show her trying to steer family members into direc-tions she thought good for them and her disappointment when they didn't follow her lead. Increasingly over time, letters show that family members depended on her advice and turned to her with family problems. On the other hand, she used the talents of family members to help her out when necessary; for example, to gather information or provide feedback.

Her letters also reveal, consciously and unconsciously, expectations Mead had of other family members. They show she could not depend on her father to keep promises. They show her relationship to her mother to have been closer than her autobiography, Blackberry Winter *(William Morrow, 1972), suggests. They*

confirm her respect and admiration for her Grandmother Mead, and her grow-
ing adult relationships with her sisters and brother.

To her mother, Emily Fogg Mead, December 9, 1919. Mead was in her first
semester as a freshman at DePauw University in Indiana. She was deeply un-
happy, and her parents were concerned. Her father tried to manipulate her into
leaving DePauw by telling her he was willing to pay for her to come home for
Christmas, but not for the return trip. As he had a pattern of doing, he was try-
ing to use money to control his daughter's behavior, this time to keep her near
home. [h]

Mother dear,

I am writing this on the counter in the office, so that I can send it when
the postman comes. I know that it wasn't two weeks that you didn't [hear]
from me. I wrote several members of my numerous family over the weekend before Thanksgiving and mailed them, on Monday and Tuesday. I
wrote Dadda on Wednesday, and I wrote to you the Tuesday, or Wednesday
after I came back. Since I knew that I couldn't come home, I haven't known
exactly what to say in letters.

I wish that you wouldn't tell me all the bribes that Dadda concocts, because you know that I can't accept them.[1] It would be a moral defeat for me
to give up and come home. I am going to finish this year out. It wouldn't be
worth anything to me if I didn't.

From your letter I gather that in your official position of Mother you say
don't come home, but incidentally you hope that I will give in, even tho you
tell me not to. But I shan't. The prospect which Dadda proposes is much
more alluring than six straight months of De Pauw *but*—it's six straight
months of De Pauw for me. That doesn't mean that I'm not homesick for
you all, and that I'm not dreadfully disappointed that I can't go home. I am,
but I just can't give in now.

I don't know what questions remain unanswered. I like my dress a lot, it
is mightily comfortable. I am well and I haven't any cold at all. As I told
Dadda I bought a little everyday gray hat to go with my coat. My school
work itself is going quite smoothly. There is a rumor that we will get out
the Friday before Christmas, the coal supply is so meagre. You need not get

the poinsettia for Dorothy. I am making her something. But will you get Richard's[2] present? I can't find a nice one here. I wish somebody would send me a couple of dollars to fix a Christmas tree for my wash woman's seven youngsters (under twelve). The poor people out here can't have trees because none grow wild. Would you let me have that instead of a birthday present? I am making Aunt Beth and Aunt Mabel collar and cuff sets but I don't know what to give Aunt Fanny.[3] I am making Madam an apron for school.

Will you send my Christmas present[s] c/o Mrs. Charles Greenfield, Paris House Annex, Paris, Ill. please.[4] I would rather have them on Christmas Day. Is there anything that the children particularly want, if not I have planned something for them. I had gotten *The Young Visiters* for you as a belated birthday gift, but you wrote me you had read it, so I'll send it to Alice West instead. Does Bessie have my room?[5]

With a great deal of love from

Margaret

To her father, Edward Sherwood Mead, January 4, 1920, from DePauw. This is a thank you note, what she calls a "bread and butter" letter. As he was a professor of finance, she joked of herself as goods shipped to DePauw. Her father had relented and promised the return ticket, and she came home for Christmas after all. She had been desperately homesick, but had ultimately won. Her letter was especially warm, as she acknowledged her father's control over family decisions, even as she managed to subvert it. [h]

Dadda dearest,

This is just a note to apprise you that the goods you shipped arrived safely at Greencastle this afternoon at two fifteen, none the worse for wear. I can never thank you enough for letting me come home nor tell you how much it meant to me. If De Pauw does nothing else it will have made me appreciate my home and family more than I ever would have done while in the midst of them, and it has made my father want to get a little bit acquainted with me at least. And I really like him a lot, beside loving him a great deal, and I do enjoy knowing him so much.

Why one's father shouldn't deserve a bread and butter letter when he supports one from a distance and gives one a wonderful time, I don't see, so please consider this a bread and butter *"and applesauce"* letter.

Lovingly,

Mar[6]

To her mother, March 6, 1920, from DePauw University in Indiana. Mead wanted her mother to come visit. She also planned an educational summer for her brother and sisters, with her mother's cooperation. [h]

Mother dear,

I have all my lessons for tomorrow and some for Tuesday and as it's only nine thirty I think I can indulge in writing to you. This has been a rather queer weekend. Friday afternoon and evening I studied like mad in the library, and then I went to bed at eight thirty Friday night. I find that ten or eleven hours sleep one night a week and eight the rest keeps one going very nicely. Saturday morning I went to breakfast but came home and slept. Saturday afternoon I knit and read. I stayed with Alice last night and we made fudge. Then this morning we cooked our own breakfast: cocoa, raisin bread, grapefruit and sausage. After which we took some foolish pictures. If they are any good I will send them home. I am enclosing two foolish pictures of Alice, needless to say it is not a real cigarette. Will you please return them at once.

I sent the *Crescent Moon*[7] on Saturday. I would have sent it Special Delivery but I know that is rather useless there. I hope that it arrives in time to be of some use. It's just as you wish about my taffeta dress. If you were going to get one for yourself, get it and I'll keep this one. If you weren't, get me a new one and I'll send you this. Voyez-vous?

Dadda says he will send you out to visit me in the Spring and you just have to come. I've planned for weeks what I would do when you came and you mustn't wreck my little air castles. I want to show you off to my friends and I want to show my friends off to you, so get some pretty clothes and come along. You've just got to, that's all. I'm tired of having other people's Mothers around, I want my own.

I'm planning a peacefully pastoral existence next summer, tea under the trees to refresh one after desperate labor at upholstery, or canning, or some other equally pastoral employment. I think we might take next summer to educate the children[8] in the art of living with all the quiet graces pertaining thereto. Then if such were the program perhaps the family finances could afford a new book a week to be read aloud in the evening. We may indulge in a French meal or two, and if you'll be looking up some Modern French plays we can read them aloud with Richard. If you will have the children thoroughly reviewed in their Greek and Norse mythology by next summer so that they will be well grounded in it for life, we might plunge them into some Egyptian & Babylonian & Persian myths. They are equally fascinating, and ofcourse, much less known.

You know I'm a bit like you whether you admit it or not. With all the love in the world

Margaret

To her mother, May 4, 1920. Mead had written and was directing the yearly school-wide pageant for all the DePauw women students, part of DePauw's annual May celebration. This was quite a triumph for a freshman, and Mead wanted her mother to come out and share the moment with her. [h]

Dearest Mother,

The check arrived OK today. I'm sorry that you had all that trouble, but I was almost penniless.

I do wish that you would come May Day.[9] If Dadda comes, he will have to stay at the hotel because the Delta U.[10] house is quarantined for smallpox—has been for some time. It would be almost impossible for me to see anything of him. But if you come, you can stay in the Dorm with me and be here Friday night for the "Senior swing out" and everything. Won't you come, Mother darling? I've been counting on it so. I want you to see all the girls and see just why I want to come back next year and everything.[11] I want to show you off. Mothers are at such a premium out here. Please please come. Oh do say you will. I've been planning for it for weeks. You know in the end Dadda will back out and I won't have anyone. And I haven't had a single bit of company this year. Please, Mother, please. You just must come. I won't enjoy May Day a bit if you don't. You just trot around and get some pretty clothes—you won't need many—and come along. *Please*!!!!! Now remember, I'm counting on your coming.

Later—Katharine came up and we went for a walk. It's a lovely night and I haven't much work to do. Last Saturday (May Day)[12] I had the loveliest time. Friday afternoon Hilda and I went out in the country and gathered flowers for May baskets. I made three May baskets one [each] for the Dean, Miss Steese and Katharine. Then May Day morning Katharine and I spent out in the woods gathering flowers. Saturday night second floor gave a party—kimono dance and stunt, and I had a spread and slumber party afterwards. Mildred was away, so Katharine and Mary Mutschler slept in her bed, and Christine and I in mine. Christine is an English girl—a Senior. She is very interesting and wonderfully well read. She and her mother live here in town. Katharine and I go out there every once in awhile.[13]

We had the Russian Symphony Orchestra here this afternoon. It was lovely. Your angel child has got to stop writing. Have those blouses made by some modern fashion book pattern please—the newer the better. And write and say you're coming.

Lovingly,
Margaret

Her mother did end up coming out for the pageant, to Mead's pride and happiness. As she wrote in a letter to her mother on May 31, 1920, "All the girls are so crazy about you, and oh, it was perfectly scrumptious to have you here."

In spite of her success during her first year at DePauw, over the summer she decided not to return and to attend Barnard College in New York City instead. Her future husband, Luther Cressman, was attending a theological seminary in New York. Mead also thought New York City was close enough, yet far enough, from her home in Pennsylvania, and she looked forward to the city's exciting cultural events.

To her father, October 9, 1921, in her first semester at Barnard College in New York City. Her father had a pattern of saying he would do something and at the last minute, would not, and would send a check instead. Here, she responded to this pattern. [h]

Dearest Dadda,

Thank you very much for the check which arrived on Friday morning but did not by any means compensate me for not seeing you Thursday night. I got all dressed up in my grandest clothes and waited for you for an hour and a half and you never came at all! I then went down to Grace's and she fed me an orange and some raisins for I hadn't any money left to get supper with.

I had been looking forward to that dinner so much. Mother says you are very tired. Isn't their being in town going to make it any easier for you?

I am continuing to sit as if a straight [sic] jacket surrounded me and my cheeks are very rosy.

> Lovingly,
> Mar

To her nineteen-year-old brother Richard, July 23, 1923. For a time, Richard thought about going into a musical career and Mead strongly encouraged him to try it. She also tried in this letter to enlist him to help her move her mother toward finishing her Ph.D.

Dear Dick,

I am visiting Léonie[14] and I am taking the opportunity to practice up my typewriting which is sadly in need of repair. Your letter was duly appreciated even if I have not had time to answer it. But I wanted to wait until I had time to tell you all about Louise[15] and my visit to Mrs. Osterlink and what she said about your voice. She has studied music very intensively in New York for years you know, and her opinion is well worth having. She thinks there is no doubt about your having a pure tenor voice. She says the quality of it is entirely tenor and she thinks that you haven't begun to fall [into] your range yet. She is very anxious to talk to you and see whether you have what she calls "musical intelligence" or not. I think that she means by that not so [much] knowledge and cultivated appreciation of music, as simply a good brain which is capable of fully appreciating and understanding the more intellectual aspects of music, like harmony and counter point, I suppose. . . .

We have been having a very gay and argumentative life at home since you left. Dadda has been feeling very poor indeed, with the usual upsetting results. Once he decided that I couldn't have a wedding at all. Mother brought to the fight for a wedding, in which I was not allowed to take part, all the accumulated rancour of twenty years ago over her disappointment at not having a wedding. However, that blew over. I am leaving it to you to see to it that Dadda lets me have the car. Uncle Lockwood and Aunt Mabel[16] are going to England on the eleventh of August and not returning until the fifth of September, so we can't have the cottage until the eighth or ninth. We think we can manage by taking the trip very slowly and going a little longer way. But you see we would be quite lost without the car. It is virtually impossible and also frightfully expensive to spend five days going somewhere by train. So please. I'm depending on your potent influence to get the car.

By the way, Dadda doesn't know about this new plan of Uncle Lockwood and Mother doesn't want him told as he is sufficiently upset as it is just now. His plant is the only one running. He is proud of that fact, and also it gives him a little more coal as the plants up above are all shut down and so of-course he gets all the coal that there is; but still there isn't much and he is existing mainly by cutting the pay roll lower and lower.[17] All this hectic uncertainty is having its usual nervous effect upon Mother. Dadda spends all the time he is at home arguing about what we are to do next year. His plan when I left home on Saturday was to leave the children and Grandma and a reliable woman on the farm, send the children either to Doylestown (which Mother won't hear of) or to New Hope (which I am going to suggest when I get home), have Mother go to town four days a week, staying in [town] two nights, on which nights he would come out so as never to leave them alone on the farm, and that he and you should get a room together if you so desired with a piano, or let you room where you wished and hire you a studio in which to practice, while Mother would stay at the club when she was in town and so be freed from all domestic cares four days a week.

I haven't had time to think it over very carefully and it will be an elephantine task to convince Mother that she ought to do anything by which she herself will be benefited in any way, but still I think it might be done. I have gotten Dadda more interested in getting Mother to take her degree by suggesting that if she did there might be a most convenient and desirable opening for her in the new girls' college at Penn. Mother isn't the sort of person who will ever be content to settle either to knitting and jam making or to women's clubs and dabbling philanthropy. She has a very fine mind and it is a crime not to use it. Also she will be frightfully unhappy and difficult to live with unless she does do something of the sort.

You have already reacted very hard against the nervous wear and strain of living with her in her present disorganized mental state. If the two children are not to come to look at her and treat [her] in the same way that you do, she will have to be jacked up by being taken neck and heels out of domesticity and flung back into some work that will keep her from getting old. I think perhaps if you would take the trouble to do it diplomatically it would be very helpful in your next letter to Mother if you should say, anent[18] what we are to do next winter, that you don't want her to plan to do anything which will interfere with her studying for her degree, because she has thought of us long enough and it [is] just about time that she should begin to think about herself a little. Don't lay it on too thick or it won't sound natural. But it will, if you are careful, because you have been taking such a preternatural interest in our next winter's plan.

I came over here to get an apartment in response to a telephone call from Luther on Saturday. We have gotten a lovely apartment with two big rooms on 119th Street, which is right over on the other side of the Columbia Campus, on the other side of Amsterdam Ave. We think we are very fortunate. Luther had his article accepted by *The Journal of Social Forces*. If I find any stories which I think will be of any use to you, I'll send them. I shall be up here at Léonie's until Friday, and I don't even know when this will get mailed.

Lovingly,
Margaret

To her mother, September 6, 1923. Mead was on her honeymoon with her first husband, Luther Cressman. [h]

Mother darling,

We are having a most beautiful time. But I'm worried about you, with all the deadly grind of packing my stuff on your hands. Even tho the operation isn't salted with tears for a lost daughter, still it bears the stamp of twenty years monotony which I suspect is even worse. I do hope Dick is staying home and helping some.

It makes me unhappy to think of what a beautiful trip I am having with a wedding and clothes and everything, and then to think of you with no wedding at all, not even a breakfast. And now that the only desires for yourself are simply hangovers from long ago, it seems hard to make it up to you. Perhaps the tale of happiness and human woe is evened up in the lives of one's children—at least I have always been supremely happy, taking the best your love endowed me with, and giving, I reckon, very little in return. I love you, Mother.

We liked the Berkshires so much that we stayed here two nights—tomorrow we go to Boston. Will you forward all my mail to W[est] Yarmouth [Uncle Lockwood Fogg's cottage there] until the fifteenth and then to 419 W. 119th Street, New York.

I am happy, so happy.

Your devoted daughter,

Margaret

To her teacher and friend Ruth Benedict, July 13, 1924. While at home for the summer in Pennsylvania, Mead wrote about working to help her mother finish her Ph.D. [h]

Ruth dearest,

. . . I fought for hours last night to try to introduce some sort of rationale into Mother's attitudes towards life, via Father's, for ultimately it is his opinion that always counts. Much of the fighting was due to my unwillingness to employ arts instead of honesty. Finally, at twelve thirty, my morale was quite wearied out and I flattered and cajoled Father into consenting on Mother's finishing her thesis, incidentally promising to keep the household running smoothly so she could. Victory, but fairly savourless, and of Heaven knows how long duration. The complete lack of harmony between them affrights me. I become more amazed that any measure of spiritual adjustment is vouchsafed to blundering mortals—and also more thankful for contacts of the flesh which at least mercifully blur a lack of deeper understanding. But that particular avenue of approach has been quite thoroughly denied me as far as the family are concerned. They just seem to be a whole dissimilar divergent group of personalities who have endless possibilities of making one's heart ache and offer little promise of reconciliation. . . .

In spite of Mead's efforts, her mother never did finish her dissertation. Mead later wrote that so much had changed in sociology by the 1920s from her mother's graduate student days at the turn of the century that she felt she would have had to start all over again. Instead Mrs. Mead made the decision to move on and put that dream aside.

When Mead went to Samoa in 1925–26, she depended on letters to keep her in touch with the family and the world. To her brother Richard, twenty-one years old, October 4, 1925. [h]

Dick dear,

You had the honor of having a Samoan baby named after you the other day. I was dragged into the house of an unknown woman who proved to be the native who washes my clothes, and before I could adjust my scattered wits to that information, I was asked to name the baby—after my brother. *Richard* translated into Samoan sounds is *Likato*. So *Likato* the baby now is, and you are sponsor for a distant little brown namesake.

The male section of my family was badly represented in the last batch of mail. I hope such negligence is not likely to continue because, with Grandma away, I will get no news of you. Mother writes about the grand problems.

You must tell me how your Trenton plans are jibing with your musical ones, and why your teacher didn't visit the family and what he thinks of your plans, and where you are living, n' everything. All my news is compacted into the typed letter and strength is too much needed in this hot place to waste any on reiterations. I need all the extra energy to kill cockroaches, as big as mice and much more destructive. I keep a great stick beside my bed for the purpose.

If you write me on this boat, I'll add to this letter—if not—

Lovingly,
Margaret

Oct. 12

Your long letter was much appreciated. You will realize that my notes to each member of the family must of necessity be short. The long one will have to help out. Don't forget to call up Melville Herskovits,[19] Ant. Dep. Columbia–Morningside 1400, when you go to New York. He'll put you in touch with some interesting people.

Love,
M.M.

To the younger of her two sisters, Priscilla, almost fourteen years old, November 3, 1925, from Samoa. Priscilla was then trying out a nickname, Pam, formed from the initials of her full name, Priscilla Anne Mead. [h]

Dear "Yam," [sic]

Your long vivid letter was duly chuckled over. Mother writes you are being tremendously good and sporting about all this uncertainty. I do hope that by this time you're well housed in Fairhope[20]—and like it there. Thanks for describing all the girls so accurately—it was just enough of a jolt to my memory so that I could picture most of them. I remember I liked Sydney best.

You'd laugh to know what my party stunt is here—I who can neither dance nor sing amuse the natives by imitating the cries of all the animals. The rooster, etc., make different remarks in Samoan, and so they are highly amused.

Here I had to stop and walk out to the Dispensary[21] to get my pictures which I took at Vaitogi. I'll tuck one in for you—just before mailing this letter.

I amused the Samoans muchly by telling them the way my youngest sister used to count to see how bad a piece of chicken she'd get—according to the number at the table. They all say your picture is *manaia*—"beautiful."

By virtue of my adoption into Ufute's family you have a whole raft of Samoan relatives, very sweet ones—to your credit—and your sister is high princess of a Polynesian village. Grand thought.

Lots of love to you, Priscilla Anne (It's a pretty name)
Margaret

To her sister Elizabeth, almost eighteen years old (seven and one-half years younger than Mead), January 11, 1926, from Samoa. Mead responded after receiving a letter from her sister describing what was probably her first important brush with sexual feelings. [h]

Elizabeth dear, I've a good mind to punish you by writing back in pencil. You're a wretch to write in pencil on *pink* paper just when you're writing something very important that you particularly want me to read. Don't do it again.

I am glad you told me about the moonlight party, dear. It's the sort of thing that had to happen sometime and it might have been a great deal worse. As it was, it was a nice boy whom you like, and nothing that need worry you. There are two things I'd like to have you remember—or in fact several. The thrills you get from touching the body of another person are just as good and legitimate thrills as those you get at the opera. Only the ones which [you] get at the opera are all mixed up with your ideas of beauty and music and Life—and so they seem to you good and holy things. In the same way the best can only be had from the joys which life offers to our sense of touch (for sex is mostly a matter of the sense of touch) when we associate those joys with love and respect and understanding.

All the real tragedies of sex come from disassociation—either of the old maid who sternly refuses to think about sex at all until finally she can think about nothing else—and goes crazy—or of the man who goes from one wanton's arms to another seeking only the immediate sensation of the moment and never linking it up with other parts of his life. It is by the way in which sex—and under this I include warm demonstrative friendships with both sexes as well as love affairs proper with men—is linked with all the other parts of our lives, with our appreciation of music and our tenderness for little children, and most of all with our love for someone and the additional nearness to them which expression of love gives us, that sex itself is given meaning.

You must realize that your body has been given you as an instrument of joy—and tho you should choose most rigorously whose touch may make that instrument thrill and sing a thousand beautiful songs—you must never think it wrong of it to sing. For your body was made to sing to another's touch and the flesh itself is not wise to choose. It is the spirit within the body which must be stern and say—"No, you can not play on this my precious instrument. True it would sing for you. Your fingers are very clever at play-ing on such instruments—but I do not love you, nor respect you—and I will not have my body singing a tune which my soul cannot sing also." If you re-member this, you will never be filled with disgust of any sort. *Any* touch may set the delicate chords humming—but it is your right to choose who shall really play a tune—and be very very sure of your choices first. To have given a kiss where only a handshake was justified by the love behind it—*that* is likely to leave a bad taste in your mouth.

And for the other part—about being boy crazy. Try to think of boys as people, some nice, some indifferent—not as a class. You are[n't] *girl*-crazy are you?[22] Then why should you be *boy* crazy? If a boy is [an] interesting *person*, why, like him. If he isn't, don't. Think of him as an individual first and as a boy second. What kind of a person he is is a great deal more im-portant than that he belongs to the other sex—after all so do some hundred million other individuals.

I am very proud of the way you are able to think thru the problems which life brings you—and of the way you meet them. And I consider it a great privilege to have you tell me about them. I'm so glad you are happy dear.

> Very lovingly,
> Margaret

***

To her brother Richard, March 14, 1926, from Ofu Island, Manu'a, American Samoa. [*h*]

Dear Dick,

I'm off the island of Ta'u, for the first time in four months. Think on it. To have stayed on an island 13 by 8 [miles] or something like that, for four whole months. It appalls me now. The Lanes are over here at Ofu temporar-ily, so I came over to see them and do a week or so of field work. It's comfort-ing to be with gentlefolk again. I'm getting to be a hide bound aristocrat again after months of knocking about with people who have no manners at all.[23]

Manners, really good ones, make it possible to live with almost anyone, gracefully and pleasantly, but without them—one must pick one's friends with terrific discrimination. Remember that. I think the best society in the world is one based on brilliance and talent and personality, such as one once had in the famous *salons* in Paris. But next to that is a society which partly makes up for its defects in the matter of brains and talent by the perfection of the breeding. As long as one stays in a well-known milieu it is not so apparent, but the minute one tries to travel that fact stands out glaringly. You don't have time to find out what is in people's souls, but if they are well bred you can trust their breeding. You know what they will do under given circumstances—a prediction which one can only make of members of one's class and of one's intimate friends.

I hope to goodness when you marry you'll remember how hard it is to live with someone of different manners. I'd be the last person to try to influence your choice of a wife. I hope I like her. But if I don't, I hope we speak the same language so that we can dislike each other charmingly. This homily is all because I'm dining with a lady and gentleman for the first time in four months. In due time a bulletin will account for the events of the last week.

I'm very anxious to hear something definite about your plans.

Lovingly,

Margaret

To Martha Ramsay Mead, her paternal grandmother, May 20, 1926, from the Pacific. Mead's grandmother was a former teacher who had been largely responsible for home-schooling Mead and her siblings during their early years. [h]

Sweet little Grandma,

I've just been reading one of your *Atlantic Monthly's* with a big penciled cross on the table of contents—which is just like a very special command from you to a curly haired little girl who *wouldn't* remember how to spell "*dew*." The pencil mark was against Kuble's article on "Participating in the Grand Adventure." And I thought how really wonderful it is to have a grandmother of over 80 who would mark an article like that. Probably that is one of the reasons that my life has been such a serene untroubled development. I not only had no quarrel with the essential thought of my parents, but I also had none with that of my grandmother—a very rare privilege in as changing an age as ours. All the energy which most of my contemporaries had to put into reconciling affection for their elders with honest revolt against their teachings, I could conserve to use for my own development. It's very much owing to you that I've wasted so little time in life, made so few false moves, chased so few chimeras, extraneous to my personality.

The Samoans always thought that the most remarkable thing about me was that my father's mother, over 80, was still living. And, put into less material terms, I think they were very right.

Lovingly,
Margaret

When her field trip was over, Mead went home by way of Europe, then returned to New York in the fall of 1926 to begin her job as an assistant curator at the American Museum of Natural History in New York City.

To Grandma Mead, May 24, 1927, only weeks before her grandmother's death. At the time her grandmother died, Mead was in Europe, studying Pacific collections at European museums, particularly in Germany. While the trip was sanctioned by the American Museum of Natural History, Mead used the opportunity to spend time with her future second husband, Reo Fortune, as well. She had met Fortune on shipboard returning from American Samoa in 1926. Here, Mead talked of getting ready for this second European trip.

Grandma dear,

I am looking forward to that longer treatise on education of children which you have promised for a steamer letter. It's a point which bothers me a lot and which I feel quite inadequate to solve. You are the wisest educator I know and therefore I look to you for help, even tho I shall never have any children of my own to experiment on.[24] So be sure and write that letter.

Then when you come over to visit me in August we can discuss the matter at more length. I would really like to have you see that the subject which I have chosen is not all a matter of collecting barbarous and curious customs and delving into a queer set of human customs, but rather that it has very real bearing upon the ways in which we live our lives here and now, upon what educational policies we adopt for those who come after us, upon how much we can build a civilization which will be a base on which individuals can build important personalities and real art. But I'll tell you all this when you come over to visit me. You are coming you know. I won't be slighted. If you can gallivant down to Alabama and out to Columbus, why New York is just a very little way to go.

I have discovered that the way to get anything done in this Museum is just to go to Europe. All the requisitions which I have had in for two months are now being filled like lightning. The carpenters are carpentering, the photographers are photographing, etc. all chasing each other around getting ready to put on a special exhibit on two new collections and to finish up the illustrations of the guide leaflet on the New Zealanders which should be printed while I am away.

Marie,[25] as usual, is a godsend, mending, shortening, dying, planning for my trip. She has taken away all the clothes which I am going to wear to Europe for fear I will damage them before I start and as a result my wardrobe is very much denuded. I have plenty of things. I have had to get nothing new at all which is a great help.

Have a good time in Ohio and collect a great many stories to tell me when I come back. And keep that NEW YORK VISIT IN MIND, dearest and sweetest of little grandmothers,

Lovingly,
Margaret

To her mother, Emily Fogg Mead, August 16, 1927, after her grandmother's death. Mead and Cressman were preparing to divorce, and Luther was getting ready to sail to Europe to visit the woman he would later marry. Margaret had just returned to New York from her own trip to Europe.

Mother dear,

Your letter to the boat missed me so that it was very nice to find a letter here. Luther sails at noon on Saturday so that Friday will be his last night at home. Then I shall be alone in the apartment and shall love to have you stay with me. Could you stay with Bertha Friday night? Life is so complicated just now that I think it would be better if Luther and I had the last breakfast, etc. to ourselves. Then plan to stay several days. I have a great many things which I want to talk over with you. I was in Paris when I heard of the De-Pauw offer and Ogburn considered Luther very wise not to take it in view of the developments at City College. I hope Dadda will understand that. Be sure to plan to stay. You might bring some work along and bring it down and work in my office, or you could have great privacy in the apartment. I wish you could stay several days. Grandma's death gives me a very uprooted feeling. Please Mother. Love to you all. Come right up to the Museum Friday morning.

    Lovingly,

    Margaret

To her mother, September 7, 1927, from New York, suggesting how she should talk to Luther's mother about their upcoming divorce. [*h*]

Mother dear,

   . . . I had a long sweet letter from Mrs. Cressman this morning. Luther decided not to tell them I was in love with someone else. So I said nothing of it in answering her, but simply emphasized the fact that the way I cared about Luther wasn't right for marriage. I didn't talk about my arms.[26] I don't think he did, but I don't know. He wrote them from the boat. I think perhaps Mrs. Cressman will try to talk it over with you. If she does, remember these things. Say nothing of Ray.[27] Emphasize how *abnormally* I've always behaved—not taken Luther's name, gone off for 15 months at a stretch, never had a real home, etc. Be sure to tell her how fond you all are of Luther—that it is in no sense his fault, but just bad luck. I was too young when I was engaged, engaged too long, etc. They can't quite believe there is no hope. Elizabeth says Grandma thought this was coming. We have a lovely airy apartment on W. 124th Street.

    Lovingly,

    Margaret

   ***

To her mother, Emily Fogg Mead, December 20, 1927, from New York. Richard decided to follow in his father's footsteps and get his Ph.D. in economics. Eventually he would end up in an academic job—first at the University of Pennsylvania, then in California—but it meant putting music aside. Mead felt he was making the wrong choice.

Dear Mother,

. . . I have never made any criticism of my brother's brains. I still think that for someone who never reads a book unless he has to, to teach in a University is a decided misfit, which may be attended with dire consequences for himself, even more than for his students. University competition is very keen these days, research is demanded of everyone, only those who feel a particular set towards such things should go into it. I still have to read more for my workers education teaching, after years of much more intense specialization than Richard has indulged in, than he reads in weeks. If he enjoyed teaching even it would be different, for sometimes very unscholarly people make good teachers better than the scholarly. But he doesn't.

My judgment on my brother is purely that I find him temperamentally unexciting, much less exciting than either of the girls. I don't doubt that I would have found many of the great minds of history unexciting. Richard has better brains than many of my friends; I don't pick my friends for brains, I pick them for temperament. I dare say it's very irksome occasionally that I look at my family as impersonally as I look at people to whom I am unrelated, but you might remember also that my favorable judgments are also impersonal, and not based on any sentimental theories about blood being thicker than water. I am very fond of my family, as individuals, and that is, I venture to say, a more flattering affection than Dadda's feeling for Uncle Alfred.

I know that you have always resented the fact that I haven't cared about my family on the purely accidental count of blood relationship. Unfortunately, I fail to see that blood relationship entitles anyone to special affection. It is fortunate that I happen to feel a good deal of special affection for you all, so much in fact, that it is only upon occasions when my judgment happens to clash with yours, that you bring up the purely extraneous matter of "being nice about my brother." Being nice about anyone on such a count I would regard as pure dishonesty, and on the whole so would you. As far as my feeling for what you and Dadda have done for me, that also is a recognition of the affection and care you have given me as myself; inasmuch as it is given because I happen to be your daughter, I value it that much less. Anyway I don't regard the truth about any member of my immediate family as of such a discreditable nature that I have to "be nice" and suppress it.

I am delighted to see that your letter breathes such a strong sense of family solidarity, of complete acceptance of Dadda's choices. That no doubt means that we will have a very happy Christmas, one in which you needn't regard me as an alien, for I refuse to be one. I am just the catalytic agent which precipitates a large number of irrational loyalties, to the peace and well-being of all of you, I trust.

I am planning to go to the Christmas meetings and visit Aunt Fanny, probably. You asked nagging for this letter, so don't take it to heart.

Affectionately,

Margaret

[Handwritten postscript] "Love and kindness" have nothing to do with matters like memory spans and interest in abstract thinking!

To Ruth Benedict, February 21, 1929. Mead was in the Admiralty Islands on a field trip to Manus with her second husband, Reo Fortune. She commented on her feelings about her father at a time when Priscilla was butting heads with him. [h]

. . . Father wrote me at last—all innocent. I suppose I found it easier to grow up because (a) I knew what I wanted—exactly and (b) because my emotional ties were always outside the home, things home couldn't touch. Then I suppose I intrigued more—I made Father pay me to come to New York [Barnard] you know, a $20 a month raise. Also I gave the impression of being good and unselfish and thoughtful while Priscilla does her best to create the opposite impression even when she is good. And I hardly ever wanted things which involved money. A change in my college allowance on coming to Barnard was the one financial point I ever tried to make and that was attended as a bribe not demanded by me as a right. I never wanted clothes, nor trips, no marcel waves, nor manicures, nor jewelry, nor a fine coat—instead I wanted to buy "good books" and was content to buy them at Lury's. In other words, I was in the family tradition and Priscilla unfortunately hits them on their sorest points—she hits father's stinginess about luxuries and objection to demanding females, and mother's intellectual snobbishness. Elizabeth gets much more of what she wants by a natural harmoniousness plus a kind of innocent craft which Priscilla lacks completely. Priscilla showed her fondness for telling unpleasant truths about her motives from the time she was a baby. . . .

To her father, October 16, 1930, from New York. Mead was anonymously financing her husband's portion of their next Pacific field trip with her savings by getting her father to donate her own money to the Department of Anthropology at Columbia University. She wanted Reo to think he was getting a fellowship from the department.

Dadda dear, I meant to write to you sooner with thanks but Reo has been having a bad time with his teeth and the scene got complicated. Thanks so very much for getting that letter through and in time. Papa Franz was able to present it to the Council on Monday and it was passed on, so the whole matter is assured. In case of any question you will always be the anonymous donor which will make no complications. But anyway there won't be any questions. Papa Franz doesn't look gift horses in the mouth. And I am really not being so rash as you think. You see during that two years I will have $5000 salary for which I need give no accounting but which will get expended on the expenses of the trip, none the less. By substituting three thousand in this way, it improves the trip, gives Reo the Columbia backing which they are powerless to give alone, not having any money—this is outside doubling by a foundation—gives us $3000 more than we would have had otherwise and leaves us no poorer at the end. It's all very convenient and I do appreciate your doing it. When are you coming over to talk about the book?

Lovingly,
Margaret

To her mother, Emily Fogg Mead, May 28, 1932, from Alitoa, New Guinea, where Mead and her second husband, Reo Fortune, were working among the Arapesh people.

Mother dear, Once and so often I have to write you a bread and butter letter for my education and this [is] one of the occasions when such a letter is due. Now it is the question of how seriously you took the theory that children should be taught all sorts of arts and crafts. There are probably many kinds of lives in which I should have been only a little less clumsy and a little more casually skilled, and the benefits might have ended there, but for this special job which I spend my life on, nothing could have been better preparation. Whether it's shades of Mr. Stringer's carpentry which enables me to make stretchers for drying prints, or memories of a little gray book called "Construction work" which makes me make envelopes when I have none the right size for negatives, or whether I want to copy bark paintings which won't keep, or draw a sago working apparatus—that training is always there, making me twice as efficient as I would have been otherwise. And if you hadn't taken the business of educating your children so seriously, and hadn't utilized every passing craftsman in the places we lived in, I'd be without it all now. So many thanks, sweet little mother.

Your letters are delightful, only when you occasionally refer to yourself as an "old lady" it sounds so ridiculous, sandwiched in between an interest in everything in the world.

This typing looks as if it had the measles but it has been only by endless pains that I have gotten this blasted typewriter to write as black as this.

I was delighted to hear of Richard's marriage,[28] I'll write him and her both for this mail. Aunt Fanny wrote of how sweet and dignified you were not to resent the way it was done, which made me realise as I mentally framed an answer to her, how extraordinarily free you are from stereotyped, traditional attitudes whenever anything important is at stake. You get right to the heart of the matter without any posing, or posturing. It is one of the refreshing things about you. The point being that Richard appears to be happily married and other matters are after all mere superfluous trimmings. We probably couldn't run a world on such direct and simple lines—we need a fair amount of the flumduddle of traditions and prejudice to make the nasty natured people behave with a modicum of decency—but that doesn't make the way you take things stand out any the less finely. Of course it was the same thing in the way you and Aunt Fanny treated Aunt Mabel. Aunt Fanny will be lovely, if she feels inclined or if all the cards of etiquette require it, but let someone make a mis-deal, and she will take advantage of the fact. All of which needn't be quoted.

Incidentally Aunt Fanny writes with great enthusiasm of Priscilla, so that little impasse is settled. I wonder if she guesses that it was her criticisms of Priscilla which made me stay in Madison last spring instead of stopping over in Chicago. Anyway that battle seems to be won. . . .

June 7, 1932

. . . I had no letters from any of my family this mail, except you, and one from Dadda written ages ago and forgotten, so I've a mind to discipline them. . . .

Goodbye for this mail, and I do hope you are having a nice summer. Reo sends his love.

Affectionately,

Margaret

Mead returned home from New Guinea in the fall of 1933 and did no field work again until 1936. That 1931–33 New Guinea trip also saw the breakdown of her marriage with Reo Fortune and the beginning of her relationship with anthropologist Gregory Bateson, who would become her third husband. These events are explored more fully in Chapters 2 and 3.

In the summer of 1934, Mead participated in the Hanover Seminar on Human Relations, a month-long interdisciplinary conference held in Hanover, New Hampshire. She wrote to her mother, May 5, 1934, from New York, and asked for her help in preparing research materials for the seminar.

Dear Mother,

You are an angel to suggest making so many plans for me. What I want is this. The Hanover conference is a conference to prepare materials for use in adolescent education on Human relations. I am responsible for anthropological material. I [would] like to get a group of girls, or boys, but I think I prefer girls, however if I had several groups, I'd like some of each sex. I want to get as wide a sampling as possible, different social status, different intelligence, etc. Here I have had children from the big private schools, Dalton, Ethical Culture, Birch Wathen, and the extreme opposite, very poor Jewish public high school. I'd like some in between these two if possible, white collar class. I prefer the most intelligent groups and one of two schemes, either that I give a lecture first, 20 to 30 minutes and then have questions afterwards, or a pretty small group, not more than fifteen, for discussion only. If you could get me several, for Monday morning, Monday afternoon, Monday evening, Tuesday morning, Tuesday after-noon and Tuesday evening—and leave one vacancy there for your group that you wanted me to talk to, why I could stay over then until Wednesday morning. If you can't get that many, concentrate close to Monday. Make any sort of deal you like with the schools, an assembly talk with slides in re-turn for a class hour later with questions, etc. You have far better contacts for this sort of thing in Philadelphia than I have in New York and I would very much appreciate it.

I am sorry Dadda was worried about the Bank. Richard has just paid me back the money he borrowed and I won't need it for a couple of months and I was just considering whether it was wise to put it in Doylestown or whether I'd better start an account in a bigger bank in New York. That was all. Richard paid me back promptly the first of May, on the dot.

We'll not be home until very late on the night of the 11th if at all that night. Elizabeth is coming with me. I have to speak at Briarcliff College that evening.

> Lovingly,
> Margaret

Mead seemed to be asking the impossible, but she knew her mother. In her re-ply three days later, dated May 8, 1934, Emily Fogg Mead had two groups al-ready arranged and four others as possible, plus her daughter's talk to her Shakespeare Group and a possible luncheon.

Mead saw herself as a keystone in dealing with family problems. Here, in parts of two consecutive post-Thanksgiving letters to Gregory Bateson, she spoke of this role.

To Gregory Bateson, December 3, 1934.

My darling,

. . . I went home for the weekend following Thanksgiving, with a horrid lot of work to do, and then got much less done than I had intended, because the family needed so much attention. Dick is sleep walking again and I had to work on that; I've set up a three-fold diary system of reporting on his interviews with his father and the results, for that is undoubtedly the crux of the matter. Then we had a family dinner which bored Father stiff and Elizabeth took advantage of that to put over a trip for Father and Mother to the West Indies; more politics. Then father needed a lot of encouragement about a book that he is thinking of writing on obsolescence. . . .

November 29, 1935.

My darling,

. . . Thanksgiving at home was lovely, the family in better shape than they have been in the last ten years. Dick looking happy and well, all the asymmetry gone from his face and it like his 16-year-old face, Priscilla placidly married, Elizabeth radiantly excited about life, both Father and Mother contented and peaceful. I couldn't have hoped to have left them all in such good shape, or at a time when they seemed to need me less. Even though I had cramps, Helene had had an umbrella drip down her neck for an hour at the game, and Elizabeth was half asleep from having danced all the night before—it was lovely. . . .

In this letter to her parents, written on the eve of sailing from New York to Bali in early 1936, Mead deliberately lied about her intent to meet and marry Gregory Bateson, en route. This letter is directed at her father, as her mother already knew about Bateson. Mead feared that her father would react badly and so held off telling him until after the marriage.

Dear Dadda and Mother,

I am writing this rather than saying it because I do not want to talk about it. It is just to say that Reo and I are now divorced and I will be working alone in the field this time. It is too bad that it didn't work out better, but the difference in our backgrounds, the depression and my fortuitous fame, were too much for any marriage to stand quietly. You don't need to worry about me, Bali is a safe, tame place, with doctors and motor buses and I will be able to work quite well there. I am sorry to have to tell you things which will upset you, but this seemed the simplest way to do it.

I love you both very much.

Lovingly,
Margaret

Her mother wrote to her on February 6, 1936, that her father was not upset or shocked by the subsequent news of her marriage and that all was well.

To her father, August 27, 1936 from Bajoeng Gedé, Bali. An ongoing family conflict in the 1930s existed between her brother Richard and their father over the farm their father owned and that Richard managed. Many of Mead's letters in the next few years to her parents and Richard included advice on how to deal with the problems relating to the farm.

Dear Dadda,

I got your two letters almost together, just as I was busy composing an answer to the first one. I am sorry your summer was so stormy. It isn't often that a member of your sex has a chance to find out what it feels like to be treated like a *mother*-in-law. Any time you want to blow up, blow up to me, I won't tell. I am in fact the safest confidant in the world at present.

You should have seen me this afternoon, perched on a bamboo bed in a dark little house, while some ten people squatted at my feet, discussing such weighty questions as did Gregory and I belong to the same caste. Gregory has already explained that he belongs to the caste whose chief concern is with the making of books. So I told them, yes, and our fathers before us, which of course is the necessary point. I had gone to visit the house because one of the daughters thereof has just eloped, theoretically been abducted. If only the boy wants to marry the girl, it costs one cow; if the girl is suspected of being willing, it costs two.

I know that balancing the farm and Richard on the head of a needle like two angels is difficult, but on the whole it has been worthwhile, hasn't it? Richard is so much better and so much happier.

I hope to get a bulletin[29] done soon, but I am not a very ready correspondent these days. There is so much typing to be done everyday as part of the work, that I find myself in anything but a typing mood by evening. And there are always all the variegated struggles with Javanese doctors, Balinese officials, Dutch shops, etc. to fill up the time allotted to letter writing.

This is the most comfortable fieldwork I have ever done, no malaria, no heat, no mosquitoes, good fresh food, and no need to push anyone. These people simply work all the time, rather like breathing, and without realising that they are supposed to be doing something unpleasant. Instead of having to chase and bribe informants, they come and say: "Now about that temple land that is distributed to old people. You said three days ago that you were going to write something about that and you haven't." We had some days of a driving hot wind which made some of the pillars of our house crack, and we have a few little earthquakes every evening, but the house still stands. The people do nothing exciting, but they do a lot of things.

The house has been familiarly full of proof of Gregory's book *Naven* which should be out in a month or so. I was reminded of conversations heard in my early infancy: "Sherwood, you need five more words in this paragraph." "Well, write them." Next week I will have to forget the Balinese myself for a bit and work on my Arapesh mss. which Wissler[30] has reversed the order of.

Now that Mother is home again and it is autumn, everything should be going well again with you.

> Lovingly,
> Margaret

P.S. Tell Mother NOT to try to publish any of my letters. I'll write her in a day or so and explain why.[31]

To her mother from Bali, December 10, 1936. As Mead approached her thirty-fifth birthday, she reflected upon her resemblance to her mother. In an earlier letter, October 8, 1934, she had written, "I am feeling very much like you in several ways, I don't know why, perhaps just feeling a little grown up and remembering your gestures when I was a child." In this letter, her analysis was much more trenchant.

Dear Mother,

You know you have often said that sometimes you reminded yourself of your mother and how you disliked it and that you hoped I never felt the same way. To which I didn't say anything because it was true that sometimes in the last few years I have caught a note in my voice which I disliked and which at the same time reminded me of you. But just recently I have noticed again that my voice was reminding me of you and that I did not dislike the reminder. So I set to work to analyze it. And I think the point is not that one dislikes being reminded of one's resemblance to one's mother, but that one dislikes being reminded of attitudes in one's parent's adjustment to life which one dislikes. The tones in my voice which I used to dislike were ones of frustration, when I would complain, in a semi-martyred tone, that ofcourse I would give up this informant too, if Reo wanted him, but really I had to have some kind of informant to do my work, etc. etc. Now, the tones which I recognize, are cheerful, "Yes dear, I'll be there in a minute," "Just a minute and I'll get it for you darling," which represent an adjustment in my marriage which is similar to part of your adjustment, but which is happy.

It seems to me that Dadda watching me now would say even more distinctly that I was like you—in these respects—and I like being like you. But if you have had a sort of nightmare fear that your daughters would feel as you do about your mother, this may relieve your mind a little. When we detect notes of failure in our lives which happen to repeat notes of failure or frustration in yours, we will probably dislike them, and rightly enough, although it is on the whole a superficial matter then to emphasize the resemblance. But when we detect notes of adjustment and content and enthusiasm which echo yours, then we will be happy, and again make the identification that we "sound like mother" or "look like mother."

. . . Norah[32] went home five days ago—flying to England—and now we are back in Bajoeng and hard at work on photographs, texts, collecting, feasts, medicine and what have you again. It is very good to be home again; our palace was very handsome but not meant to work in, and tearing about all over Bali can be very tiresome. Gregory's mother is still here, quite contented to sit in the dining room and embroider, with the occasional diversion of someone come to sell puppets or carvings. She is easy to please, and it is only necessary to convince her that we are doing important work and that everything else in life is subordinate to it and she is quite contented, for her whole interest is that Gregory should represent his father as much as possible.

So she accepts the disappearance of every wash basin because they are being used for developing photographs with great equanimity. There is not a great deal of conversation—she has a mind rather like Priscilla's but with less ambition, and she doesn't really understand our shop, although she did understand her husband's shop quite well. But you can imagine what a pickle you would be in in terms of vocabulary alone, if you had a child who became an experimental biologist, for instance. She lacks both intensity and enthusiasm, but has a brisk good humour which is a help. Gregory and I find that it adds a little spice to our lives to have someone to define the difference between time to ourselves and not, because with the natives it is almost equivalent to being alone, and we work together so much that life hardly holds a break, unless another white person is here.

I have put further particulars of the costumes, etc. on the enclosed sheet for your guidance.

Lovingly,
Margaret

To her youngest sibling, Priscilla, May 16, 1938, from Tambunum Village along the Sepik River in New Guinea, where Mead and Bateson had traveled from Bali. Priscilla had just had her first child, Philip, and had asked her sister's advice as an expert who had studied childbirth and child raising, although Margaret would not have her own daughter for over another year. This was Mead's response.

Priscilla dear, This mail brought the first news that I have had since the cable. I am glad you told me the names before hand, so that I could immediately name him in my mind, and the picture makes him seem quite real. I have no affectivity at all about Caesarians, but I do think they may make it easier for the baby! Is it really true that you can only have three children? I know I used to believe that and was corrected by a New York doctor who said he knew of a woman who had had seven! But likely three will be a good nursery full anyway. And I agree that it is much much better for the eldest child to be a boy, probably anywhere, certainly IOC.[33]

. . . About weaning, the great points are, I think, that it should be gradual, not strictly progressive, and without guilt on your part. The gradualness is produced by gradually decreased nursings, and increased bottles, but bottles given holding the child. Even sometimes I think it might be worth while to give both together, half a feeding from each medium, as one breast, and then turn to the bottle rather than to the other breast. (Or does your doctor make you nurse only from one breast at a feeding? Remember my experience is so much of natural man, that I fail to take into account current pediatric prejudices.) Then the routine should not be iron absolute; say you have got him half on a bottle and he is teething or has a cold, then more nursing is indicated. Once he gets through the little crisis and is healthy, more bottles. I know this is harder on the mother and doesn't permit the doctor to give such nice cut and dried rules, but it means treating the child like a living and adaptable creature, not a machine.

All the way through the child's development you will see periods of regression, of a falling off of gains before the strain of acquiring a different skill, or developing a new sense, or before teething or illness. These periods should be met in the half weaned baby with more suckling; in the older child, who is already eating a lot of food, with more bottles.

One of our sternest moral attitudes towards children is that *they must grow up* and this attitude fails to take into account the discontinuous and uneven character of development, the fact that the child must regress in one way, while he consolidates new gains in another. These periods of temporary and non-significant regression may be met with the various accompaniments of suckling; suckling, holding, a bottle later, and I think, in many cases, with giving the child the dry breast occasionally as comfort, after the mother no longer has any milk. . . .

I have rather stolen the time to write this, as I am supposed to be writing up notes. But Gregory is away watching the division of a pig, and I find I need his notes to make the write-ups of yesterday's events, so I am meanwhile playing hooky. We are camping in the government rest house, while our new houses are being built, and living inside a big mosquito net. I've written a long bulletin which Mother will send you which will provide you with all the current facts. Do keep on sending pictures, so that I can feel that I know something about Philip as he waxes and thrives.

Before Philip was born, I gather that Dadda behaved more like a husband than a grandfather, getting a little too sulky with *Mother's* preoccupation with the new baby! I don't think that too many people are hard on a baby, IF they all handle him in the same tone of voice. Ten people who use the same tone of voice and same muscular adjustment will do much less harm, than two people who use radically different systems. Crowds of-course, or too much play, especially play with adults when they try to wring too much emotional satisfaction from the child, can be bad, but not merely a group of accustomed people.

I am writing Lee [Leo Rosten] separately.

Love to you all,

To her sister Elizabeth, June 16, 1938, from New Guinea. Elizabeth had asked her advice on whether she and her husband, well-known artist and car-toonist Bill Steig, should be psychoanalyzed, and her advice on how to bring their father and her husband together.

Elizabeth darling, today is your birthday, your twenty-ninth isn't it, and how good it is to be as young as you are at twenty-nine, with the world still ahead of you. Next year, when you are thirty, you will find it correct to have very staid and sober thoughts—at thirty I decided I would never do any-thing better or different from what I was doing, and what a bore that was—but this year, you are still in your twenties, the very best year of them because with the joy of the twenties you almost hit the wisdom of the thir-ties. So be as happy as you can, my love, and don't live too much under the shadow of coming events.

Coming events, seen in retrospect, have never been very hot. What if you had been a bride in 1860, or a mother in 1935; what if you had been a Euro-pean bride in 1914, or an English mother in 1895. You have read *Joan and Peter*,[34] haven't you, in which Wells describes the beautiful safe world, in which all the nurseries had rounded corners and science was going to solve every problem, science and humanitarianism. Just the age in which to be hopeful and happy, to breed freely for the great new world. And the boys who were born into those English nurseries so lighted with hope, are dead.

We can't know, anymore than the parents of the 1890s knew, what the fu-ture of twenty-five years from now holds. Which isn't an argument for your having children, if you don't want them, and if Bill doesn't want them.

There is something to be said for a cross-racial marriage[35] avoiding children, if the husband and wife have rich enough ties, not to need them. And Bill has such heavy responsibilities now. But don't dramatize too heavily the evil to come, even as a dark cloud against which the present looks so golden. You may stampede yourself into a position on politics which has actually nothing to do with them at all, but is simply an enriching personal device for living more intensely and doing better work.

As for analysis, I haven't really anything new to say. If Bill wants to risk his whole creative gift so as to have his family and you get on his nerves a little less, I don't suppose anything can be done about [it]. It is about as hazardous a course as it would be for him to go into an occupation where the casualties for the loss of the right hand were 75%, or a singer with a beautiful voice who deliberately went out without a coat at 10 below zero. Contentment can be bought at a price that one can not possibly pay.

As for you, it is not so serious, I don't suppose it would do you a great deal of good, but neither need it do you a great deal of harm. You are far enough from having hit your stride artistically, to make analysis possibly a releasing experience, as falling in love, or a religious conversion, or going temporarily insane might be a releasing experience. From that angle they are all about on a par.

It is only to the artist who has reached some part of his stature, who has definitely started to grow in a direction fixed enough so that to break it, may break him, that I feel analysis is so dangerous. Arthur[36] was much younger than Bill; he had not any art that really mattered yet. . . . [orig] And for you, yes, I would trust Horney[37] more than anyone else; she is wise and kind and warm and intelligent. One can't ask for more. I can't conceive of her hurting a patient.

As for Dadda and Bill's family. Darling, are you too nearsighted, too close to it, not to see the analogy between you and Dick in this case? Both of you have married into an ethos which is alien to your own family ethos;[38] with part of yourselves you are still loyal to that ethos, and to make your marriages wholehearted you have to beat down that loyalty to a position where it doesn't threaten the wholeheartedness with which you want to embrace your new life. There is no reason on earth why Bill and Dadda shouldn't like and enjoy each other, as much as Lee and Dadda, or Reo and Dadda enjoyed each other, except in yourself.

There is however one way in which your position differs from Dick's. Dick has married into an ethos which he was taught was *lower* than his own family, so he is defensive and his family tend to resent his embracing it, and so to resent Helene. You, on the other hand, have married into an ethos which you have been taught to regard as *higher*, or at least much rarer and more precious, than the ethos of your own family, therefore you are inclined to feel superior, defensive of your family rather than of your husband, and your parents instead of resenting Bill, are enviously and wistfully anxious to partake a little of the rare atmosphere into which you have married.

Priscilla and Lee, Gregory and I, are not in that position, as the ethoses are so very similar. But with Luther, I was in that position. When I wanted to embrace the religious life I tried to do it by making Luther out to be as different from and as incompatible with Dadda as possible—there could be nothing in common for them, I insisted, and ofcourse they never got on, never learned to respect each other.

There is no doubt that Reo and Gregory would have got on perfectly well, if not too congenially, if they hadn't been representative, each one, of a conflict in me which I was present to exaggerate. I can think of dozens of points on which Dadda should be able to get on with all of your in-laws. But you don't feel safe in letting him do it. And it doesn't matter really, if only everyone understands what the reasons are and so doesn't feel guilty. Bill probably doesn't care at all, I doubt if he's sentimental about knowing his father-in-law, he already has plenty of relatives; Dadda doesn't represent a world to which he wants to belong, as he does to Lee, or a world to which he does belong, as to Gregory. . . .

Only if you fuss and worry, get annoyed with Dadda, or hurt with Bill, or guilty about yourself, need it matter at all. Especially as long as you have no children; Priscilla has provided a perfectly good grandchild and in all her sentimentality about home ties, she will let his grandfather enjoy him.

The other contributor to the schismogenesis[39] between Bill's family and ours, is ofcourse Mother. She consciously deprecates the fact that she gets on with Bill better than your father, but secretly she is proud, because there has always been a side to her which wanted what she would call "the finer things of life," for which she feels Bill and your whole way of life stand.

It has always been a point against Dadda that he had no educated taste in the arts, although he has a more intense appreciation of great prose than mother has of any art, and she is, without knowing it, using your new life to make the old point over again. That can't be helped either. I am willing to badger them about Dick because I think he is too vulnerable to stand on his own feet, and because they live in the same town. But after all they are both over sixty and it's no use bothering them too much.

I have explained to Mother the conflicting ethos point, and how it is natural to expect that Priscilla and I qua our husbands, will take a different course than you and Richard. Get it thought out yourself and get Bill to see the point so that he can remind you of it when you have a fit and forget. And then forget it.

You're probably right about letting yourself go and talking out depressions at home and that is a little hard on them, because they only have the consciousness of your happiness to go [on], a consciousness which I always get through all the ups and downs, ends of the world and moments of Creation, in your letters. And they don't understand the artistic temperament nor the role that intense feeling plays in keeping creativeness alive. . . .

My dearest love to you both.

To her mother, July 10, 1938, from New Guinea. Mead wrote about problems with her brother Richard.

Mother dear,

. . . I think you would feel less annoyance over expenditures on Richard if you compared it with how you would have felt if he had been made an invalid, say, in a motor accident as a child, and been completely dependent. Then you would have counted in the expenses of wheel chair and nurse, of no plan made which he couldn't be fitted into, as a normal expenditure which you would have made perfectly happily. The present expenditures are made because he is to some extent psychically invalided, always a less obvious call on sympathy than physical invalidism.

But ofcourse actually what you are resenting is another expenditure which used to [be] pigeon-holed under headings as follows; briquettes, the creek, Trenton, the Farm—i.e. an interest of Dadda's which you don't share, and which somehow can be made responsible for the fact that Dadda and you don't take trips together, which again is a symbol for the fact that you and he have separate interests and different ideas of how to enjoy yourselves. But the creek and the farm and Trenton had no feelings and Richard and Helene have nothing but FEELINGS, and therefore the accidental labeling of an annoying fact with which you have lived for 38 years, with their names, might do some harm.

Well, this is preachy, I'd better get on to more appropriate subjects, but after all, parents and children in our family have always had a reciprocal rather than a complementary relationship. I know I forgot to put the house plan in the bulletin; I'll try to get it done for this mail. But Gregory has a bad cold and the boys have colds and all the village has colds and we've had an epidemic of Europeans again, 4 in twenty-four hours, and we are trying to get ready to go and hunt crocodiles.

I think I shall have to write another book very soon, or *Sex and Temperament* will be used for very wrong uses.[40] As education is in no way adjusted to the kind of differences which I discuss as between individuals rather than between the sexes, the conclusions of *Sex and Temperament* are entirely irrelevant to the issue. Our educational system is at present concerned with character structure not with elaborations of temperament, and men and women do have very different potentialities for character structure. I think it is possible that if education were very very differentiated for little children, were almost completely different for boys and girls under five or six, then it might make sense to have identical educational systems in some of the more abstract subjects later. But the assumption that men and women are essentially alike in all respects, or even in the most important ones, is a damaging one, as damaging as the assumption that they are different in ways in which they aren't different, perhaps more so. . . .

All the magazines and newspapers are arriving faithfully, and I am beginning to be a little up to date again. Keep the *NY Times* summaries up, specially I find them very helpful in giving a quick orientation.

This somehow sounds like a cross letter, because it is so controversial. I'm sorry. It's late and frightfully hot and sticky, and I am worried because Gregory has been sick so much since we got here, two bad attacks of fever, a whole succession of small infections, and now this cold, which always leaves him with a trying cough. But I don't mean to be cross, Mother darling, and I love your letters and I like arguing with you, and after all we are both interested in the same things, but from different temperamental and experiential biases.

No, don't send newspapers and packages care BPs;[41] nothing sent by post needs to be sent care BPs, only freight, which you aren't likely to be sending. We have no trouble with paying duty on postal packages which come direct.

I hope you *have* had a nice summer,

Lovingly,

[P.S.] Love to Grandma Fogg.

To her sister, Priscilla, June 12, 1939, written from New York City. Mead wrote about her own pregnancy and the difficulties she had getting pregnant.

Priscilla dear,

I have felt really deprived over not having had a chance to write you a long letter, but I am taking it as easily as possible, and that combined with unavoidable appointments, a little writing, etc. and evening and evening of different kinds of specialists who want to look at our movies, and pictures and carvings, makes a pretty full life, especially as I am doing my own cooking.

The movies of Philip came and we are delighted with them. Mother saw them all, and we showed Ruth[42] the bath-tub reel the other night which she enjoyed very much. I find myself almost talking to Philip. I suppose you had them copied before you sent us on the originals?

. . . We lead a sort of Mad Hatter life over this big apartment, abandoning first one devastated tea table and then another, with films everywhere.

Now to answer some of your questions. The Museum had already given me a leave of absence from July first, and Wissler is just informally letting me work at home until then. It was good of him for it makes almost three hundred dollars difference. We are sure that we will stay here until autumn, then if the war cloud lifted more over Europe and the doctors thought it was safe, we may go to England for the winter. There is now a New York Board of Health Law against having the baby in the room with the mother! It is supposed to promote diarrhea in the nursery, where the baby would never go at all! So after finding an obstetrician who was willing to fall in with all funny ideas, he himself found he was powerless.

You will be amused by the way in which I chose my obstetrician. I started backside first and chose a pediatrician, Dr. Benjamin Spock,[43] who has been working on various nursery school projects I've known about, and also taken care of the children of people in the Adolescent Study.[44] Then I went to him, and outlined the points I wanted made about which he would have to have cooperation from the obstetrician from the start, including his presence at the delivery. He said he would be glad to be present but no one had ever asked him before. I think I got that idea from you, didn't I? I think it's a very good one, like having the father's relatives to concentrate on the newborn and the mother's relatives to concentrate on the mother, in primitive society.

He selected Heaton, Clyde Heaton, the author of *Modern Motherhood*, as the most flexible-minded obstetrician in New York, who also has a very high reputation as an obstetrician. Perhaps you've seen the book. He is very moderate about anesthetics, wants to give as little as possible, and is writing a history of medicine in America and wants me to criticize the Indian section, etc. He figures that a baby should cost about ten per cent of the year's income, that therefore the whole thing is less than $500, should be $400. I am to go to the French Hospital, to a special small more expensive maternity floor, with a nursery on the same floor, and usually only two or three babies. It's the best we've been able to work out so far, as I think home delivery would be silly.

He is giving me Vitamin D, calcium phosphate, and thyroid, and I was taking some Vitamin E already at Eddie's advice. A quart of milk a day, but no other dietary rules. He allows ten cigarettes a day, but I find I smoke very little. I am feeling so well that if I didn't have the word of the rats and two local examinations, I wouldn't believe I was pregnant. I can see now how mothers with unwanted babies just drift along hoping it isn't true until it is too late for an abortion. The first striking changes in the alteration of the breasts no longer claim one's attention, one could easily think it was imagination; lassitude I would ofcourse—not knowing—put down to a disordered period. I daresay that if I had to get up in [a] hustle in the morning to care for an older child, get a husband off to work, or go to work myself, I might have morning sickness and dizziness, etc. I certainly feel less well than I do at night. But Gregory gets me tea, and fries the eggs and I get up just in time to put the coffee cups on the table. He is working partly at home and partly at the Museum, enough out so he doesn't feel calabooshed.

The major dangers of miscarriage seem to have been surmounted; the uterus has turned up alright in spite of the retroversion, and I have passed my third period. I wonder if I don't owe carrying this baby through the first six weeks to the fact that my periodicity was disturbed by that miscarriage just a month or so before, so that my body couldn't quite make up its mind when the menstrual period was—a month from the old menstrual time, or a month from the miscarriage—gave it a nice bit of leeway and me a headache and backache for two months, but it may have saved the baby. Ask David what he thinks, and tell him that the examination he gave me and the history he took were better than anyone else has done since.

However, I am still going nowhere, not riding in trams or buses or trains, and moving as slowly as I can remember. This rather special cocoon sort of life is necessary to make me conscious enough to slow down my naturally abrupt quick pace—for I have no bodily symptoms to keep me aware of the need for caution. But in my dreams I am always pregnant, just as a known condition of life, a matter as taken for granted as my sex.

I have the notes Mother kept before my birth, "prenatal influences"; "Felt pressure of *Encyclopedia Britannica* articles." It seems that, although I had always been told I was starved as [a] baby for mother had too little milk and they had to feed me, Mother took four months to wean me, gradually reducing feedings and substituting a cup, and when she finally weaned me, no trouble at all. That makes more sense as a background for my character.

Tonight, my obstetrician, Tao's[45] obstetricians and some others, are coming to see our baby pictures from New Guinea. I am going to—I don't know what, as Dick just arrived for supper. I'll have to send this off and not wait for a chance to add more. You don't know what a pleasure your furniture is being to me—a comfortable bed, a lovely easy chair and the first real desk that I have ever had at home in my life.

My letters to you will make a sort of diary, for I find although I am willing to put down all the physical things, faithfully, I haven't the necessary introspection to write a diary of moods. —Oh I know what I was going to say—I am going [to have] a movie made of the baby from moment of delivery on—in colour. The colour shows up the points where the circulation is weak; also after 11 months Tao's baby still lies in the position it took as it was lowered to the delivery shelf. She had a colour film of the whole birth, Psyche made it (Myrtle McGraw),[46] a high delivery after turning the baby from the anterior position.

Lovingly,

A few weeks later, to her mother, July 5, 1939, from New York. Her father did not want to go to California to see his daughter Priscilla, her husband, Leo Rosten, and his new grandson, Philip. Here, Mead analyzed differences between her mother and father.

Mother dear,

The indexes arrived safely and I am enclosing a check for $50, which you will then have if you need something extra for Leland.[47]

As for Dadda, you know well enough that he can't be mean and uncooperative without making himself a little sick, just to ease his conscience. He doesn't want to go to California; he's afraid of every strange contact and strange situation. But he hasn't the brass just to say he won't go so he gives himself mild palpitations of the heart so he won't have to. I think it would be just a waste of time for Elizabeth to go and see Dr. Small, for I don't doubt that Dadda's symptoms are genuine, but neither do I doubt that they are purely functional. If Dr. Small told him it was safe for him to travel, then he'd have to get much sicker in order to get some other doctor to say he was too sick to travel. On the whole, he has chosen a very mild disease— just a heart which gives no trouble as long as it is coddled—and I'd be glad he hadn't thought it necessary to get stomach ulcers, and let it go at that. It means, I think, that he doesn't feel very guilty about not going, as long as he has sent you. If he felt guiltier he'd have to get a more active and unpleasant illness. And for Priscilla, the main thing was your going out. So I'd leave it at that.

You know, you ought to know, that you can't make him do something that he doesn't want to do, all you can do is to make him sick. I'd even sympathize with him, mildly, tell him to be sure to be very careful, etc. because otherwise he may make himself sicker in order to get you to come home, or take more interest in him. So I'd pretend to worry, but I wouldn't worry really. It's no good comparing him with Grandma. Grandma wasn't afraid of life; he is. You go ahead and have a good time, and try, just for a change, accepting your husband as he is. You are always interested in new experiences, and that would certainly give your life an entirely new flavor, like falling in love with a different man.

Priscilla's letter came while I was writing this.

Richard was over just before Dadda, and we had a very pleasant time. Gregory likes him, and it all went off very well. I hope he gets the Indiana job.

Lovingly,

P.S. I am very well, gently swelling a little, but otherwise perfectly normal.

To her brother Richard, August 19, 1941, from New York. Mead sought his views about the family farm he managed for their father.

Dear Dick,

I gather that with Mother away, the farm is occupying a central position in Dadda's anxiety system at the moment. As you are one of the people most likely to be affected by his decisions, I'd like very much to know what you think about it, what you think ought to be done, how much more you yourself want to go on with it. I can imagine that for the time and effort and fuss which goes into the farm now, you could very easily do something as or more remunerative which you would enjoy more. As I remember one of the original points about your participation in the farm was how good it was for [you] to do some outdoor work in summer as an escape from the petty worries of the University. But they aren't as worrisome any more, and it doesn't make sense ofcourse for Helene to think of living on a farm or anywhere where life is so hard, physically. In what terms then does it still make sense to you? I know that you are genuinely concerned that Dadda should save some money for his post-retirement years, and that the question is how is this best to be done. Would you take a moment and write me please, confidentially, what you and Helene think about the matter? I will probably see Dadda in New York before I have a chance to come to Philadelphia and see you, that is the reason I am asking you to write.

It seems a long time since I have heard any news of you or Helene. With Mother away we don't get news.

It's been a pretty hectic summer. I have had a grueling, full-time job to do in the day time, organizing a study in medicine and social work which turned out to be much more time consuming than I had bargained for. In the evenings we've worked at selecting pictures for a monograph on Bali and Gregory has spent all day printing photographs. Teaching the baby to feed herself and preserving a rather unstable equilibrium have about completed my time scale. And at that, I have just managed to pay for the summer.

I tried to lure Dadda up here but he wouldn't be lured.

Love to Helene and Roma,[48]

Lovingly,

To her mother, August 11, 1945, from New Hampshire. Mead enlisted her parents to do research for her on attitudes about the atomic bomb.

Mother dear,

Cathy,[49] Marie [Eichelberger] and I all safely up here and the weather is gorgeous. I have decided not to try to finish my book this summer for I want to know how much difference the atomic bomb is going to make in people's consciousness. Could you do as much interviewing for me as possible, just get anyone you are talking to to talk about it, and write down the key things they say? First write down what you thought yourself and what Dadda thought. I need this badly. . . .

Lovingly,

Margaret

To her friend Rhoda Metraux, January 23, 1949, from Philadelphia.[50] [*h*]

Rhoda darling,

. . . I am writing this in a hospital room in Philadelphia sitting by my mother's bed. She had a stroke on Friday—right arm and speech were affected, but seems to be coming out of it, pretty well. She's quite rational, a little extra mischievous—and humourous about herself—and very sweet. There are of-course terrific problems to face—what is to be done when she comes out of the hospital—generally dealing with father, etc. Father is at present just burst-ing with pride in himself for "not minding" the attack, for being able to "bear to do anything for the dear little woman"—so sweet—and yet, somehow—so unbelievably insensitive. He doesn't understand a word she says—and it is very difficult—for none of the consonants are back yet and one has to be ter-ribly attentive—and guess at a lot and ofcourse be willing to guess and be wrong if necessary.

And I feel the exclusion of the other children so. He definitely didn't want any of the others here—and he only wanted Elizabeth told. My brother, who went through a long cancer illness and death of his wife, is not to come be-cause he must—as always—be protected—as weak. Priscilla because he is terrified of her managing him, and Elizabeth because she would be upset. And finally he doesn't want them to see mother while her face is still dis-torted and her speech thick, doesn't trust any of them—except me. And in the end I go ahead and make decisions—decide Priscilla is not to come now, but later when she can stay longer—and Richard is to remain in readiness but not come yet—because both of their arrivals would alarm Mother. And Elizabeth is to come on Monday while I go back to New York and meet Ge-offrey.[51] Why does Geoffrey always arrive in such difficult times?

I sit here marking papers—in the intervals while Mother sleeps—or trying to talk with her—with my head full of images of me as a child, of Cathy, of Mother as a child—of each of us being both mother and child to the other— of Cathy's saying one night when she came up and found Gregory and me both crying and I took her down [to bed] and said "I'm crying because Daddy is going away so soon"—and she turned to me with a beautiful smile and said, "You always have me, Mummy." She has my mother's eyes and smile.

It's curious to realize—watching the fleeting expressions on Mother's face—that she is as confused as Daniel[52] as to what is inside and what is out-side. For a contraction in her gut, or a bit of gas, she looks out the window— or into the corner just the way a baby does—and then looks so miserably anxious because she can't localize it.

It's very reassuring to watch the return of familiarity, though—her face smoothing out—more control so she can drink—the enormous comfort of getting back control over elimination, and all the little disorientations to think of—no glasses, they were broken. No plate yet back in her mouth—and she fusses about that. The nurses putting her slight deafness down to her condition because no one had told them she was a little deaf. It really seems to me a major miracle that anyone does get well in a hospital—just from the terrible unfamiliarity—the lack of anyone who really knows who you are.

But perhaps very few people care about other people knowing who they are—and feel quite safe if there is respect, or competence, or function in other people's voices.

. . . The next few days are necessarily very indeterminate. I must go to California on Wednesday if possible, if only for two days. I promised them a year ago. But it isn't like leaving Ruth to give that lecture, because then it was only my feeling[s] to be considered.[53] Here there is no one to take my place and be with father if mother should die and California—so far away.

Also I'm afraid this will reactivate all Geoffrey's anxiety about leaving his mother.

A lot of things have become clearer these last two or three weeks. . . . I do tend to disregard neurotic manifestations as unimportant—just distortion—not the "real thing." I want to react to what was in the unconscious at birth—not to the miscellaneous truck moved in after birth.

So, I think I said, watching my father: Here is an impaired human being—his aggression is reactive to a sense of weakness and sexual inadequacy and he is a *destructive* father—I don't want to marry a man who is so organized. I want to marry a man who is gentle and virile and constructive. I wasn't fixated—I didn't lose my feeling for men—or for what he might have been—but I did become libidinally withdrawn from the special reaction formation of the first phallic phase. And whenever I encounter that in a man—who is or may be a father—I withdraw. Probably that is very important about Reo because I only consented to marry him when I had been told I could have no children. But its appearance turns me from a wife who is to be a mother into something between an older sister and a mother. It all came out sickening really in my relation to Edward,[54] in which I said "He will try to destroy his children and I will have to fight for them and it would be hopeless." I suppose ultimately too I am most moved by the man whom I see as a father of a child.

Now if you like it is neurotic in me—in the sense of an over-statement, an excessive investment of libido at a particular phase so that I shrivel from phallic fixated men and also phallic identified women—but it can't provide any motivation for analysis because I went on quite happily to acceptance of genital sexuality. . . .

Now suppose that I react positively to gentle considerate tenderness— and you know how I did—because it makes me feel like myself—like a woman and like a mother. . . . And I think it is very significant that I can see this sort of behavior quite well when it is directed against *someone else* and isn't that an indication that I fear—not what may happen to me—but to the child—or potential child?

. . . Father will be here in a moment to relieve me while I go out to lunch. So I'll say goodbye, my darling, while Mother lies sleeping—as quietly as a child. Are we right to resent it when the old become sweet children? . . .

To her friend Joan Erikson, in March 1950. Mead wrote about the death of her mother.

Dear Joan,

. . . It's bitter cold here. My mother died last week just four months be- fore their golden wedding, but leaving everyone else whole and able to manage without her. A year of coming back from a stroke, learning to do everything again, and at the same time letting Father learn, through looking after her, how to look after himself—a few weeks of being herself, and set- ting her house gently in order—and then another stroke from which she never recovered consciousness. . . .

To Elizabeth, August 27, 1953, during her first return field trip to the Manus of Papua New Guinea since her field work among them with Reo For- tune in 1928–29. Mead reflected on relationships, seeing herself, at the age of fifty-one, in a grandmother role to the Manus and to her young anthropological colleagues. She had chosen Ted Schwartz, a graduate student from the Univer-

sity of Pennsylvania, and his wife, Lenora, a dancer, to accompany her to
Manus on what was their first field trip. They worked in a nearby village while
Mead worked in Pere.

Elizabeth darling, It was wonderful to get such a happy letter and the news of the license[55] all at once. So now you really are set, aren't you. And I'm so glad the summer turned out so well. Have you done anything more with any report on the Puerto Ricans? If you wrote it up Larry[56] could see that it got to someone who could use it.

I think I have a clue to how to handle the difficulty which you described as due to the fact that I ought to be a queen without a consort, of a sort of distance between my ability and most other people's. I have no fancy for being a queen but I think perhaps instead of acting like everyone's mother, I might move to the grandmother position with the accompaniment of more indulgence and patience—being what grandma was to me, rather than what she was to Dadda and Mother.

Here in this village I am a grandmother, my children's children are adolescent and married in some cases, and I find a great gentleness available for some hot-headed violent boy whom I knew as a baby. The other night I had to find and disarm a man who had run away with the intention of committing suicide and then lead him back to the village—as a loving grandmother it was a very simple role. And I am loving them, there are more of them who are real living intense parts of my life than has ever been the case before. My own particular grandchildren—Ted and Lenora—present problems but they are very good stuff and will come out alive if they set me to tossing about on a sleepless pillow sometimes. Kilipak—the one of our little boys whom I loved best—is a pure northerner,[57] with all the zest and gaity which a Northern can have and ought to have, and so like Ray in many ways that it's uncanny.

I am so glad that everything is going well darling.

Lovingly,

To her father, November 15, 1953, toward the end of her return field work
to the Manus. Her mother had been dead for three years at this point.

Dear Dadda,

This is the last letter that I will be able to get off before leaving Manus, or at least it looks that way at present. There is a flu epidemic of the kind that isn't dangerous to Europeans but is fatal to natives,[58] raging on the New Guinea mainland, so they have imposed a territory-wide ban on any movement from one sub-district to another. This means all the normal movement of small ships, mission, trade and government, and the movements of natives by native canoes, has been interrupted between here and Lorengau. We are not dangerously isolated; we have our own outboard engine which can take us into Lorengau in three hours, but we aren't supposed to go in at present. There is said to be a ship coming through in four or five days, so I am tidying up all my mail and getting letters off on that with the expectation that I won't be writing again before I leave.

I am not going to get home for Christmas, as I am going straight to the WHO[59] Conference which has been moved from Geneva to England and set up for the seventh of January. This means that I will probably get home somewhere around the twentieth of January. . . [orig] well before the New School term begins. Then you and I will start our teaching there. You really had better plan to stay over in New York the nights that you teach. I am sure we can make you comfortable. And I plan to come to some of your lectures—I haven't heard you lecture since I was ten years old, and you gave a lecture in the La Salle Hotel in Chicago—with charts.

Work here is practically done. I am busy making lists and tidying up bits, and rereading all my old notes and papers on Manus to make sure that nothing has been left out or left unchecked. People make speeches about my being like a turtle who is going away off into the deep sea and who knows whether I will ever come back. The sun has shifted so that my little house wind[60] on the beach is now sunny at eleven in the morning instead of at four in the afternoon and has to be taken down and moved; the babies that were born when I came are on the edge of learning to crawl. The job is done, and six months was a good allowance of time for it.

See you before your birthday,

Lovingly,

Margaret

Mead's father died in 1956. She wrote to a friend and anthropological colleague, Ray Birdwhistell, on September 10, 1956, "I am only back for two weeks, a good part of which was taken up with my father's funeral. He died peacefully and painlessly at exactly the right moment in everybody's lives."

To her brother Richard, February 17, 1962, a hurried note with family news. The family had grown smaller: Her youngest sister Priscilla had died in 1959.

Dear Dick,

I've enjoyed your long letters and I'm sorry I've been such a poor answerer. Janet[61] says you will be back in June. Will you be coming though New York? If you are do let me know, as I may be away a lot and I wouldn't want to miss you.

General news not much. I had a profile in the *New Yorker* and that means a lot of mail. Also the newspapers have been rather generously misquoting me this fall, not with malice but with stupidity. Interest in work on peace has stepped up and that means a lot of work.

Family news: Elizabeth got off for Europe with Aunt Beth.[62] Jemmy got married but the marriage didn't last: he grew up, got a job and became responsible, but she didn't. I have her working for me so I can look after [her] a little. Lucy has an apartment and a job.[63] Elizabeth rented her apartment which she will have to give up in July, and will move to her studio. She sold a picture just before she left. I got an honorary degree from Temple (!) University last week, and Ellen and Lockwood Jr. came. Aunt Mabel[64] didn't make it but is as lively as ever.

I go to California in late March–early April and will see all the young and younger ones then.

I'm so delighted that you are gathering material for [an] article when you get back. We'll have a lot to talk about.

Love to Jess,[65] and to you? [sic]

To Aunt Beth, her mother's sister, July 27, 1962, about her own sister Elizabeth.

Dear Aunt Beth,

I am now ready to give you what I think is a well-based opinion about Elizabeth's immediate future. Rhoda[66] has been watching over her very alertly, and I have now been home for two weeks. We couldn't make any definite judgments until there was some idea how all the complex factors, (her acute pain, Jemmy's operation, Lucy's marriage, her reaction to her brief stay in Florence alone) worked out. Yesterday she saw her doctor again, a gynecologist in whom we have very great confidence. He thinks the acute condition has cleared up, will not need an operation although she will have to eat lightly and with care for some time. Jemmy is on his feet well enough to come in and out of town. Bill has definitely and apparently with pleasure taken over the main responsibility for Jemmy and at this point Jemmy needs a father more than a mother. But it is very hard on Elizabeth to have to know he is in town and not know where he is or have any way of doing anything much for him. Lucy is comfortably married and they plan to move to Milan, where he is setting up an office. Elizabeth has explored teachers' agencies again and found those who were negative three years ago are more receptive now that she had had even this few months abroad. It has made her into a person who could do college teaching—in agency minds that is. Ofcourse she would be very good at it. I believe she should teach in a teachers' college, and impart her enthusiasm and resourcefulness to generations of teachers who will in turn affect many school generations of children.

But she is not well enough to start teaching this fall. She is painting with enthusiasm but far from ready to do a day's work in an exacting teaching situation. Her painting is going very well and she might—with a few more months—have enough pictures for a show which would clinch the teachers college teaching and—in any event—give her all the necessary sense of herself as a competent painter ready to teach adults.

It looks as if the best thing for her to do would be to go straight back to Florence for several months. . . [orig] and the sooner the better. She should not fly for any distances ever again, the doctor thinks.

So—for what it is worth—this is my estimate of the situation.

I do hope you are very happy. Sometime, if you are going west, I'd like to have you stop and see the Retirement and Gerontology Center we are developing at Topeka. It's a sort of child guidance center in reverse, where children can come to understand their parents, older people can get a reliable assessment of their powers and limitations, etc. While you have Fanny's problems to think of, this might interest you to look at. I don't mean that Fanny needs it but that she makes some of the things they are trying to do for less sophisticated older people, more vivid.

My regards to Will.[67]

Affectionately,

Mead continued her pattern of helping her siblings and their children in times of trouble, but increasingly letters gave way to telephone conversations. Her brother Richard died about two years before she did of the same disease, pancreatic cancer. Her sister Elizabeth outlived her. Mead herself died in 1978, a few weeks before her seventy-seventh birthday.

2

Husbands: Starved for Likemindedness

❖

Margaret Mead was married three times. Her first husband was Luther Cressman, a young man who was the brother of her science teacher in high school, who became first an Episcopal priest, then a sociologist, and finally a prominent archaeologist of the Pacific Northwest. Mead's second marriage was to Reo Fortune, a New Zealander studying psychology who became a cultural anthropologist. Finally, she married Englishman Gregory Bateson, who was also a cultural anthropologist, but after their divorce worked on communication and problems of ecology.

The differing relationships with each husband are set out in the letters in this chapter. Luther Cressman she referred to later in life as her "student-husband." They went through graduate school together, although in different fields, and tried to have a marriage based on the most avant-garde principles of free love. Margaret met Reo Fortune on shipboard on the way back from her first field trip to Samoa. After marrying, they worked together as an anthropological team, doing field work in Nebraska, and then the Admiralty Islands and New Guinea in the Pacific. She met Gregory Bateson on her third field trip with Reo, on which they were studying three different New Guinea peoples and Gregory was studying a fourth. He offered her gentleness at a time when Reo seemed harsh, and she felt freed from the restrictions she had put upon herself at Reo's desire. After their marriage they worked together doing field work in Bali and with him Margaret had her only child, Mary Catherine Bateson. Their marriage floundered in the aftermath of World War II, and they were divorced in 1950.

Unfortunately, the Margaret Mead Papers at the Library of Congress do not include early letters to her first husband. In the 1920s, when she was married to Cressman, Mead wrote most of her letters by hand. While Mead retrieved many of her letters from family and friends, her letters to Cressman were not among them. All we have are a few snippets about him in letters to family, and reflections upon him and their relationship in letters to Ruth Benedict and others from that time. In the later part of their lives they reestablished a friendly relationship, and some of those letters are included in this chapter. Below are some of the snippets she included in letters to her family.

In letters home to her mother and her paternal grandmother, Martha Ramsay Mead, Margaret occasionally mentioned her then fiancé and future first husband, Luther Cressman. [h]

January 3, 1919
Dearest Mother,

. . . I have had a most wonderful visit [with Luther's family], and am happier and more in love than ever before. I am quite properly convinced that Luther is the most wonderful thing that ever happened or could happen to me.

September 27, 1919
Dear Grandma,

. . . Luther got his scholarship for two hundred, so he will have a much more comfortable time this winter, than he expected. He is in love with the seminary.

February 13, 1920
Dear Grandma,

. . . Luther has been home ill, and so he didn't know that I had had "the flu" until he got back to the Seminary yesterday. I got a telegram from him last night with an urgent request to be careful and take care of myself. I've practically gotten my strength back now however.

March 7, 1921

Dear Mother,

. . . You should see how far Luther has swerved towards the left wing since last winter. He no longer boasts an inflexible moral code nor a single reactionary principle, thank heaven.

December 7, 1923

Dear Grandma,

. . . P.S. The Cressmans weren't a bit shocked at my keeping my own name. Mrs. Cressman told Luther it was our own business and she was perfectly willing to call me Margaret Mead.

A letter with no date, possibly from their honeymoon (circa early fall 1923).

Dear Grandma,

. . . Luther has gotten breakfast all the time we were up here. It was delightful to come down to a beautifully cooked and served meal. By the way, did you ever eat swordfish? Oh, boy, it's good!

To anthropologist Ruth Benedict, fourth day at sea on the S.S. Sonoma, *en route to Pago Pago, American Samoa, after having stopped for a few weeks at Hawaii, in August, 1925. Luther and Mead had an open marriage based on her theory that a person could love more than one person at a time. Mead felt free to love other men and women and Luther also had the freedom to love other women (but was not interested in other men), with the understanding that there was a special bond between them in marriage. This letter reflects that philosophy. [h]*

Ruth, dear heart,

. . . The mail which I got just before leaving Honolulu and in my steamer mail could not have been better chosen. Five letters from you—and, oh, I hope you may often feel me near you as you did—resting so softly and sweetly in your arms. Whenever I am weary and sick with longing for you I can always go back and recapture that afternoon out at Bedford Hills this spring, when your kisses were rained down on my face, and that memory ends always in peace, beloved.

Then Edward's letters were all in his later mood of gentleness and love.[1] For a whole month I shall know no other mood of his, and because I'm so easily convinced of joy, I shall be happy.

Then I wrote Luther from the train after you left me—telling him more about Edward than I had cared to say while my mind was in such a turmoil of uncertainty. His answer reached me in Honolulu with his first thought— "The poor darlings, they had so little time"—and ofcourse I mustn't worry. I will be able to make Edward happy, and much more in the same vein. He's one of the people who really know what love is. . . .

Mead's unconventional marital arrangement that allowed her to be away from her husband for a long period of time did not go without notice among the European community in Pago Pago, however. Mead wrote Ruth Benedict on September 11, 1925: "And this sweet little group of gossips are just seething with speculation as to why I 'left my husband'. Of course, they are sure I have. And I know I oughtn't to mind but it's so depressing to be greeted with suspicious unfriendly glances."

To Ruth Benedict, December 7–9, 1925, from Ta'u village, Manu'a, American Samoa, on her relationship with her first husband, Luther Cressman. There came times in each of her field trips when she hit rock bottom emotionally and afterward would start to climb back up. This letter reflects one of those points during her Samoan field trip. [h]

Ruth darling,

. . . Léonie[2] never fails to convict me of having been awfully wrong somehow. She said you and she both agreed that Luther was the chief sufferer, and that he had been miserable all last spring and everyone reported that he was very unhappy and depressed when he left New York for abroad. Do you see what that comes down to? I'm not able to make him happy either. His unhappiness is not over Edward. Last winter he didn't know about Edward. And it isn't a fear that I am going to leave him because I wrote him quite positively that I was not—and he got that letter before he left home. If he was unhappy it was because of the one thing which can't possibly be remedied, that no playing by Léonie's rules or anyone else's could possibly remedy—because he knows that he isn't wine to me and never will be.[3] Happenings, events, don't matter to him any more than they do to me.

He has said that he would rather be bread than wine if he couldn't be both—I thought he meant it. But I have no confidence in myself now, except as a general marplot.[4] But if it's true, all that Léonie says, then all I'm doing is hurting everyone all round. Ofcourse she's silly enough to think that if I stay with Luther, it makes all the difference. But it doesn't really. You know too how my not trying the impossible with Edward is a great deal out of obligation to Luther—not an obligation of wife to husband or between lovers—but something more than that. There's that and the fact that Edward would never be content unless I changed my whole attitude towards Luther—and I can't do that.

But if Léonie is right and Luther is fundamentally unhappy now—then everything is immediately changed. I'm too horribly mixed to understand anything. With my head I know perfectly well that Léonie distorts things, perverts them, makes them up just to fit her prepossessions. I also think as I've thought all along when people reported that Luther was depressed before leaving that it was because he was lonely, worried about my health and my being miserable and horribly depressed over his visit to Ethel.[5] It doesn't make anything clearer or easier that she is in love with him. He cares too much for her to take at all lightly the fact that he knows he can't give her the least glimmer of happiness.

I'm just unmitigatedly miserable and my head aches so I can't really think. It's the hottest day we've had this year. The room is full of flies, alive and dead. I have to beat off hordes of visiting Samoans. I don't dare cry— and I cry all the time.

Ruth, what have I done that is wrong? What have I done? It is very truth that your love is keeping me alive. I could only face life for you, now. I love you, always.

I've just cabled you "Is Léonie right?" That's inexcusable, I know. You'll be awfully worried because you won't know what I mean and even after you get hold of Léonie maybe you won't know. But I'll die if I don't do something. The last stakes are knocked from under me now. I did think Luther was happy with me—that I could make him happy. But nothing is changed between us by my decision in his particular case. Léonie doesn't seem to realize that. Oh, do you agree with her? I feel absolutely baffled and helpless. If I were to write Luther he'd maybe get the letter by the middle of February. The middle of February! And it's now the 7th of December. I feel so hopelessly beaten that I could almost decide to come home and give it all up. Fighting the climate and fighting my arms took all my energy. I didn't half hope to be successful anyhow. And now—you can't get people's inmost secrets out of them if you're on the verge of bursting into tears any moment. Tonight there were just mobs of Samoans around and I could hardly think of a word of Samoan to say to them. I guess Edward is right in that there is certainly something fundamentally wrong with me. Just because I haven't done anything permanently dreadful to you doesn't say I won't soon.

All Edward's letters this time were more horribly convincing than ever. It would be the impossible, we're a million miles apart. But that's cold comfort.

I know I'm selfish to heap all this on your head. If I could wait a few days perhaps I'd be less selfish. But I'm so cut off here—no chance of getting a letter off before the 6th of January—a whole month. And it's hopeless to try to write Luther. It would only add to his unhappiness without helping me any. But oh, Ruth, you know, you do know how I've builded on Luther and my relationship. It's always been there, to keep the world from rocking, and if that's wrong?

I think this is the worst day I have ever spent.

December 8, night

I didn't send the cable. I was afraid it would upset Léonie too much. I reread all your letters this afternoon. I'll manage somehow and your love is an armor to my need. Goodnight, darling.

December 9, night

I've just reread this hysterical outburst. Sooner or later, you'll give me up as hopelessly morbid and childish. . . . I feel very strangely calm at present. I'm almost tempted to believe that maybe I've been dodging facing the idea that Luther was unhappy. I don't half believe it now, but after the first shock of the idea, I feel suspiciously peaceful. And peaceful is certainly not what I should feel this week. . . .

Darling, your picture is at the foot of my bed where I can feast my eyes on it—and it's infinite delight to have it there. I just lie and look at you when it's too hot to sleep—and quite forget the heat in the joy of it. You'd be an angel to send the other picture too. I read your letters the first time as a drowning man—they stabilized the universe. Yesterday I reread them all again—not for salvation alone, but for joy—so much joy—tho it was pretty hard to get by the picture of you raking leaves. Do you remember—or perhaps you didn't know—the spring a year ago when you weeded while I rested and I lay and watched you—beautiful against the hillside.

I love you—I love you—Kisses on your lovely face—

Margaret

When Mead went to Samoa, she and Luther had been married for two years, but had known each other since 1917. Mead had an emotionally lonely time in Samoa, and on the boat taking her to Europe to meet both Luther and Ruth Benedict, she met and fell in love with the man who would become her second husband, New Zealander Reo Fortune.

To Ruth Benedict, July 15, 1926, from Paris, France, explaining how meeting her future second husband, Reo Fortune, on the ship to Europe, made her question her own philosophy of free love. [h]

Ruth darling,

I'm going to try to write out as clearly as possible just what all the fireworks and ambiguity are about. I despair of making anything but a mess of it. However—

If you got my letters from Australia, you know in what a state of depression I started the voyage. Nothing seemed real to me. I looked at your picture but you seemed unbelievably removed from me. I'd lost all sense of being in communication with you. And that was the only thing which buoyed me up this year. And with that came a general sense of the uselessness of coming back. I couldn't make myself believe it mattered to you. It seemed like a ritual act performed years afterwards in memoriam. This feeling explains my not writing during the voyage and also the pitch of the Australian letters. I can't explain it but it gave me a sense of panic. Perhaps you will never realize how hard I've clutched at your reality this year.

(There are a whole series of different things which I have to tell you related only in terms of my general mood and by chronology. I fail to find any other credible relationship.)

Then I had Luther's letters filled with an unclouded joy over my return and at the same time a general depression because he hadn't a job, because the Church had gone quite by the boards this year and there was no work which promised to even hold his interest. And I knew that I had no comparable exaltations to match with this; I could say grace but I could not chant canticles of praise for my daily bread. It seemed so rottenly unfair that I couldn't match his mood and also I feared that this inability would make me powerless to help him out of these other difficulties. My whole general escape pattern grew stronger than ever. I finally decided and started writing in letters and telling people that I was going back to the South Seas in March 1928. That helped some, and the fact that it did was the most dispiriting part of it.

Then, there may be added to this the very material fact that I was tired and seasick.

Also, altho I refused to recognize their validity in his case, all the various accusations and deductions which Edward made were still fermenting in my mind. From them was bred a tremendous distrust of myself, with Luther and Marie [Eichelberger] to point the moral. In trying to be all things to several people wasn't I simply failing all round? And as the other side of the proof—wasn't I quite miserable? And you had suddenly gone very far away.

Then I'd planned to work and I found I couldn't. Any attempt to work meant a violent headache. I had another major attack of conjunctivitis so that I couldn't read, and I had to lie for hours with compresses on my eyes.

I wonder if I'm telling you all this first to make you discount the next announcement or to disarm you. I don't quite know—anyway, I fell in love with Reo Fortune, the young New Zealander whom I wrote you about. And then all Edward's dicta bore fruit. For Ray,[6] without being either pathologically jealous, or possessive, or demanding, is essentially a monogamous person. (I'm not going to try to tell you about him—That will have to wait till I see you. He is a very clean cut, essentially simple person—far less temperamentally complex than anyone I have ever known. In some ways he has the effect of a clean-cutting knife on all my involved inter-explanation.) Because I cared so and because I know how great a lack there was in my whole unexpectancy in meeting Luther, I finally worked myself into an acceptance of his point of view. I felt as if I were a different person. I was entirely skeptical of its lasting, but there it was. And with the idea firmly rooted in my mind that ecstasy and ecstasy alone justified demonstrative affection, I met Luther in Marseilles.

Luther was far more unhappy than I knew and I was powerless except to add to it by announcing this new attitude. I wrote you those dreadful letters out of those first days together. All that we had ever had seemed to be falling to bits and I had no other explanation to offer except that I was a complete failure. The real difficulty was, I think, that I was acting in a way which was extraneous to me and that unintentional but complete dishonesty was acting as a corrosive agent. Anyway I switched about again to my former belief that one can love several people and that demonstrative affection has its place in different types of relationship.

And yet I can't get away from the seed that Edward sowed in my mind. I vacillate between thinking I'm being greedy and arrogant and thinking that I am incapable of meeting the demand of a single hearted devotion. As I'm gradually getting my strength back and my wisdom tooth has stopped aching, I'm getting a more confident attitude again. I'm again beginning to feel as if I could manage anything. Maybe I need even more protracted ill health to induce a decent humility.

As to practical consequences, there are very few for the present. Ray is going to Cambridge to get his doctor's degree in Psychology. He has a two years' fellowship. After that he thinks he will want to go on a field trip with me. I am going to spend these two or three weeks between the time Luther leaves (the 22nd of July) and the time I join you, with Ray. It's very little to build on against two years' separation. As far as he and I are concerned I am unqualifiedly happy. It's my punishment that I can never have one pure emotion that is not qualified by several others. And I've such a sense of having no right to this happiness that I fight it off—with both hands—and yet at times on the boat I surrendered to it. I shall for these few days this summer. What is to come after I do not know. But I do know that leaving Luther could not possibly have the horrible disintegrating effect on our relationship that this temporary extraneous attitude of mine had. And that was the thing which shattered my confidence so completely.

I don't know what kind of a job I've made of this. It's desperate to have to tell you like this and in such a contest of extraneous despairs. And you will see why I couldn't write anything illuminating. I did want to arrive at some sort of conclusion in my thinking before I wrote to you about it. The whole voyage I thought I'd be able to see you and talk to you at once—and when I found it was to be six weeks it seemed impossible to write you anything which would make sense. . . .

I'm still smarting under the fact that I was too obsessed with my own problems to keep from hurting you. Ruth, Ruth, you'll never doubt that I love you, love you, love you? Soon I'll make you believe it.

Your Margaret

After a tumultuous visit to Europe, Mead returned to the United States to be with Luther Cressman. But this was the beginning of the end of their marriage. During his year in Europe while Margaret was in Samoa, Luther had also met a woman he was attracted to, Dorothy Cecelia Loch, a Scotswoman who had grown up in England. Mead and Cressman slowly moved toward divorce. Meanwhile Fortune was doing graduate work at Cambridge University in England. Mead managed to get to Europe once to see him in 1927.

To Reo Fortune, on shipboard returning from Europe, August 12–13, 1927. Mead was returning home after an eleven-week trip to Europe to visit other museums and learn more about Pacific material culture, as well as to rendezvous with him. Again she questioned her free love philosophy, which she named "polygamy," a term used by the avant-garde in the 1920s. [h]

Reo, my sweetheart,

It was so good to get your steamer letter, darling. I hadn't expected it at all. I thought you'd be too rushed and anyway mightn't remember the boat. So I stepped on any little hopes which I had—to be really surprised in the end.

I'm writing on deck, for the Social Room is horribly crowded and makes me sick at once, but it's so glary and windy and shaky out here that I can't write well. I'm just dully miserable not really believing that you're gone but still lonely for your hand on mine, and then by turns come sharp stabs of realization that it will be a year. And with it all I'm a thousand times happier than I was last year sailing back. For then all the things which are right now were wrong. You know, you must know, that all my endless argument and violence grew out of my own lack of ease with the formulas I made up with which to deal with a very complicated situation. And now that I don't need them any more, now that I can remember that song in the "The Star and the Plough"—"When you said you loved only me, Norm, and I said I loved only you."—and not find it all mixed up with a lot of silly formulas about polygamy. I feel as if I were shriven.

I saw Ogburn[7] in Paris and told him the whole tale—or such parts as I deemed good for him. He was immensely pleased for he has always suspected me of lack of femininity. He's going to work hard to get me my fellowship and he thinks there is very little chance of a hitch. He's going to buy your book. I had a beautiful time talking about you. . . .

I had several long letters from Luther in Paris and he seems to think a London divorce may be the only possibility. That's 3 months residence there. It will be a question of his going out there I guess. I never could get 3 months more on top of the fellowship and all. Luther is sailing to England on the *Leviathan* return trip, without, he says, any intention of marrying Dorothy. But that doesn't seem quite clear. . . .

Aug. 13

I read all your letters over after all, and I'm glad I did. Now I'll never think of them again as cruel or unfair. Reading over the phrases which had seemed like the end of the world at the time I only remembered "This was the phrase that made me cry so on the way to lunch" but all the bitterness, the hurt, the blind miserable fear is gone. What a tragedy of errors, and lost letters and horrible impasses it all was. But, oh, I'm glad I read them dear. Against a background of misunderstanding there were times when I thought I'd be ashamed for anyone to see those letters and now I only wonder anew at your love which survived such provocation. We did get thru all that winter this summer didn't we? It took that last afternoon in Strasburg to wipe out the invidiousness of that one remaining phrase which bothered me. "I feel clearer now." I think I've taken over your feeling about that particular thing too.

I'm sick with cramps by the way—which is all to the good. I didn't quite trust that German B.C.[8] and somehow I don't know how I could have born to have violently interfered with something which in other times we will want so much. . . .

I mean to get an English cook book and learn to make Yorkshire pudding and lots of other English things. You might tell me the names of the ones which you particularly like. This is always assuming there will be an answer to this letter. It all seems so impossibly fantastic to realize that there mightn't be. . . .

Oh, and because you didn't like some of my friends' comments I must quote Marie. She writes, "So you have your caveman and like mindedness all in one after all," and "I think of you as perfectly happy, as having entered into the possession of Paradise without suffering the pangs of death."

This trip has been pretty awful so if this letter is scrappy, you must blame that fact. The vibrations are horrible; they've loosened every bone in my head. We're on [14][9] deck below the water line—not a porthole on the deck, and the air—unspeakable. I keep wakening and believing I'm suffocating. I've tried to write in the noisy hot social room but it made me sick. So mostly, when it isn't raining, I've stayed on deck and simply sat and wrapped myself in the peace and beauty of your love. I am so unspeakably happy, darling. To have given up all the arguing and reservations forever and simply fallen into your arms, to be held there, safe, infinitely blessed, for always.

I love you, Reo, I love you.

Your Margaret

To Reo Fortune, August 21, 1927, from New York City, showing that Margaret had taken her mother into her confidence about her marital situation. [h]

Sweetheart,

Mother is here with me and I've been talking, talking, talking about you. It's such a joy. It's a great help to have the picture.

This week has been pretty awful. It's so funny to have people think that they can't mention Luther's name in my presence—and all such silly tomfoolery. Luther himself has been as considerate as always—and one thing I've realized at last is that he never believed the things I said about our marriage. I said them enough but in some fashion he never believed them or he wouldn't have gone on with it. But I couldn't know. It took my actual final decision to leave to convince him.

One of his principal conflicts now is whether you'll let him see me as a friend when we come back. It's so hard for me to know whether you will or not.

Mother is going to see about the possibilities of a divorce in Pennsylvania as soon as she gets back to Philadelphia. I'm afraid they are pretty slim. . . . So consider this—divorce in Nevada—if necessary—means either not joining you till March or April or giving up Samoa. Consider this carefully, dear. I'll have the best lawyer in Penn. seeing what he can do. . . .

Goddard[10] isn't back yet. I'm going to tell him about Luther at once. I decided any other course would be impossible, involving everybody in strings of lies. . . .

Part of the time I am so miserable I don't see how I'll stand this winter, but mostly I can look forward far enough to manage. Reo, Reo, sweet, I love you so. I love you.

    Your Margaret

Mead ended up getting her divorce in Mexico. She prepared to leave for the Pacific to marry Fortune and to go on her first field trip with him to the Manus of the Admiralty Islands.

To Ruth Benedict, September 1, 1928. The week before she left to meet and marry Fortune, she was swept into a brief infatuation with Morris Crawford, an internationally known expert on fabrics and costumes who consulted for the American Museum of Natural History. It made her question the future she had chosen.

. . . Ruth, I'm wretched about it all. I'm not exactly unhappy but enormously puzzled as to the rights of it. If I tell Ray that my refusal to let Morris make love to me was formal and legalistic, that its only foundation was a respect for his wishes, and that as far as my own emotions were concerned I gave far greater and more fundamental assent to Morris than he would ever be able to stand, I don't know what he'll say. It oughtn't to matter to him what I did or didn't do but only how I felt. And I wouldn't give two figs for such an acquiescence to my wishes as that. And if I don't tell him, is that fair? And anyway what am I running into? I'd felt so safe, so settled, so "conservative." And now, I feel completely unsafe and at sea. It isn't that I have to make a choice, of course. Nor that I can't bear to give Morris up. I don't want to have my cake and eat it two [sic]. If he'd wanted a casual relationship I'd have rejected it and as he wants an important one, I'd reject that too, because you can only have one of the sort and I'm quite given to Ray.

But—the fact remains that Ray couldn't bear my caring this much for Morris. I know he couldn't. And when I started out, I thought I could be honest because there would be nothing unbearable, impossible to face which I'd ever have to reveal. Now what. Oh, please write to me air mail so it will catch the boat and tell me what you think, Darling. I feel hopelessly at sea and as if it must all be wrong. Ray should fill my field of attention and exclude other men as you do other women. I don't have to be afraid of meeting the loveliest women God ever made. I'm sealed and bound to you. And why can't it be that way here too?

. . . Maybe there is something fundamentally wrong with a relationship in which one person will constitutionally, inherently do things which the other person can neither understand nor forgive. Isn't that what is so right with us, that we understand each other well enough so that no act which is germane to our personalities is incomprehensible nor unacceptable? Or am I running after an impossible high standard? I know you and I don't happen but once. But I'm doubtful whether less perfect things ought to happen, granted that such perfection can. Marie asked me at the train if I had to choose between you and marriage which would it be. And you know I haven't any doubt. Ought one to do things about which there is doubt? It just boils down to this. If I tell Ray the truth, I'll wager he'll not want to marry me, and yet I have done nothing alien to my own personality nor which I in any way reject. The nearest thing I come to rejecting is being enough bound to one person to make you act insincerely towards another. Oh, sweetheart, I need you so. You should always cross the continent with me, you know you should. Things become crystal clear in the light of your understanding when they are all murky to me. Of us I am so sure, so sure and of everything else, doubtful. . . .

Same letter, later, on the train after Washington, D.C.

Sweetheart,

I didn't realize that we'd get to Washington so soon, or I would not have eaten any dinner. I'm afraid it was awfully disconnected. I simply dashed through that last page. Perhaps you'll write to me and tell me I'm crazy. My feeling for Ray isn't a bit shaken or questioned. But Morris has moved me very much, more than I have been moved for over two years—and more than Edward moved me I think. If I didn't have such heavenly patterns I'd think I was head over heels in love with him. It's all so completely crazy. He left me saying the point was to care, the point was that I was in the universe, the point was that there was something as meaningful to desire, that that was more important than any possession, and that come what may he was exultant over having known me. And no plea to have me change my stand now, complete respect for my doing what I had said I would do, but a crazy, blind optimism that someday I'd come back to him, want to come back to him. I'm trying to write as I'd talk. I want your judgment, as you would be able to give it if you were here and could listen to me talk. I want the moon, don't I, beloved? I can't understand that I've passed beyond our perfect circle where all questions are answered and all is made clear.

I'm telling you all this because I want you to see just why I'm worried. I haven't any idea of how much this really means to him. But as it stands as it was presented to me, there is nothing which I did not respond to. They are all my own points you know. And made with absolute honesty. And yet I can't tell Ray all that. And what is the rest of my life going to be like? Is it going to [be] chock full of feelings which I must hide and dissemble unless I shut myself up in a country cottage with three children and don't see anyone but the milkman and the butcher? Heaven knows the details of this scene are simple enough. I met him professionally and accidentally in response to a request of Sydney Bunzel's! I have lunch with him taking Eda Lou along, I go to see his collection and he showers gifts, a la Santa Claus.

He left his film pack in the taxi that day; it was found and he came up and got it, stayed and talked awhile and said he'd give me a camera.[11] I went to get the camera; he made love to me, I explained that he mustn't. He took me to dinner and I explained further and in detail, he accepted with good grace and I came home at nine thirty and spent the evening with Marie. You'll think this array of incident is ridiculous. But what worries me is that for the rest of my life I'll have just such accidental encounters with people, and run the risk of outraging Ray or lying to him. I can't see either course clearly. Please write me all you think. I'm trying to be perfectly honest.

. . . Oh, my lovely one, my lovely one. I love you so,

Margaret

Mead did end up telling Fortune about Morris. She wrote to Ruth Benedict, October 13, 1928, from Sydney, Australia, "I told him about Morris. He said, 'Well, it's a good thing you didn't live with him, because you're likely to get tied up with people then and you mightn't have wanted to marry me.'—all quite simply." Fortune and Mead were married and went to do field work together among the Manus of the Admiralty Islands in the Pacific.

In a letter December 27–30, 1928, Mead wrote to Benedict her early reflections on her relationship with Fortune. [h]

Darling,

. . . I'm a novice yet at understanding Reo. I still can't tell hurt from stubbornness. Maybe they are the same thing and it's a false distinction. We squabbled so over interruptions that I finally said "Let me have Banyalo[12] to myself for an hour a day on the language and you can have all the rest of him and the other informants." He kicked at the time, said it was beastly unsocial of me, indecent, that I wished to throw him out of the language, etc. But I stuck to it—for about ten days—and Reo worked on legends which he had said was the way he preferred to work on language. I typed all my results—all organized under headings making grammatical points. At the end of ten days it devolved that he'd refused to look at one of these, that he'd refused to analyze his legends grammatically, that he refused to talk Manus and was spending all his time on photography. I had "booted him out of the language." I don't pretend to know whether this is excellent, tho unconscious technique, or real hurt. He certainly has an enormous faculty for spreading affect about.

Anyway ofcourse he has me beaten, because I think the work is so much more important than any personal issue. It's hard to know how to tone down the inevitable clash between people who care *how* a thing is done and people who don't. With you I've solved the problem, where it arose, by adoring you too much to permit myself to criticize. With Léonie I've solved it, but not so well, by the dogma of her complete importance because of her gift. But where the other person is conceived as an equal—that was the case with Louise Rosenblatt—I seem given to useless squabbling. The impossibility of retaliation *in kind* is maddening. If one could organize a punitive expedition upon the other person's bureau drawers, letter piles, boxes, notebooks, etc. it would be simple. But Léonie and Reo are equally invulnerable to such attacks.

So now Reo "listens" to my working on the grammar, listening consisting in his deciding it would be amusing to get the personal names of all the village pigs when I am in the middle of trying to work out the provenence of the term only-one-throat-to-the-two as a term for behavior. If I disapproved of his results, it would be another matter, but I don't. They are very good. He has a grudge against me for not emphasizing my technique of personal relations more last summer so that he could do more in Dobu. It's curious how different this is from the complex against being taught which we anticipated. . . .

December 30 (?) [sic]

. . . I have such a sense of not knowing my way about in my relationship to Reo. It's so shifting, so charged with completely unknown potentialities. And it's all so sharp. Nothing is blunted yet nor blurred. It's a matter of wandering in a forest full of rare unknown flowers, of strange flycatching, poisonous plants, of vines which threaten to choke me, little springs of clear water which are without landmarks, so that I can never be sure of where I found them nor of finding them again, resting places on new fallen leaves, which tomorrow will be dry and harsh, long trails over sharp little stones, among mosquitoes and stinging flies. Nothing is known, nothing is sure to be re-encountered. I have to be continually on my guard and continuously brushing away the scars of fresh brambles. And if you were here to hear all this you'd say "You like it. You don't like things to be too simple!" N'est-ce pas?

Mead wrote again to Ruth over the next few weeks, reflecting about herself and Reo.

To Ruth Benedict, January 2 and 13, 1929, from Manus. [h]

Darling,

. . . I am happy, sometimes piercingly so, but so lost as to generalizations. I know one thing, I'd much rather be the one in the wrong who has to do the apologizing. He's far better humoured than I am, but if he is bad tempered he's so miserable about it, that life isn't worth living. I'm not satisfied with my own temper at all—and still he's satisfied with the cause of a lot of it, which I know well enough is partly due to lack of time for each other and piled up tension. However, I suppose this is an inevitable sort of situation with this newness and isolation.

We have a good time talking over the problems altho the conversation usually starts with Reo's damning me, anthropology, and Manus in particular—as having no problems worth tackling and being generally something nothing.[13] After I've been goaded into a furious defense, then he finally admits perhaps there is something worth doing and we have an amicable and profitable séance over the stuff. Perhaps it's a game to do it that way, and maybe I don't recognize these games quickly enough altho you say I play them. . . .

January 13, 1929

. . . I drive Reo into stages of desperation which are similar to the kind of mood in which I had that whole stream of automatic imagery about you last winter. It seems to be exceedingly similar—and I've got only the slightest clue to what sets it off—an apparently equal pitched battle in which small sharp missiles are being hurled about will suddenly change into a flash of passionate outrage. It's more likely to happen when I'm tired and collapse into weary misery, I know that. It's a circle that can be managed in time, I guess. I've only got to realize that in working that [sic: as] he does is Reo ever consciously selfish or disagreeable—that the things which I fuss about are in the nature of the beast—so that I might as well give over fussing. It accomplishes nothing anyway in the way of change and he feels that he is mysteriously armed with some weapons for hurting me—weapons which have been smuggled on his person and which carry no safety catches so that they go off without his volition. This reconstructs his whole dreaded home pattern and he goes to pieces frighteningly. However, my respect for abnormal sets[14] being so much greater than my respect for the normal idiosyncrasies of the masculine mind, I can probably handle this much better with more knowledge. . . .

Margaret

To Ruth Benedict, February 21–March 2, 1929, on Fortune and their working relationship. [h]

Darling,

. . . I think Reo will probably do better all-round field work with me along, but I doubt if he'll ever enjoy it so much. I've had to hit and hit at the idea that because something was hard to get it was therefore valuable. Which ofcourse is true, but he doesn't find field work difficult enough, no sufficiently puzzling and arresting problems in the material to challenge all his attention, and he gets bored. Our methods of work are astonishingly different. He gets a clue of some new aspect of the culture . . . then he follows it up eagerly, relentlessly, gets the point settled, writes up the results— looks them over and decides they are something nothing after all—. . . .He simply can't understand my method of working in terms of time, of putting in just the same amt. of time when I'm bored as when I'm thrilled—time that is long enough to cover a go at new material, if it comes up—without showing any special new display of attention. He varies between accusing me of taking no material at all and of being an unaccountable grind. . . .

Feb. 28

. . . I've got a little more insight into what kind of scene it is that I treat you to occasionally when we begin by discovering that I have hurt you and go on in a fashion which if pushed to its full conclusion would mean your comforting me because I have hurt you. It doesn't get quite there because you get stern and insist you need sleep and leave me to my images of misery and self abasement. With characteristic variation that['s] the same type as the worst kind that Ray and I have—he does something which makes me miserable, then he gets perfectly furious with me for being miserable, then finally the evening winds up with him in a state when I have to comfort him and beg his pardon for being miserable. It's a funny twist. Why do we do it?

Have you any theory as to why I take particular kinds of small disappointments, interruption of work, disregard of my routines, so hard. In special cases I'm as bad as Marie [Eichelberger]—yet I don't know why. Ofcourse they may be one of a series of sets—broken luncheon dates out of former unpopularity—interrupted work here out of a fear Ray won't respect my work. But why should I be sick with misery because a dog breaks into the cookhouse and steals the one chicken we've had since we came, a chicken I spent an hour cooking myself? . . .

Mar. 2

. . . I don't know what sort of idea I really give you about my life with Reo. It's fundamentally just as I would have it in intensity, in delight, in congeniality. All the friction is surface adjustment but I'm afraid I stress that too much in writing to you.

Sweetheart, I kiss your eyes—

Your Margaret

Mead and Fortune returned to New York in the fall of 1929, where Fortune became a graduate student in anthropology at Columbia under Franz Boas and Mead worked as an assistant curator at the American Museum of Natural History. They spent a summer among the Omaha in Nebraska doing field work, Fortune on religion and Mead on the effects of social change on the women, work which both of them found trying.

She wrote to her mother, August 22, 1930, from Macy, Nebraska, "I am sorry I have sounded so blue. It's just been a reflection of Reo's dislike of this problem.[15] *He's getting more of what he wants now and feels much more cheer-*

ful. He has the most deadly intensity though and when he's worried he sucks the bystanders, innocent or guilty into his moods, just as Léonie does. However I like that kind of temperament even if I do get drawn under occasionally."

Their next field trip was a long trip back to the Pacific, this time to the mainland of New Guinea. Here, from 1931 to 1933, they studied three different peoples: the Arapesh, the Mundugumor, and the Tchambuli. On the third and final stage of this trip, Mead met the man who would become her third husband, English anthropologist Gregory Bateson.

To Ruth Benedict, December 26, 1932–January 1, 1933, after meeting Bateson for the first time. [h]

Darling,

. . . I've got a lot to tell you and as there is no need for you to have unnecessary worry, even for a letter page—I'll say at the start that I don't think there's going to be any gun powder in the situation—and I don't think that it's just my inevitable rationalization either. It's Gregory Bateson—ofcourse. Nothing that I had heard or read prepared me for him. We'd been thinking of Christmas in terms of meeting him primarily, but mostly as someone who would talk a piece of our language, and deal in familiar ethnology and familiar gossip—and hoping that he wasn't going to be hostile. Reo's friendships are so ambiguously constructed and Gregory had a reputation for being so casual—that I've never known whether they had ever really been friends or not. Reo always just insisted that "Gregory is alright." I know now they never had been friends although they are in a fair way to be at the moment. Then Gregory's Sepik article—did you ever look at it—is incredibly bad— and you know how much I am swayed by work. All the tales about him have missed every single point—

He's six feet four and yet has all the slender unplaced grace of the most complete fragility. You've no idea how moving six feet four of vulnerable beauty is. He gets all the points, is extraordinarily sensitive to people. For all his utter at sea-ness, he has the kind of authority which people have who are absolutely faithful to their own temperament—I'm more moved than I've been since I met Reo—

I don't know whether I told you that he started up here with a woman named Bett,[16] who had been twice married in the territory—a warm, tough, little pagan, of about 38–40. He left England with her. She's been in the territory for 15 years or so, is a licensed engineer and lives on boats—in spite of his very cold, stiff mother's violent opposition. In Brisbane he met a girl he'd been in love with in London—I think. I got the whole story from Bett about a month ago and it's not too clear—and fell in love with her. Bett who has the sure animal instinct of her type faded out of the picture. She came to the Territory—but got ill en route, stopped for an operation and finally went off with another man—of her own sort. She's been living on his pinnace for a year and they were married about three weeks ago.

Meanwhile, just before we came to the Sepik, Steve,[17] the younger girl, came up. Gregory had sent her all his papers and wrote out the language for her—and when she arrived she hadn't looked at them. She was ill and miserable and everything went wrong and she went down on the next boat—but she's coming back. And I'm not clear yet about her. He's not at all sure he wants her. But he is quite sure that Bett was a mother and although they got on perfectly, that the relationship was frustrating and that Steve stood for his own unplaced age group. She apparently needs to be given a few black eyes, and he has no genius for aggression. Anyway, once the affair with Steve went wrong—a mother ofcourse was in order again. I obviously could be nothing else. And I think it's clearly written down that I'd be rotten bad for him in any other capacity. So if we can keep sex out which will be comparatively simple if Steve comes back—it should be valuable without great danger. Reo feels in no way threatened and I'm not making my old mistake of explicitness. His instincts are perfectly correct—and a lot of conversation and explanation might upset them. And I do think I've learned once for all that sex need not be permitted to spoil things. And ofcourse it's a situation in which I hold all the cards. And there is the practical situation of teaching him to do anthropology which is of immediate importance. He knows he has to learn or get out.

His difficulties are all of temperament and bad sets—that much I had figured out from his work. His flesh creeps at thrusting himself forward into the natives' lives and then there is a bad tie-up about his father—and what is science. He'll obviously get Cambridge—and Reo sees that without resentment—still as an aspect of a stable and orderly social system. And his mind is fine—if only all these other things can be overcome. He's docile as a child and sorry for his sins—and he's going to bring his stuff up to us—replan and reorganize it. I'm preaching the busy work theory hard—he's quite enough like you to make it completely relevant. He hasn't a shadow of drive and I see no chance of his acquiring any. His father's image might be revamped into one, but I'm dubious. I'll know better when I've read his father's essays.[18]

I don't know how definite this picture is. I want it to be as clear as possible because it will be such an important element in my thinking—probably for the rest of this trip. We're probably going to a place about three hours away from him—so he can come over for weekends. Ofcourse part of this was planned before he knew Steve was coming back. How she'll work in heaven knows. I'll have to make her like me somehow.

. . . Gregory says that the English reaction to my last book—and that includes Malinowski—is that it's impossible [utterly],[19] that either my methods are incredibly better than theirs or else I'm a liar. Malinowski believed that the Samoan book was possible but not the New Guinean book. Now that he has talked with us he does believe me—but that was his own feeling before. I believe it at last—tho I'm still in a dilemma as to how to counteract it. I've never talked so much about method to anyone as to Gregory because he's so pathetically in need of suggestions. . . .

Dec. 30, 1932

. . . On the night after Christmas Reo did what he has never done before—he got drunk. I had about four cocktails which hit me [like] one of those sleep giving thunderclaps and I went to bed. The next day Reo was still drunk. We'd plan[ned] to take the pinnace up to Washkuk, a place about 6 hours away which we wanted to see—and Gregory and I knew we'd have to get away from Ambunti or go mad—so we packed and left, Reo mainly listening to music and just not there at all. And I was dreadfully repelled—he became generic and it was almost as bad as insanity in the effect it has on me. We got to a village called Jambon and decided it was no good going further. That evening—with Reo still not himself quite—the natives warned us to sit up all night in fear of an attack—by ghosts they said but it seemed serious. They hung lamps all around the house and sat up themselves. So we spent the night with loaded revolvers sitting in a big net with blazing lights and sentinels out. Reo dozed off and on. Gregory and I smoked and he woke up once just as Gregory was handing me a cigarette—That was all—but he got a sort of vision out of it, brought it up the next morning in the canoe. I thought I was telling him how things stood, but he still wasn't sober enough to really quite know what was going on—it was a combination of sleepiness and the aftereffects of alcohol. Then came Washkuk—the three of us in a big net—on top of a mountain—and things gradually getting clearer—with some pretty violent scenes along the way. And Gregory loathes violence as much as Max[20] does. And now we are back here and going with Gregory to his place and then for a search expedition in his motor canoe. Ofcourse it *would* be all mixed up with anthropology again—

It was rotten, rotten luck Reo's getting drunk, and I'm still having to work over the results of that. Then his misunderstanding of Gregory means a re-documenting of the degree to which he misunderstands me. It is always Gregory and me on one side and Reo on the other—which is ofcourse the reason I can marry Reo and couldn't ever make a permanent picture with Gregory, but he doesn't see that. And meanwhile, I'm starved for likemindedness—just starved. And I make all the mistakes I made with him about you—I scold him for being bored because I feel guilty to have talked to Gregory—or something. Still we've been managing under terrific conditions and so we should be able to manage when they improve a little. But ofcourse Reo identifies Gregory with himself—and it's the South of France and Paris all over again.

Seeing Gregory in the field there these days has changed my estimate of his work considerably. What he's after is your kind of thing only worked out from the individual native rather than from the structure of the society. His touch is immensely delicate and sure—and if he can get over this notion that it mustn't be literature—he should do good work, very good.

Jan. 1, 1933

This mail must go off. We all leave tomorrow for Kanganaman, Gregory's village. Things are going much better—and will go increasingly so, I think. I'm terribly happy. I'd forgotten just how happy I could be. . . .

Oh, I am so happy—and I do think it is going to work. It's gorgeous to be speaking one's own language again—and I think I'm convinced enough that permanency lies in a relation built on other points. . . .

    I love you, Ruth,

    Margaret

To Ruth Benedict, February 14, 1933, among the Tchambuli in New Guinea, on talks she and Fortune were having on renewing her free love philosophy. [h]

Ruth darling,

    Still no mail. It's now over two months since I've heard from you. I've sent off a mail as far as Kanganaman—but no boat has passed there to take it down. Robbie[21]—the D.O.—is due here in the next fortnight and may have mail. He's [inspecting][22] up the next river at present—

    Gregory was here this weekend—and it's the best time we've had since Christmas—obtained by my succeeding in asking for nothing in the way of time alone with him, except the casual pickings of life in a house. But it has made Reo more peaceful—he's beginning to take up his original attitude towards Gregory of complete acceptance of him—and as his affection and trust come back—I can talk more freely to him.

    The whole thing has given us a frightful shaking up, and tho I don't think even my rationalizing will ever make me cease to resent the bad accidents we've been subjected to—still, I think it can be turned into gain all round. The big battle I've been fighting for the last six weeks is with the knowledge that Reo was emphasizing the lack of balance in the thing—and planning to "back" it.[23] I had—I still have a little—the feeling that his attention span for people is so narrow that if he tried to pay much attention to someone else, he'd quite stop paying any attention at all to me—i.e. that he has an essential genius for monogamy which I lack and that if he tried to do what he calls "take over my, (M.M.) culture" [sic]—he'd either outrage his own temperament or else I'd lose him altogether.

    And yet to give up Gregory meant to smash him possibly—at least as far as this field trip's work was concerned—and would have been a kind of cheating on my part—a backing down from a path I voluntarily took up.

But now, under the greater ease of this weekend—Reo is phrasing it all in a way that I think I can manage—that he primarily identifies with me—if I have an experience which he can neither share nor duplicate, he feels that identification is shattered—and so if he can develop a side of himself which responds to other people—he can feel it as done—for us—and so feel no treachery, but rather weave it in as further extension of the possibilities of identification and common swiveling to meet experience.

Because—in the last year I've tried monogamy pretty hard—and I did have the sense that a part of me was going numb—it was that sense of re-turned self-consciousness which I had when he went away in Aliatoa. So I do feel that I've given monogamy—in an absolute sense—a pretty fair trial—and found it wanting and now it's fair for him to try my culture for a change—if he can do so without violence to his own temperament. . . .

To Ruth Benedict, February 23–28, 1933.

. . . He [Reo] simply can't stand my talking to Gregory alone. It all got so bad that I began to think we'd have to go away from here—and just desert Gregory—to his own devices, which ofcourse might not be suicide but merely lethargy and no work—at all—and gradually getting further and further from reality. If it were necessary—it may still be—tho I've got life smoothed out at present but by such desperate measures—taking out the last wish of all the fairy gave me, as it were—I'd feel a cheat of the worst order, someone who had promised something which I was too weak to perform.

I'd worked it out that it's because I'm such a bread and butter person— that all the trouble comes. If I were a heady wine, people could wear my colours in their hair and be secretly thankful that they didn't have to have me around all the time. It's part of the point that men always want to marry me—it's the only way I'd be any good to them. As I'm bread and butter to Reo too and he insists on eating me for every meal—I have no right to make any other commitment in the world—because I'm likely to fail them. I had the desperate sense that you couldn't *trust* me to do anything against his obstreperous documentation of hurt.

Feb. 25

. . . This has been—I think—about the most frightful two months I've ever spent—mostly due to the peculiar local conditions. I've faced as difficult emotional situations before, but never where there was no way, no place of escape—not even a bathroom door to lock—for the local version swarming with mosquitos is no substitute. The only minutes I've not spent with Reo in a high state of tension of some sort—good or bad—I've spent with Gregory—and not even his ability to relax and be happy could make them a relaxation when I knew at what a price they were purchased, and always had the haunting feeling that each hour might be the last. And Reo and I go to bed to argue, wake in the night to argue, get up in the morning to argue. Then if all goes smoothly, why that blessed state must be savoured to the full. I've only had one attack of fever out of it, so I think I can vote myself a horse. . . .

Gregory's comments on Reo at Cambridge worry me too. He says everyone interpreted Reo's painful eagerness and worry over personal relationships as a social inferiority complex—the N.Z. outsider determined that he was going to be slighted. It seems to me they ought to have had more sense—and known it's personal affront he fears, not social slight of which he has no sense at all. But as they didn't, I think his living in any English country is definitely out. He'd be open to the same misinterpretation again and I should know it and it would be unbearable, besides continually interfering with his relations with people in the worst possible way. By the grace of God, the American social inferiority complex towards everything British is just the antidote which is needed for his peculiar vulnerability. They—Americans—don't expect social inferiority—and so have some chance of reading his painful shyness and anxiety correctly. At least it reconciles me to taking him to America for good. It's the only possible solution. Then if I can stage manage the environment enough he may come to feel secure enough in it, so that he will lose his nervousness about people and be able to meet the people whose opinion he values in England, easily.

. . . If Steve had only come and made Gregory happy—I think I'd have said for us to go away from here—and then I'd pray whatever gods there be that I might never fall in love again. . . .

To Ruth Benedict, March 29–April 9, 1933, on the situation between herself and her future third husband, Gregory Bateson. While separated, Mead and Bateson stayed in communication by sending bundles of small note slips back and forth across the lake by messenger. [h]

Ruth darling,

I shall send this off by the next mail canoe and hope very earnestly that it catches up with my last two letters—then, altho you will have a bewildering sequence of moods to disentangle—it won't be as bad as if you are left with only my ink[ed]-in P.S. on my last letter. If you'll bear in mind the very strange and difficult conditions under which we have been living, the lack of privacy—lack of any time to think, etc. you'll perhaps understand it all most surely. During the two weeks when Gregory was at Aibom, and over here for dinner and breakfast every day, I only saw him alone twice—and for the rest the three of us ate meals, and talked, or he and Reo played chess inside this tiny cubicle "room wire."[24] And that is the particular kind of frustration from which I always get a bad "kick back." I got an attack of feeling that I'd made it all up—and it was out of that that I wrote the end of my long letter. Gregory came over this weekend and for the first time for weeks we really had decent conditions—I have my own mosquito net "office" in the House Kiap[25] and we had a chance to talk this whole morning without any strain or fear of interruption—as Reo was working happily in full possession of the "room wire." It's the best time we've had and it left us both very happy.

I've less a sense of his youth than I've had before—and a more than ever vivid sense of his having the full authority of his temperament—"and of such is the kingdom of heaven"—I've also a more definite picture than I've had before of the enormous contribution he's made to working this out for all of us—Hardly anyone would have seen with such clarity that it would only have been a hurt and mangled me which he could have won completely—and so in a way by losing me he has kept me—in the only way in which, given the ordering of my life—it could have been done. If we hadn't had such absolute trust [in] each other through most of it—what fits I've had, ofcourse, I've not visited on him, nor him me—and hadn't such a similar philosophy of love, it could never have been done. As it is we have a kind of peace which is final, even tho all the intermediate steps which usually lead to it have to be omitted. When all our communication has to be anthropology and arguing about mails, collections, pinnaces, etc.—it gets badly muddled but an hour of talk puts it right—Reo's hurt has been mainly in a redocumentation of how frightfully different we are emotionally—but that was fate—once I knew he thought I was something I wasn't, I had to cark ofcourse.

April 9—Monday in Holy Week

And then the world began to spin again—and I've not had a minute to write since. I found the mail had gone so I didn't send this. And meanwhile I've been too obsessed or perhaps enthralled is the better word, to write coherently. For Reo and I are happier than we have ever been, even in France—we have all the delight of those first months and something far more—for we've carved out a vocabulary at last for communicating our so contrasting and hitherto incommunicable temperaments—and intellectualization has done what would otherwise have been impossible. It doesn't mean ofcourse that we are less different, but now we can capitalize that difference completely, where before it was a source of heat and separation as well as of excitement.

I can't tell how much of what has been happening I have conveyed to you, but Gregory swung me back very sharply to my original self, the self I was before I married Reo and tried to lay aside all the things about which I couldn't talk to him. I don't know how much you've realised also how completely I'd pared down my life—putting away all expression of religion—poetry—and all my old philosophy of personal relations. Given the frightful pace at which we have worked at anthropology, I ofcourse, succeeded in leaving very few empty spaces for retrospective glances. I did so terribly want to make my marriage a success—and I was so convinced of the essential truth of you.

"What have they cherished behind lock and key
One half so precious as this injury
That leaves us still together"

April 9, p. 2

that I was willing to make any sacrifice to keep this precious thing—you were the only sacrifice I would not make—and yet, because so much which you and I had had together was being suppressed, you lost. Particularly did I let the "stained glass window" darken over with false simplicities of explanation in terms of Stanley and Tom[26]—and I accepted them the more easily because it was so difficult to keep such an important faith which Reo found unintelligible. And ofcourse I had to repudiate a good part of the good which Luther and I had had—and rereading it in Reo's terror—lose it. Léonie too had ceased to be the anchor of my spirit which she once was, because I had found no terms in which to accept the Edmund[27] affair— And so when I said in a previous letter that I had not developed—except anthropologically—in these 4 years—it was only too true. Having laid aside so much of myself, in response to what I mistakenly believed was the necessity of my marriage I had no room for emotional development.

Those weeks alone in Alitoa paved the way for change—I knew there was something uncanny in the way I felt I had gotten back my identity there. And meeting Gregory, it all—the built up picture of behaviour appropriate to an opposition of temperament—went down like a house of cards. That has been the real problem all along—not that Gregory was taking me from Reo—that he was making as strong an effort not to do—as I could have done—but that he was shaking my unreal picture of myself to the depths. But it has all come out like a beautiful pattern to which God held the clue to the finished design. All that Reo never believed about Luther and about me, he has seen demonstrated in crystal clarity in Gregory—and we have found terms to phrase all the points in, so that we will never have to face misunderstanding again—and the crossness and snarl are gone from my voice—and I am happier than I have been for many years.

While I write this, Reo is writing to Luther—so you can see how the world has changed. Gregory is coming for Easter and we are all to be happy together. This will not mean any essential change in my relationship to Gregory—I think we can keep it as it is and not run any risks of the complication which would be introduced by demonstrativeness.

. . . .We've talked a lot of nonsense about sex—we've mixed up an interest in turning sex into a form so specific it is like a fight with ~~manliness~~ being well-sexed—I don't believe any of it, anymore. I think specific sex is a form which is primarily suited to a clash of temperaments—between people who can't communicate with each other—and is utterly and absolutely inappropriate between people who do understand each other. The kind of feeling which you have classified as "homosexual" and "heterosexual" is really "sex adapted to like or understood temperaments" versus "sex adapted to a relationship of strangeness and distance"—To think one goes with man-woman relationships, or that if it's within a sex it's because one person belongs in the other sex, is a fundamental fallacy. I believe every person of ordinary sex endowment has a capacity for diffuse "homosexual" sex expression, and specific climax—according to the temperamental situation. To call men who prefer the diffuse expression "feminine"—or women who seek only the specific, "masculine," or both "mixed types" is a lot of obfuscation.

This is a climax of the work I've done this last year—a combination of anthropology and reviewed biography. I think, for instance, it explains Gladys & Goddard—a heterosexual situation between two contrasting temperaments. And I think the way to bear Gladys is to treat her like a man—I believe she'd be quite bearable. Reo says that's what he's always felt. The fact that sexually she was likely "passive" and "feminine" to Goddard is intensely irrelevant. Similarly, Brown should be treated like a man, not like a mixed type.[28]

I'm inclined to think all this talk of "mixed types" is a nonsense as the Papa says. You're not a "mixed type" at all. You're a perfectly good woman—who prefers different temperamental types and responds differently to them.

We'll have acres and acres of things to talk out when I get home—but I think this should be enough so you'll feel you are following events. I've got back my "stained glass window" and my belief that all love is good and essentially compatible. I feel as if my whole life had been gathered up in a golden basket and given back to me. I'm mentally alert as I haven't been for ages. I'm on tiptoes with happiness and a sense of the goodness of God—and all three of us feel that it has been worth it—even Reo who has suffered most feels the rewards are greater than the price paid—far greater.

And oh the joy of being able to get—as intelligibly as if you and I were saying it—the reaction of a really different temperament on all the points upon which conversation seemed forever barred out—It's a discovery of a great magnitude. And you'll have it all too—for it will be an easily learned language once one has the key[29]—One might almost talk to Edward in it—think of that.

Ah, my darling, it is so good to really be all myself to love you again—

I am telling no one about Gregory—nor about this change between Reo and me—so if a sense of a new mood gets into my letters, you need not think anyone has any data. I doubt if I can keep the mood out and still write honestly. I feel so very much closer to Léonie—Gregory is simply a living embodiment of her fairest moods.

The moon is full and the lake lies still and lovely—this place is like Heaven—and I am in love with life. Goodnight, darling,

To Ruth Benedict, June 16, 1933, from Sydney, Australia, bringing her up-to-date on the twists and turns of their developing triangle. [h]

Ruth darling,

This is to be a real letter—if you knew the number which I have written and then had to discard as no longer relevant. At Easter we finished thinking out the hypothesis which I think is going to be so important—and we—Gregory and Reo and I—all decided to come down to Sydney. We were all in a fine fervor of pure intellectual emotion—Gregory was to try things over again with Steve—Reo and I were to come home slowly—via N. Z., Honolulu, etc. There was a large amount of religion in all our hearts—and everything seemed clear. I left the equipment for a comfortable house in the field—in case Gregory took Steve back. Then Gregory went to get a pinnace. Reo got a bad attack of fever, I got badly bitten by a scorpion and was invalided for a couple of weeks—the schooner trip down to Madang took five days of broken engines, practically no [fruit][30] and a bad S. E.—and all the peace and friendliness was pretty well done in.

Gregory and I had reached such a pitch of understanding that we had convinced Reo that we had and could make sex irrelevant—but that was all lost again. The days in Rabaul and then the steamer voyage down were ghastly. It got to a point that I simply gave up talking to Gregory at all—except formal chatter at meals—and life ahead looked as if it were going to be an endless mass of reproaches over sins which Reo was convinced I had wanted to com-mit even if I hadn't committed them. And then—the uneven humour of the gods intervened again—and on the boat Reo met a woman he'd known years ago—whom I have named Mira as we do not like her real name. When we reached Sydney I knew that all the banns would have to come off and that Gregory and I could be happy for a little while—as Reo had decided to stay some time in Sydney—and now my hands were tied—for Steve was on the dock to meet Gregory and I had to give that relationship every chance to come to life again—if it could—so I said nothing to Gregory.

Steve carried him off—I spent the next few days laying out tête-a-tête luncheons and then vanishing off the scene—and going about with Carrie and Timothy[31] and seeing Gregory only with Steve in groups. Then fortunately it became sufficiently clear that he wasn't in love with Steve, nor had she any clarity about him. And now life has a layout which seems nothing short of miraculous. Mira and her daughters have taken a flat next door—we have a flat on the 3rd floor, Gregory on the second. I am now down in Gregory's flat writing to you—waiting for him to come home to lunch. R. and M. are lunching upstairs. I am extra-ordinarily happy—altho I've no idea how long it will last. I don't think that it means any fundamental rearrangement of my life—tho it's hard to know yet. I have very little sense of elation. Gregory and I have been buffeted by Reo's passions and mores so long that it seems the natural order of the day. We neither of us feel that we can possibly will anything to be different from the way he wills it—he is too precious to both of us. It's a strange [diagram of a triangle]—Reo and Gregory—in love with each other—and hopelessly blocked ofcourse in any expression of it except through me.

I have made friends with Steve—she's a combination of Lee Newton[32] and my sister, Elizabeth—but more gifted—really very gifted I think—utterly unplaced and childlike. She expects me to give her Gregory on a platter—when she cries for him—and I'd be moved to grant her prayers if I saw a way. But his objection to clash and violence is enormous—and the perfect peace of his relationship to me seems his ideal at present. Meanwhile he's spending a good deal of time with Steve—and we're trying to work things out for her. She's completely immature—all her experiments with sex have been fiascoes—and there's lots of work to be done to straighten her out. Meanwhile Steve and Carrie and Mira all dislike each other and [are] jealous of each other—

which is horribly complicating. Even with doing nothing but see people from morning until night there just isn't enough time to deal with life. Reo hopes to get the bones of his theory written up here—he has a little more time than I have but not much.

The happiness which Gregory and I have is closer to your and my relationship than to any other happiness I've ever known. There is the same balance of activity and passivity, the same completeness of understanding and absolute integration of feeling—and the same timelessness. Ofcourse our terms are all different—all very clear cognitive science rather than art or literature. That and the difference in sex make a sufficient contrast so that I feel no substitution—but—again the feeling I had when I wrote "The Gift"—that this after all is what life is about. It brings me ofcourse very close to you—whereas Reo always came between us.

Letters worry me dreadfully. . . . I couldn't write you until I could at least speak of a happiness which seems likely to have a few weeks duration—and it is only in the last ten days that such promise was given. I feel reprieved— if not from death which was very near us often—at least from carrying always a stigma of being evil because I could love more than one person. Reo at last knows that such a course is possible in all innocence—and that means a revision of all his past judgments—which were at heart almost as condemnatory as Edward's—and the process which began as soon as I met Gregory of being given back my real self, is now complete.

I don't think anyone, no matter how forceful and how mad, can take it away again—but perhaps that's an idle boast. Reo has shattered all my convictions to bits so many times. I do not know how much understanding you've really gotten from my letters—they have been desperately incoherent I know, but these last months have all had the quality of near madness— sometimes great lucidity, sometimes ecstasy—often despair. Gregory and I fully expected that the most possible outcome would be that Reo would shoot me, then Gregory, then himself—and there was nothing we could do—except try to hold on patiently from day to day.

I knew that if Reo attacked Gregory—that could pass—for Gregory was willing to take his blows unreturned and quiet—he loves him quite enough for that. But I can't help fighting back. I can't achieve an invariable disarming docility, and there was risk of my maddening him when I reached an end of patient submissiveness complicated by lack of sleep. There were weeks without sleep—and without privacy. I only kept alive by drinking seven and eight quarts of milk a day and the scorpion bite left a great hacked sore the size of two silver dollars—a beastly thing to cure up.

But I'm alright now, darling, really all right. We're in Sydney, among people. Gregory has more right to look after me—and I would not now hesitate to appeal to him, I think. It's a great comfort having him in the same house. Mira and I can talk easily—she's really a darling—and she is very helpful when Reo gets difficult. All her interests lie in protecting my happiness with Gregory—and she doesn't quite realize that I would be happy for their happiness anyhow. I do not think you need worry about my health or safety—and about the future I can say nothing definite myself. . . .

To Ruth Benedict, July 19, 1933, from Sydney, in the midst of the tensions with her husband Reo Fortune and future husband Gregory Bateson. [*h*]

Darling,

A letter. The first letter since you'd heard about Christmas. It gives me an extraordinarily solid feeling just to have a letter written out of knowledge. And then you sound detailedly and specifically well—which is so much more convincing than the cable.

First I want to explain in more detail about letters so you can quash that association with the time I left Samoa and didn't write. Just before Easter and directly following the letters which you answered in the first letter to Sydney—I wrote you a very long letter—but there was no mail between that time and our decision to leave—so I brought it out with me. In that interval things all changed again—and on the boat down I wrote a new long explanatory letter to be mailed when we reached Sydney and precede us as we intended staying over along the way. Then as I wrote you by the time we reached Sydney all that had blown up and there were a few days there which were so mad that writing was hopeless. . . .

You seem to have the feeling that I'm telling you very little but it's only that I've not wanted to give you nightmares, my darling.

Life is still a very unstable and expensive affair. I've been staying out at Carrie's for a week while she gets a new house in order and left Gregory and Reo together. Half the complications is Reo's crazy ambivalent attitude towards Gregory—also it's wretched that Gregory is probably the best friend that he has in the world at the moment. Anyway it was a good thing to leave them together and let them talk. I'd rather have Gregory acting entirely from genuine emotion about Reo than from rules. I find I can't get up much enthusiasm for rules. As long as Gregory and I are combined in our attitude toward Reo and that our happiness must be made secondary to his, we can both manage. We have the same kind of timeless relationship that you and I have and he has the additional prop of loving Reo. He's an extraordinarily different person from what he was at Christmas—then he seemed the weaker and Reo the stronger. Now the positions are reversed— he is far stronger and surer—and it is Reo who needs care. . . .

There is no possibility of attempting to build up again a complete and explicit relationship with Reo. We have decided to regard our work as a child to which we are both bound until it is completed—which will be a year or so. It is quite clear that life in America does not fit his picture very well. I think it is likely that he will come back to Australia and do some Australian field work under Elkin[33]—say next June or so. I think I shall very possibly apply for a Guggenheim for that year. I shall never go into the field again with Reo under any conceivable circumstances.

He has elected for a life of complete reserve and complete freedom in which each person makes an effort to keep up appearances and does exactly as they like. It means nothing to me at all—in fact it doesn't seem to be a relationship at all. Gregory says he feels exactly as if he had been asked to stop calling him "Reo" and to substitute "Mr. Fortune." However, he has proved pretty conclusively that he can't stand the conventions of a truth-telling world—which was ofcourse what you said in Paris seven years ago. I am equally convinced that the world of surface living and polite evasions means nothing to me. Meanwhile—there is our common work—which is interlocked in a thousand ways and cannot be sacrificed to anybody's private life.

You say several times that you don't understand Gregory's problem. Up to the trip down it was one of complete self-abnegation based on a desire to preserve my marriage which I had assured him was the most important value in my life. Also, he was still tied by an obligation to Steve so that neither of us ever indulged in any fantasies, or even in daydreams to ourselves. But Reo had one consuming urge—revenge—and he took his first opportunity—I didn't see it as such then—I thought he'd learn that it was possible to love two people and so I would be cleared of all the abuse which I'd had heaped on me for my wicked feelings. Also, through his eyes I saw M[ira] quite differently at first. Well, he's had his revenge, and to get it, he threw Gregory and me together—without doing that he'd have been unhappy over this disequilibrium. And ofcourse that has meant the unleashing of our rigorously controlled nerves and imaginations.

Gregory no longer has as a prop the belief that my marriage is a beautiful thing which should be preserved at any cost. He'd marry me tomorrow if it were possible, or he'd take me to England if it weren't—and stand the racket of it. But he does agree with me about the importance of the work and the danger of leaving Reo adrift. M[ira] is no tie. Her picture is one affair after another—all gay and light. She won't make a slight sacrifice of any sort. About three weeks ago—a week after I got Steve left—I decided that the one chance of saving anything from the wreckage was for Gregory to go home. He was willing to go. I asked M[ira] to cooperate with me—I told her Reo was wearing himself out and I couldn't stand his worry and that both Gregory and I thought it would be better for him to go. She promised to help me and then she threw all her weight into persuading Reo that Gregory shouldn't go! Ofcourse it's far more convenient for her for him to stay and take up my time. Reo made a point of affection to Gregory out of his request to stay so we decided to try it—and now what little was left is torn up. However, she has things organized even more to suit her. I'm out of the way, Gregory is there to occupy Reo when she has more important fish to fry—and she can summon Reo when she wants him.

And that will have to go on until the 25th of August—I won't go back to live in that flat—it's too hard on Gregory's nerves. Carrie and I have a flat in another part of Sydney which Reo knows nothing about and we are staying in town from Tuesday to Thursday. I have some reason for coming to town occasionally and meet Reo for lunch or tea—and he comes out to Castle Hill occasionally. I took him to the Consul's and arranged his visa last Tuesday. Today I speak at a luncheon and meet him for tea. They have a maid to look after them—who I found two years ago—has been Carrie's ever since—and she is devoted and silent.

I'll try to get another letter on the *Monterey*, but it's hard to get mail in from the country. You'll only get this about a month before I get home— My darling—Oh, it will be heaven to be in your arms again.

Margaret

Mead wrote at least three letters that do not seem to have been sent to Fortune after their separation that show her unsettled state of mind. Next is one of these, dated September 10, 1933, on a ship home from New Guinea by herself. [h]

Reo darling,

This is going to be one of the letters which I shall probably not send you. But I keep on writing them because then if I should die suddenly you could have them—at least that is the reason that I keep them. I write them because I can't help it. I've been reading Marie's proof all morning and it makes me ache with homesickness for you. All I know is that you were in Sydney when I left Honolulu, for Gregory cabled me you were well. Oh, I wonder, I wonder, what you are going to do. If only this terrible carking doubt as to whether you love me at all, or want me at all. As to whether I can ever make you happy hadn't entered in, I'd be so much clearer. For it is really true that I care more about your happiness and success than about anything else in the world. And if I'm not really the right person for you, then I mustn't make any positive moves to influence you to come home.

Oh, dear—I wish you hadn't hit me where it would show that night over there. I wouldn't have had to go away from you.[34]

[End of letter]

This is a curiously unfinished letter. Mead late in life revealed that Fortune not only knocked her down in Tchambuli, but also that she was pregnant at the time. She blamed him for her miscarriage at a time when she had made the conscious decision to try to have a baby, in spite of having been told by a doctor that she could probably not carry a baby to term.[35] They had been engaged for over six weeks in what Bateson called "soul rakings," intense thinking together about anthropology and about themselves, that resulted in the idea of the squares.

Mead later wrote to Bateson, June 12, 1934, that they had been in "an extraordinarily abnormal state. . . . It was of course a form of madness, a rate of thinking and feeling that one couldn't keep up for more than a week or so without snapping into something too remote to matter. . . ." and their tenuously balanced relationship did snap. Mead seemed to accept Fortune's right to punish her within some drawn limits—not to hit her where it showed. But she used the story of his hitting her later, privately among friends, as justification for the end of her marriage. She seems never to have directly referred to the miscarriage in letters to friends or family, even those to Benedict or Bateson; nor in letters to Fortune.

To Gregory Bateson in England, October 8, 1933. At loose ends in New York, Mead once again found herself questioning her philosophy of free love with another potential husband.

[no salutation]

. . . I feel singularly free and lonelier than I ever have in my life, but that without any unpleasant affect. I have always suspected that it wasn't that I needed people and was afraid to be alone, but only that I felt people needed me and so I dashed around after them. Now everyone is happy and adultly adjusted in their own terms; I have been away so long that everyone has learned to get along without me. Even the people who want to see me most because I am the only person whom they talk to about their troubles, have learned to curb any urgency about it. During the two years here with Reo I had to be so inaccessible and make such a point of keeping other people from making moves that it really amounts to a five year break in the old life, the old life which now seems so vivid and is really so very far gone.

I hadn't realized how I had trained everyone to regard me as hopelessly busy and to only come to me in major emergencies. Now even the way in which they handle major emergencies has been reorganized. And so much of my life was organized in terms of other people's needs, not who I wanted to see but who wanted most to see me. And so in a curious free fashion I am wandering about, seeing whom I like, without any sense of responsibility to anyone, not even to Ruth. Of course in the two years I was here in New York before Ruth and Marie both needed me. Now Ruth is all settled with Nat, Marie is in Pittsburgh, my little sisters are in Chicago, Louise and Léonie and Paula[36] are married and caught up in their own patterns.

And don't think I am making the point that I don't feel plenty of love and affection about me. I am perhaps more conscious of it than ever. But a universe in which I make the choices seems so strange as to be almost fantastic. I don't even mind sleeping in my big flat all alone, although I have hardly spent a dozen nights alone—in civilization—in as many years. The element of making my own plans has ofcourse never bothered me, but being free to make them everyday is thoroughly unreal.

The more I think about it the more I think that Reo is pretty right in his point about one relationship impinging badly upon another. I don't believe now that I am capable, or ever have been, of handling two love relationships. From the minute I began to fall in love with Ruth, I fell out of love with Luther, and so on. But I wasn't sufficiently placed to assert my own temperament against western attitudes.[37] And ofcourse Luther always felt the same way. While asserting that one could love several people—which was the standard attitude about—the one time he has a chance to demonstrate it—it was his year in Europe and he had met a girl who [had] gone to Vienna and hadn't the money to go there, and I sent him $100 for a Christmas present, and he decided not to go—he baulked. Meeting Steve in Brisbane only got you into a muddle. We probably don't ever have to handle them quite as the northerners do with suppression and failure to face the facts at all, but now I should be inclined to say that faced with Ambunti, I should get out. I mean if I ever again have to face the fact that I am falling in love with someone other than the person to whom I want to give full allegiance.

As one failure to see clearly begets another, we are all caught now in this tangle out of which heaven only knows if we will ever come. But the primary reason why I left Sydney when I did was that I couldn't handle the conflict; I have got to get out of my relationship with Reo before I can really be rightly happy with you. And if that blessed outcome ever arrives, I am all for cherishing my allegiance to you and for my part, doing no experimenting at all on any theory of the possibility of several contemporaneous relationships. Don't darling, think that this is meant as counsel to you now. I am just talking about my feeling about myself. You know for yourself that one love excludes another, anyway, only you explained it on the point of inhibitors or something which I think isn't enough.

The image of you holding my spirit in your hands is so very present with me. No "Closed Doors," no battering onslaught, at all. I can just lay it quietly in your hands and know it is held completely and for always. Whatever happens to us and however much we have to lose because I have made such a twisted pattern of my life, at least we shall always have that. Oh, Gregory, I love you so.

To Gregory Bateson, November 7, 1933, from New York.

. . . I had a difficult evening with Ruth last night, because she said that I had thrown a monkey wrench into Edward's life. I find it very difficult now that I have to go back over my life and see what a dominant role I have played, while not meaning to do so, in the lives of less vigorous or less intelligent people. It's of the nature of a sin to take a role for which one has no temperamental preference, and ofcourse one deserts it in the end, and lets the other person down. That I think is where Luther felt, and quite justly, a sense of betrayal. I had dominated him, he had accepted my domination with whole-hearted devotion and then I walked off and said "I don't want to do this." This is not ofcourse by way of repentance. I very seldom saw myself as dominating and when I did I attempted to get out in every relationship where I thought it would matter.

I was willing to boss Pelham and little Louise and David[38] because they are all or were all, either children who needed bringing up, or weak people who needed looking after. But now that I know ofcourse, I have a frightful sense of needing to be very careful, because it is so easy for me to give direction to other people's lives just by the very vigour with which I can picture one course of action as opposed to another. And about Steve, I'm worried, worried as to whether I fundamentally am enough involved for it to be right for me to become more involved and so involve her more, and afraid that I will dominate her and then inevitably be driven to desert such an uncongenial role.

If I could consistently see her as a child who has to be brought up, I could manage, but sometimes I am not sure that I can bring her up, and if I can't, if it isn't to be a definitely cherishing relationship which she will outgrow, then I don't know that I have any right to get into it. I know that when I tend to go over and over my past relationships to people and try to reckon up the harm that I have done, it is always a sign that I am uncertain about a present step—like my obsessive interest in the whereabouts of notes when there is a hurricane in the air. I am not comfortable about the whole thing. I see her coming over here and making a temporary personal heaven or hell out of her relationship to me and that is about all. And what she should be doing is working out some relationship to life and getting some kind of work. I wish I knew whether she could write or not. . . .

To Gregory Bateson, November 28, 1933, as she fantasized about being his wife. In this letter, Mead discussed at length the issues of balance and reciprocity in relationships, not just as they involved two people, but also as loyalties to third parties were invoked. Her analysis, as is typical of her in this period, used the language of the squares, as well as the Pidgin concept of bekim (which she translates as "backing"). As this letter illustrates, Mead's intimate relationships were often stoked by intense intellectual discussion, whether in person or on paper.

. . . My crawling-inside day dreams, have now got more content and structure. Inside has become your house, in which I live, under a complete cloak of your name and my identity is only that of being your wife, and no one knows anything else about me at all they are just pleased when I understand what they say, these your learned friends, who sometimes come to tea. And nobody knows that I have ever written a book. I feel completely smothered by a hard glittering armour of built up and unloved identity at present. But of course to strip it off and leave no cloak, means standing all undressed in the center of a crowded place and having to make moves, which I also hate. But in your house, I could just live happily folded away.

. . . It's really not fair to talk of Reo's having a will to power, in general, or of Brown's having one. Malinowski probably has more. And it's as inaccurate to talk about southerners as having a will to be dominated, what we have is a repulsion to dominating, just as a N has a repulsion towards being dominated. But the NW enjoys his power because he can feel also, how it feels to be dominated, and the weak SW enjoys being dominated in the same way. But N and S have instead a kind of insistent individualism, the N in ego terms, the S in libido terms. And they have both a great desire to make things even in the world, as over against wanting either to be in a position of dominance or submission. I think that is the basis upon which a super-ego for N's of a decent amount of self-abnegation can be built. It was Reo's repentance when he had been unfair to us, only as he said, repentance hurt him so that he usually felt that he had done enough without making any further amends.

And the correlate of that is my willingness to sacrifice for him, not because I enjoy hurting myself or being hurt, but because I want to establish a balance, to suffer as much as he has suffered. As his repudiation of suffering is so enormous because it affects his ego, and yours and mine is correspondingly slight because of our so different organisation, we have a tendency, or at least I do and you too at times, to seem masochist, but I don't think we ever are. It's the point which I made about my day dreams that he will do something dreadful to me, and then *I will be free.*

I find that every time Ruth leaves me at ten o'clock to go home to Nat,[39] or can't make an engagement because of Nat, there is part of me that is pleased, for I think "That wipes out one of the times when you (Ruth) were lonely and I was helpless to spend time with you." But that is *backing*, not masochism, and I think Reo's picking up Myra[40] should be interpreted as backing, not sadism.

But here is a further point. N initiated situations in which they make too great demands, and so continually put S in conflict between meeting those demands, out of responsiveness—their upper libido layer—or wanting instead to assert their ego and back the over-demand. Similarly, if we initiate, we are likely to initiate a too-great concession to suit their notions, and they are torn between taking it, which their ego-upper-layer would dictate, or instead trying to back it from the libido level by a sacrifice which they find it almost impossible to make.

And then culture comes in. If they can phrase their compensatory, that is reciprocal, abnegation culturally, it doesn't hurt their egos so much, and if we can phrase our reciprocal assertiveness culturally, it doesn't hurt our libidos so much. So if I phrase it that I owe certain duties to my work, to the institutions which financed me, to the training I have had, then I get the energy in the field, to fight off Reo's excessive demands about taking informants away from me, or publication or something.

But without such cultural bulwarks, my desire to please him, and respond to his wishes was so great that it was likely to win, but at a price: underneath my ego scowled and was restless. That ofcourse is what Reo meant by talking about his over-drafts on my masochism. But I think they should be phrased as overdrafts on my emotion for him; and probably my sacrifices and efforts for him were equally overdrafts on his emotion for me, which became more and more conscious of a lack of balance, a lack of reciprocity, of always taking and never giving. That left his emotion in a bad way and my ego in a bad way. And I should think that a N who carries around a sense of an unpaid debt of feeling, would carry a corresponding fear and insecurity which would run quickly into extreme jealousy. And a S who carries around an undischarged sense of resentment will always be quick to anger given cause. . . .

Is this clear at all, some of it I am thinking out as I write, which may make for freshness of literary style but does not make for perfect coherence. But it is so good just to be talking to you, even on a typewriter.

. . . Oh, Gregory, darling, it's good just to be alive and love you, so good.

Margaret

To Gregory Bateson, January 26–27, 1934. By this time she had made the decision to leave Fortune no matter what Bateson decided to do in his own life.

Darling,

. . . I've turned a definite corner in my own thinking. Up till yesterday, I have been seeing my life as an alternative between marrying you—if that were humanly possible—or going on in some fashion with Reo, if it weren't. I've passed that now and I think for good, and I now see the alternatives as marrying you or going on alone, doing more field work alone, and putting most of my energies into my work. But I am convinced that any hope of even a modus vivendi with Reo is over forever. Whether you and I can ever marry is pretty tied up—with the divorce, with whether your mother approves and if she doesn't whether you could possibly afford it (those two things are pretty mutually inter-dependent aren't they?), with whether what position you hold or want to hold will be too badly jeopardized by marrying me. It doesn't look like a tangle which could be very readily solved. But you need no longer have as one of your nightmares my going back and trying life over with Reo.

Just how the details are to be managed I don't know, but I'll manage them somehow. I can always do field work on my salary, and the year after the Guggenheim,[41] if he were here in New York, I could go out into the field. I am practically the only person on the Museum staff who would be willing to do that. I feel very very clear about this and not as if I would change my mind. I have proved this year that I can live alone, and I could do it forever if necessary. I am physically stronger than I have been for ten years. And it doesn't look as if there were any danger of Reo's coming over here at present. But obviously I can't tell him any of this at present; he simply couldn't stand it.

Now on that basis, can't you work, darling. Can't we both work and be ready to write books happily together this summer, whatever eventuated. See Reo if you want to and think it wise,[42] and don't get any fits about your having spoiled our lives, on the basis of renewed affection for him. Our lives were pretty awful and he certainly will be much happier some other way, and so will I. And meanwhile we can hope for happiness together presently, if we both want [it] enough. And if you find you don't want it, it will be because you want somebody else more, and that will probably mean happiness too. And my work is sufficiently an anchor to keep me steady. To have loved you has been the loveliest thing which ever happened to me, and nothing in the world can take it from me.

Maybe this letter sounds harsh, if it does it's just me answering Noel[43] indirectly. But I feel very clear and lucid and straight, and going back to the Sepik and doing Meramba by myself seems more possible than ever trying to live with Reo again. My sweetheart, I love you and belong to you, and soon it will be June.

Margaret

To Reo Fortune, June 6, 1934. This letter demonstrates some of the problems Fortune and Mead were having as they moved toward a formal divorce. The work they had collaborated on was a particularly sensitive subject.

Dear Reo,

Your letter written when there had been no letter from me to disturb you came last night and was one of the nicest letters which you have written this year. In it you said that Malinowski advised you against Cairo.[44] I should think that meant you ought to take it. On your own statement, Malinowski does not advise his rivals for their own good. But that is just passing comment. The thing which I am mainly interested in is that you should do what you want, and if you want the field work, it's that much better for anthropology ofcourse, except that it would also be very good for anthropology for you to get more written up before you go.

Anything that I have ever said about cooperation in work has never been based upon criticism or repudiation of your work. Any criticisms which I have ever made of that have always been special, in relation to some small point, never of the whole. When I sorted out your Manus mss., for Jeannette,[45] I kept reading bits of it and crying, because it was so very good and because it wasn't being published quickly enough, as it should be. But as long as you continue to regard my approach as some form of insanity, it seems a little difficult to talk about cooperation.

I am continuing to find the two points: the need of recognizing the way in which society institutionalizes sex temperament as a basic point within the personal relations of that particular culture; the fruitfulness of thinking of certain aspects of human nature in terms of a fourfold polarity; very useful and stimulating ones in my use of material. I know how very difficult it is for me to work under your disapproval and contempt, even under the conditions of Omaha where you constantly heaped depreciation upon the job which I had been sent out to do.

I valued the fact that you thought any of my work good very much; I valued what work we had succeeded in doing together without the virus of competition entering in also. After the way in which you have talked about my Manus kinship paper I can hardly have much faith in that even when it was past, or seemed past. The only terms in which you can grudge me that piece of work are terms on which you would grudge me anything I could do in a culture which we studied together. No impartial judge of the facts would say otherwise about that Manus work, against the setting which I gave it in the introduction. Yet you did say otherwise. The fact that I was your wife did not prevent you as seeing me as a competitor and someone who was greedily taking what you wanted out of the culture which we were supposed to be working in common.

That is the reason why I do not see us as actually cooperating on one job under any circumstances again, although you remain one of the few people whose opinion I care anything about. And I care terribly terribly about your work and your having a chance to do it.

I am dead tired. I have not managed my planning for this Hanover Conference very well. I gave my research money to Elizabeth to work on the Arapesh material culture, and tried to do it all myself, and it's more than I bargained for, combined with everything else. I am just finishing up a brief rewrite of Junod. It's really beautiful stuff. He knew what those people were about.

The Hanover offer came just when you were writing about the importance of getting in touch with the big foundations. Well, this is my best and possibly only chance to interest the Rockefeller Education Board, and I felt it was too good to miss. The discussions which preceded your and my papers in Sydney hardly suggested that our giving papers at the same conference was a very felicitous arrangement. This Hanover thing is a piece of work which I think is worth doing—organizing material to be used at a lower educational level, high schools and normal schools, and junior colleges.

Speaking of economics, I am trying to get things arranged so that there is no danger of my ever becoming a burden to you. I am taking out heavy accident and health insurance, and my life insurance is paid up through next May.

I have to go now. Elizabeth is giving a dinner party for six so I am having dinner out.

Love,

[Handwritten P.S.] I am glad that you feel that the work is so well worth doing whatever happens.

To Reo Fortune, June 19, 1934. Mead expanded on her theory of temperament and the squares and where many people she knew fit in. It was the development of this theory that led to the meltdown on the Sepik in 1933. Fortune took it as an attack on him as he was placed as a Northerner and Mead and Bateson as Southerners.

Dear Reo,

. . . You asked about my spreading a theory of temperament which led to hostility. I don't think I am doing that, although the hardest point is the Turks. That is the point which Ruth has taken up with great enthusiasm; she pronounces the word Turk with a vigour which I have never attained. It's interesting to compare the way she sees them, as pursuing ignoble, non-spiritual *ends*, while I see them primarily in terms of their use of ignoble, contemptible *means*. I do so little really feel difference in temperament as a point of hostility—I have loved and admired so many people of different temperaments, Léonie, Ruth, Boas, Ogburn—I don't feel the point of difference in temperament as primarily a point between you and me at all. It has been back of so many of my attitudes for years, my old point about being "Martha."[46]

What people I have talked to about the point have all been interested and unhurt by feeling any hostility, even Dadda who is a good deal of a Turk, said "You say something like 60 per cent of the world are like me?" and when I said yes, he said: "That suits me." And I told Hugh and Isabel[47] about it, Hugh I would call N. and Isabel S. and Hugh was as pleased as Isabel, and felt that it was a scheme which gave him a place which he liked. The only problem which I have myself is about Turks, I am not quite as for-[bearing] of Louise Rosenblatt's snobbery as I used to be, for instance. And yet I suppose that if one had perfect understanding and kindness, one would find nothing in the Turks to forgive either. I am not sure.

As I see temperament at the moment, there are numerous sets of polarities, which are significant on different levels and for different people. The basic points beneath these various contrasts are probably innate. How many of them are associated as linked traits, there is as yet no way of knowing, perhaps a great many, perhaps very few. For instance, take Ruth's dichotomy between Dionysian and Apollonian. This fits her and Léonie; it has nothing to do with you or me, or Stanley or Papa Franz. It is possible to say that that dichotomy is a smaller one within the polarities, as I draw them, and that only N.E. will be concerned with that particular point, or it is possible to say that it is a function of something else, strength of temper, or strength of acculturated character, etc.

Certainly in my own case, the polarities which I worked out on the Sepik are so important in the understanding of other people, that it is more significant for me to see Léonie and Ruth as very similar N.E.'s with strongly developed S.W. personalities, but with Ruth writing out of her real N.E. nature and Léonie writing out of her built up acquired Apollonian day dream personality. Léonie agrees with that version of herself and Ruth. But as every theory of personality at present is to a great extent intuitive and limited by the temperament of its begetter, it is impossible to tell whether the great relevance which it has for the begetter is real objectively or merely real subjectively. For each person who is in any way a pronounced or extreme type in any given respect, the way in which others manifest or do not manifest the same or contrasting character traits is important.

So for you and me, people's attitudes to other people are very important, for Léonie, people's attitudes towards art and religion, towards the mystical points which absorb her whole personality and emotion, are far more significant. What I tried to do was to build up a picture of contrasting temperaments which would include not only the points most relevant to me, but also the points which I felt would be relevant to Marie, to Edward, to Eleanor Phillips, etc. And I had such good material about them and their attitudes that I made a stab at it. I knew at the time that as far as delimiting and describing their points of view, I was doing only the barest sketching.

When understanding others becomes a real and sophisticated part of our training of people, it may be that the emphases which I worked out will only be significant to southerners, or only to northerners and southerners, and not to others, that it is a kind of sorting of attitudes which is non-revealing to people who do not show the special n-s traits sharply. Or it may be that it will be found to be a genuine underpinning of personality to which many other things can be attached. There are many interesting points, the extent to which E and W look alike to N and S, as compared with the way N and S look alike to E and W.[48] I still believe that the point is an hereditary one and that the same or very similar distributions can be found in different present day, physically defined "Races of Men." There may be a lot of these hereditary points, which can be inherited independently, or they [may] be inherited as a complex.

But as I have never felt any race hostility, neither do I feel that a belief in temperamental differences means a war. Culture can introduce just as many difficulties. Léonie and Bill,[49] who are extraordinarily alike in many ways spend all their energies, each trying to make the other bring the "artist" a cup of coffee. It is a situation full of conflict on very different grounds. Certainly the fact that temperaments complement each other is as important as the fact that they are contrasting, and there is no telling in any individual case, which will be most important.

Love,

To Gregory Bateson, July 6, 1934, about Fortune.

My darling,

My office is all swept and bare, everything is stowed away behind green curtained glass doors with little notes on the outside. All the Sepik photographs and the B.M.[50] postal cards are put away, and altogether I really feel as if I would be leaving soon. There is no telling when you decided to start to write there. I am quite serene and only a little hungry.

Yesterday I got pages and pages from Reo, raging over the entire surface of the last year with a whole new set of explanations and rationalisations. He stayed in Sydney to keep me from going home to Papa Franz with an idea which would ruin my scientific reputation—although the plan had been to spend several months in New Zealand; he hit me because I had mentioned money in front of Mira—the time—four weeks before—when she had asked me if I'd made a lot of money on my book and I said a little and I'd saved a bit so we could drink it up in Sydney! But he is still suggesting some sort of rapprochement, a sterile sort without intimacy or insistence on fidelity or mutual confidence, or anything in particular. Then there was pages more of raving about the squares. I wrote back that I wouldn't mention the squares again, and that, while I was not repudiating the past, I did not feel that I could possibly make anything of the future with him, that I felt it was beyond my strength just as he claimed that the Sepik had taxed him beyond his strength.

After writing the letter I went through all the poems that I wrote him in those five years—they are an amazing collection and his to me are just as amazing. We neither one ever gave the other an ounce of security. As a result I was desperate and he was jealous. He had said in his letter "Divorce and re-marriage is not the only alternative." To which I answered, repeating his phrase, and adding, "No, ofcourse we can go on as we are until I, if not you, are old." I think that is as far as it is safe to go at present. Also in this letter he said: "Keep your world with Bateson apart from you[r] world with me." This was anent talking squares, which looks, for the first time, as if he thought we still had a world, heaven knows.

He says that Malinowski has lent him money, that he is more indebted to Malinowski's kindness in money than to me—I hope to heaven that isn't a literal statement of indebtedness, and also that he has told Malinowski and Firth nothing of his private affairs—which is a comfort. I was afraid he had. Altogether, I think it's hopeful. I think once it is settled, he will be steadier, and I think there is a fair chance of getting it settled. I really do.

[No ending. Continues on July 7.]

To Caroline Tennant Kelly, July 28, 1934. Mead wrote to her Australian friend Kelly on the situation with Fortune and Bateson, trying to control the gossip and keep all of their careers on track. After they separated, Mead took it upon herself to try to manage Fortune's career, even though he repudiated her help, and this is a theme that comes up regularly in letters about him.

Carrie darling,

. . . Now as to the sphere of discourse which might be called "keeping Reo's halo on." Just look at the matter realistically. I have a job, a family who are secure whom I love, hundreds of friends, a publisher for my work, a secure and walled in life. Reo has no job, more enemies than friends, no one to back him up, a family who are so poor they could give him no more than a grudging house room, no publisher for his work, and very doubtful prospects for the future. I couldn't gather from your letter just what asper-sions Ian[51] was shielding[52] me from but I gather they aren't the real ones. Or are they? If the gossip is running that I got jealous of M[ira] and walked off and left Reo, or even that I am a greedy person about money and pres-tige and refuse to share with Reo and told him he had to get a job himself, let it go. And steer it so that he will not be condemned.

After all, all the world knows how hard it is for a man to be married to an awful intellectual woman, who makes a cheap popular success and flaunts it in his face. Who can blame him for taking his fling away from it? Etc. Etc. Malinowski has undoubtedly done everything he can to wean Reo away from me this winter without knowing just what he was doing. At least Reo says that he has told Firth and Malinowski nothing. All that they know is whatever the current gossip in Australia is. And that was what I wanted you to tell me, just what version Ian has and has been giving Lucy Mair.[53]

Chin [E.P.W. Chinnery] told G[regory] that the gossip was that I was hurt about M[ira]. Remember it doesn't do me much harm either if I am accused of having walked off in a jealous huff. That is perfectly conventional behaviour. The only thing that I am concerned with is whether G is involved in all this or not. If not, then I think things are taking a very good course. As far as all the world knows, I will have been in America continuously for the last two years; G continuously in England. Reo refuses to come home to America, instead goes back off to New Guinea without me. Meanwhile it was very well documented here that I expected him back this winter. I came back, took the same big apartment which was originally taken to suit his special needs, put his name on the door and in the telephone book, expected him—as his original cable said—to be home by Christmas, and he never came. As far as America is concerned, they all think I am going to Europe to see him so they won't think there has been a break.

There is no communication between the English crowd and the American, or between the American and the Australian. You'd be surprised how cut the lines are. Reo writes to no one in this country except Ruth. If when Reo gets to Australia he hears any rumour from America that I have been abroad, you can say that I took an extra month's leave of absence and stayed away from the Museum so they would think I had gone abroad and not know there was any break between us. But that is only in case he hears a rumour which I think it is unlikely that he will hear. I really think things are working out pretty well, though damn slowly. Still the slowness is a good point in itself. The Cambridge chair doesn't come up for three years yet. I have got to get this Sex and Temperament book published under the status quo and while Reo is off the scene of action or he might attack it disastrously. If it gets a six months start of his attack, then the attack wouldn't matter. But there can't be any scandal about that.

You see I feel that in the lives of people over twenty-one, that part of our culture to which they have dedicated their attention must come first, and that I have no right to make any personal choice which will jeopardize anthropological work in the future. That has several facets: Reo must be kept sane and working and making his unique and beautiful contribution to anthropological work; I must do nothing which will injure my usefulness, either by getting involved in a scandal or wrecking my private life so that I can't work; Gregory's value to anthropology at the moment is a matter of his being the probable candidate for the Cambridge chair and the possible writer of a good book on the Sepik. From the standpoint of anthropology he is a future. Reo and I are present values.

Therefore in a clear eyed essay [sic: assay] of the whole problem from the standpoint of anthropology it reads like this. If I can get out of my marriage to Reo, quietly, without scandal and without breaking either Reo or myself, it can be done. Furthermore, if I can marry Gregory, he and I can do better work together than we could do apart, and furthermore it will mean a center of good work in England. Brown and Ruth and Lloyd Warner, (he to a lesser degree intellectually but a greater degree politically) can look after the future of anthropological thought in the U.S. with Columbia, Chicago and Harvard back of them.[54]

It's a large order to manage but I think if we keep our heads and our health we can do it. Ofcourse there will be a large element of luck too. For instance what happens to G's rivals for the Cambridge chair; whether there is a European war, etc. etc. But it's just a question of "Get the distaff ready and God will send the flax." You'll have a big role to play in Sydney in keeping Reo calm and happy, keeping him out of useless entanglements. By useless entanglements I mean ones which will lead nowhere, with women who are married already or are not marriageable, and will use up more money than they will give him peace and security, or will involve him in scandal and talk. This applies ofcourse to new entanglements. If M[ira] is there I don't expect you can do anything except divert any funds which might pass into her hands. He is going to the Mandated Territory not to [Australia so] that I don't think H.E.

will have any role to play but keep him friendly in case he has. In N.G. Chinnery and Monte[55] and the Administrator will do everything they can for him for my sake and I'll feel reasonably safe about him. Monte particularly will smooth his way. He's a southerner with tact and discretion. He knows that there is some difficulty but nothing about any of the other actors on the stage. At least he didn't. How much all N.G. knows about M[ira] I don't know now. One thing—I doubt if she will gossip because the credit she can take to herself for alienation of affection would be diminished thereby.

To sum up: I want him to stay in Sydney as short a time as possible, to equip himself well and carefully, and to have to live through as few of his old emotions and miseries there as possible. When your name was last mentioned you were A high again, and be[ing] quoted as supporting his point of view that the whole theory of the squares was just a new way of saying that one liked one's lover better than one's husband. Let him think you don't accept the squares; never use a word of square terminology in his presence. He goes perfectly wild whenever they are mentioned either by me or Gregory. Don't mention Gregory except when he does and better profess not to know much about Gregory's movements beyond a letter earlier in the year, or something like that. . . .

Mead spent 1934 and 1935 in New York, working at the American Museum of Natural History, while Bateson was at Cambridge in England. She took a late summer vacation from the museum in 1934, and met him in Ireland for several weeks.

To Gregory Bateson, November 7, 1934, after the trip to Ireland.

[No salutation]

It's getting rougher every minute and I am feeling less fond of this horrible time-killing life than ever. I have just written to Steve.[56] I wonder, if one never did a single thing to please another person which didn't spring from one's own desire—not one's desire to please them, to fit in with their different view of what is lovely or right or necessary—would all relationships be honest? Or isn't it fair to confine oneself to relationships in which that is so. For it is ofcourse with you. I have never written you a letter because I ought, or expressed a polite interest in something in which I wasn't interested, or remembered to pay attention to something which you thought was important but which I didn't think mattered a bean, or smiled when really I was bored, or looked concerned when I felt like smiling. Never once. Do you know what that means in freedom from constraint, in not having a single tiny bit of sense of duty anywhere?

And of course you do, because you feel the same way. Loving you is just like breathing, as effortless, and as lovely. You sometimes think it will lose that quality if one mentions it. But it doesn't spoil regular breathing to take a special deep breath full of apple blossoms, does it, or even to hold your breath for a moment to feel all the happier when you breathe regularly again. Self consciousness can probably ruin sex, and maybe eating and drinking, at least eating caviar and drinking wine, if not eating bread and butter and water—but it can't do much with something as fundamental as breathing, so there is no room for fear.

Don't forget that man in the BM[57] who wants your notes on the Galapagos finches, will you darling?

I am being bad about my wedding ring. I never wear it you know in America, but I feel so lonely when I take it off. I love you Gregory.

To Gregory Bateson, January 10, 1935, from Chicago.

My darling—I am in one of my happiest kinds of moods about you, the kind of mood which makes me think that I probably dreamt about you, although I can't remember the dream, but I feel as if you had just gone around the corner to get some tobacco for your pipe, and might reappear at any moment and put your hand on my hair as you crossed the room to eidos.[58] It makes my hair particularly gay and curly just to think of it.

. . . Sometimes I think that my twin figure of speech, in which each person was hunting for the other twin, had a very genuine aptness—yes, I know I didn't use it as a figure of speech, but it's a good one anyway . . . You know, I must have had a twin brother fantasy as a very small child. By the time I remember it, it had become a twin sister, but I imagine there was a twin brother period preceding it. And I think my farfetched twin theory was a final attempt to face exogamy, my most desperate effort to stick to Reo in the face of discovered endogamy.

There was a time somewhere after Easter [1933] when I woke up one night with a great sense of having found the "other twin" which must go back into very early fantasy development. I'll have to look up what the analysts have to say about twins. It's probable that it came from wishing Dick were big enough to play with me, and thinking "if he were only my twin, he would be." I can remember wishing for an older brother instead of a younger one. That whole Sepik experience was as good as an analysis except that there was no one there to take notes part of the time, and so I have to pick up the threads slowly. The daydream of wanting to be of one-mind with the person I loved had become so overlaid by my fascination with the diagonal[59] and by negative self feeling that it was almost buried. I suppose your premium on one-mindedness can be put down to your image of your two brothers, and to your image of your one-minded parents. So you have the fantasy picture of moving into the perfect relationship, and I had the fantasy that it was lost—the other child was stolen by robbers or gypsies and I had to search and find it.

I wonder if you are still thinking about being analyzed. Of course you have never worked out your early childhood experiences with any emotion, re-realised them as it were, and that is probably necessary for full freedom from their dominance. If you learned to handle your own dreams, though, you might be able to do a good deal with them, if you had somebody like me around to make you stick to an association occasionally, if you wanted to dodge it. That and taking notes and reminding people of things they want to forget are the main functions of the analyst, it seems.

My darling, be happy. and soon soon you will be coming.[60]

Margaret

To Gregory Bateson, January 3, 1936, as she planned to meet him in Bali and be married. Their plans changed a number of times before she actually departed.

My darling,

It's a lovely world. Surely God didn't mean me to go in December. The thing which made me wonder all along about it was the fact that the set of documents which had been authenticated by the State Department in Mexico had never come. David[61] thought I probably wouldn't need them, but it made me a little uneasy because the pattern wasn't working out quite right.

They came this week. You may be disgusted with such irrational behaviour but I had been feeling a little blue and distrait with all the postponements, and I even wrote you a rather fittish letter last Saturday and then tore it up. I already had a good letter on that boat. When the documents came, suddenly I was quite sure that all our choices had been right.

AND THEN Elizabeth got married yesterday, radiant and lovely, and now I can leave her without a worry in the world. As Léonie says, Elizabeth and Bill "have the same look in their eye." He's a Northern fey and they are quite close enough together for perfect happiness I think. If I had gone before Christmas, two strange girls would have had the apartment; this way Jules and Zunia[62] both feys and friends of theirs, will have part of it and Elizabeth will come in weekends and everything is just perfect for them. And Elizabeth feels more integrated and blessed to have had me here to approve all the final ceremonies. . . .

Ogburn[63] was here this week and I told him. He propped your picture up on the luncheon table and left it there all through lunch, so he [could] "get acquainted" with you. I told him quite a good deal of the story and he kept say[ing] "Excuse me Margaret, I know this isn't what I should say, but My God! What a story! *What* a story!" He had just seen John[64] and was all prepared to believe that John was in love with me, but I disabused his mind of that, and then he said, something which I had never thought of because my mind has been all on the fact that I didn't wreck John rather than that I had been any use, that he thought I was responsible for John's having pulled himself together and got down to productive work. It hadn't ever occurred to me, and it was very pleasant.

Ogburn was specially interested in why I hadn't broken completely on the Sepik and decided sagely that it was because my libido was anchored to you. And I think that's a fair way to put it; at my most essential core I was completely safe. I suppose it's the same thing that I used to say when I said that if you and I had ever distrusted each other, ever had one little lover's tiff, it would have all blown up. Because I was so safe with you. I was ultimately quite safe.

Ogburn also went into great detail about my drive, wasn't I anxious for success, else how could I have succeeded so well. I told him the story of comp and coop[65] and I think he at last was convinced. (Ruth says that the reason people think I am ambitious is because I do so many things which other people are driven to only by a consuming ambition, just because I don't mind doing them.) Finally Ogburn sat back and said: "Well Margaret you'll be alright, because you can figure and you can figure even under emotion. I can't think under emotional strain and so I have to avoid all emotion." That is a greater deal truer of you ofcourse than it is of me. I think you have a much higher capacity for dispassionate planning. . . .

Gregory, my darling, in ten days I will be sailing, to meet you for always.
Margaret

To Lawrence K. Frank, March 22, 1936, from Soerabaja, Java. Mead gives her account of her marriage to Bateson. She gave different people different versions of the event. "The official version for the world," she wrote her friend Leah Josephson Hanna, March 21, 1936, "is that G. B. knew I was coming but I didn't know he was—and that he took a fast boat, got here first and intercepted me in Batavia—the Dutch residence laws were impossible and so we flew to Singapore." This explanation for Frank stretches the truth in places: the marriage was pre-arranged; Mead had only known Bateson for a little more than three years; and Bateson was actively involved in developing some of the theory Mead refers to his knowing.[66] *[h]*

Dear Larry—

Well my version of your 1936 plan is complete—with breathtaking suddenness. When I arrived in Batavia I found Gregory Bateson—whom I've known for years—waiting here armed with a Ciné camera, a dictaphone, a Balinese informant who could speak English, a proposal for scientific cooperation and an offer of marriage. I accepted the 5th. I had a frightful taste of the non-joys of apparent spinsterhood coming out on a ship. This plan had been a potentiality someday, but—we had to fly to Singapore—much to the amazement of the Museum[67] there—as I'd just gone through there a couple of weeks before. Singapore is the Gretna Green of these parts. I do wish you'd met Gregory—he was in America for a short time last year—but he was dashing about to Chicago and Boston—and you were also on the wing. However he knows all the points—including the points of the compass[68]—and the Hanover Outline, etc.—and will be one more laborer in the ripening field. . . .

Don't broadcast this news—but I am writing the few people I'm fondest of at once. The longer the news takes to get about, the better.

I hope your quest is nearing an end, Larry dear.

Affectionately,

Margaret

To Beatrice Bateson, April 6, 1936, from Bali. This was Mead's first letter to her new mother-in-law.

Dear Mrs. Bateson,

Your air mail letter came tonight and made us both very happy. I am sorry that our marriage came with so little warning, but it really was a question of starting this work out together from the start or not at all. Now in only ten days in Bali, we have the grammar of the language worked out and are in a fair way to find our village and start house building. We are drawing a set of house plans preparatory to actually choosing a site, and these include a special room for you when you come out to visit us. Gregory feels pretty confident that you will be persuaded to come. It is an extraordinary country, a combination of real native life and good motor car roads and we should be able to make you comfortable and show you some real things at the same time. After New Guinea, it seems like a fairy tale to be able to make trips in an hour which would take a week on the Sepik, and to see more ceremonies in one day than we saw there in months.

Working together is going very well. Our minds are quite different; I do not mind the masses of concrete detail, which bore Gregory, and he introduces order and method into my rather amorphous thinking. It has been very amusing learning the language together because Gregory doesn't believe the language is real until it is spoken and I don't believe it is real until it is written down. This little house that we have at present is so small that we have to work together in it so it provides a ready made test of our ability to cooperate.

Gregory is looking very well. He has, I think, gained a little weight, and he threw off his attack of fever miraculously quickly. We came all the way by boat instead of risking the train trip, in the event of a recurrence of Gregory's fever, but it never came back and the extra leisure of the ship was used for manuscript revision. Gregory made an analytical table of contents for his

NAVEN, and I proof read the writing I did on the boat. The Captain of the little ship was very shocked because we did not go sight-seeing in Lombok.

I am posting you tomorrow a copy of my *Kinship in the Admiralty Islands*; this should have gone to you sooner but we were a little slow in getting settled and locating papers and envelopes in our boxes. The circumstance that we each have complete sets of all sorts of equipment seems to prevent our being able to locate either set in a hurry.

The night before last I had a long dream about your coming to visit us and woke up with a memory of such a pleasant night. I hope that will happen when our house is built.

Yours ever,

To Ruth Benedict, April 22, 1936, about doing field work with Bateson in comparison to Fortune. [*h*]

Ruth darling,

I know my letters must have been a trial this last month, but we have been doing everything together—including letter writing and Gregory is so vividly interested in everything I do and write. He doesn't always read my letters but he has no sense of not doing it and may come and look casually over my shoulder at any moment—and I find that a little inhibiting. This is practically the first moment I've had alone since we were married—and this is just because my mouth is too awful looking to go and call on a government officer. Once we get our house built—and each has a house—we won't be living quite so much in each other's pocket and letter writing will be easier.

Do you remember you said once you'd like to write a play in which the lover and the husband were indistinguishable? What gives me a far weirder sense is when all the external circumstances are the same but the essence is absolutely different. So much of this is a repeat. Government, informants, language, deck tennis, doctoring the natives, malaria, tropical fever, etc. etc. Seen out of the corner of my eye in the same clothes, G. might be R. But when I unconsciously prepare myself for similar behavior I never get it— and sometimes nothing seems real.

I still don't believe really that I'm not going to be abused for having forgotten my hat and got my lip in this awful state, for instance. He is so uniformly gentle and reasonable it doesn't seem possible. We are being very self-conscious and critical of the whole problem of working together. On every single cognitive point, we are direct contrasts. Where he is systematic, I am [modulated][69] and vice versa, Where I have total recall, he has partial, but for words, for instance, he has total recall and I partial, etc. It's really so pat, it doesn't seem possible. I think probably he is under more of a strain than I am; working to fuller capacity, and steadily rather than in spurts of excitement. We have had one funny episode—a long period of apathy which arose out of his dislike of Beryl de Zoete,[70] who is undoubtedly, what John would call a castrating female and a witch. It just redocuments the point I've known all along that what I have to fear is never Gregory's loving other people, but only any negative feeling that he may have toward them—it was all so funny. When he discovered that "Beryl had holes in that tough hide of hers"—and he could be rude back—if he wanted to be—the fit was over. Incredibly, Beryl, after a few bricks, behaved much better.

It is so delightful to live with continual affection—easily and happily expressed—altogether—I am very very happy. Even six days of real misery with these lips hasn't disturbed my spirit.

And I think the point is true that I worked out—was it on the *Tapaneoli*—that one perfect relationship never threatens another perfect relationship—that it is only imperfect, incomplete, or partially realized relationships that interfere one with the other. I feel no pull in myself between you and Gregory—no sense of counter or opposing systems.

One point that is a little difficult is that "being a wife" redintegrates my relationship to Reo at a dozen points—in a way that having Gregory for a lover never did. My tongue stumbles in talking to the boys about him. After all Tuan (the Malay for Master) is very close to Tuag (the Arapesh version of the Malay word) etc. And all that besides confusing me at the moment brings Reo continually into my mind, to worry and worry and worry about. If he can only meet Eileen[71] and fall in love with her again.

Your picture smiles down so serenely from the shelf—the loveliest picture I've ever had. Do you realize there has been a different one for each field trip. And this the best of all. I love you, darling. I love you.

Margaret

To Leah Josephson Hanna, July 15, 1937, from Bali.

Dear Lee,

. . . The other news is that Reo is marrying his first sweetheart, a New Zealand girl whom he went to college with, fell madly in love with after she had sprained her ankle and he carried her for miles, who refused him in 1925 and as a result he flunked his BA that year. She had never married, and, according to the newspaper clipping I was sent—she is leaving in July to marry him in Hong Kong. No one can say that I am an enemy of the home; she is the fifth unmarried woman who has got married as a result of the faith in matrimony which my wayward conduct instilled in the male breast. I really feel proud of my record. When I was in New Zealand I had his brother promise to see that he met Eileen again, and presto—here it is. Now he has a job and a wife from his country and he ought to be alright. I feel very pleased. . . .

Mead and Bateson finished their field work in Bali and came home to the United States in 1939; Mead found out that she was pregnant. Meanwhile, World War II had begun in Europe in August, and Bateson decided he needed to go home to England while Mead stayed in New York to have their baby, who was born on December 8, 1939.

To Gregory Bateson, no date, but it was letter Number 5 sent on a ship called the Dixie Clipper, *due to leave September 16, 1939.*

Darling heart,

I am making this letter with carbon paper again, as I haven't got a new ribbon yet and the bad ribbon doesn't show on this crinkly paper. I came back from Philadelphia with a cold in my head, but I have now just about scotched it; it never got really bad only exasperating. I am settling down to life here at Marie's; it is ofcourse eminently comfortable, full of little glasses of orange juice and suggestions that I take hot baths and unquestioning affection—all of these must be very good for the baby, who kicks placidly. It has never kicked as hard again as the day after you sailed. Louise is taking your parting admonitions very seriously, and makes extra precautionary remarks whenever she leaves.

. . . It's hard to convey the strength with which I resist any thought of publishing anything by myself, or of writing or thinking by myself. I realise so keenly that all my hopes for the future of anthropology are bound up with you, that without you I see no one who can lead it; no one! I think Brown has shot his bow, I know Boas has, Ruth is no leader, and God knows I'm not. It's not just because I love you; I can imagine going on with anthropology with you in the world even if you didn't love me anymore, but were there to criticize and lead. Everytime I open my mouth in a discussion, I want to turn and ask you: "Is that formulated right?" "How would you say it?" I've realized my dream so entirely of having someone whom I can wholehearted[ly] follow. No single human being deserves the luck that I have had, to combine lover, husband and father of my baby, with the most exciting mind, the most perfect cooperation in work, and the kind of intellectual leadership in my own field that I crave. Perhaps it needed a war, sent by the Gods to right such an outrageous balance, in my favour.

. . . Another way in which I have absorbed one of your attitudes towards life, I find, is style of letter writing. Do you remember in Tchambuli when I suggested you separate the personal comments from the anthropological, you said that didn't make sense to you—that life was all of a piece. I used to be able to save my most loving remarks to close a letter with, but now they just come out in the middle of it, anywhere. I love you so, my darling.

. . . The newspapers scream headlines about Nazi advances and Russian pacts with Poles, and the picture is more confusing than ever.

Goodnight, my love, my sweet sweet love, and bless you.

Margaret

During World War II, having failed to find a role in the English war effort, Bateson joined the OSS and was sent to the Far East from 1943–45. Within a year of his return, they began to have problems in their marriage, which came to a crisis point in 1947.

To Caroline Tennant Kelly, July 13, 1947, on the breakup of her marriage to Bateson.

Carrie dear,

. . . All these years of marriage have been a long uncertain battle—within Gregory—between the part of him that wanted to be a great scientist, with all that that entails, and the part of him which wanted to break out into some strange romantic gesture, away from his family and all that they stood for, and somehow, "be himself." Ofcourse I have always been part of the first course of action, a help meet in work and responsibility to the world, and the recurrent Steves[72]—and they have been pretty recurrent—have stood for the other position. Whenever we have been separated, a "Steve" has always come on the scene, and sometimes it took a good while to pick up the pieces. Nevertheless I had always counted it worth it, though sometimes life was pretty hilly, and it has been hard always to represent the sober course, and have someone, almost anyone with a beautiful face or a lovely voice, represent all the romance and excitement in life. Still it had been so good and so full, and Gregory had so many things he wanted done, in the field, in publication, and then when the war came, in war organizational work, that I simply threw myself whole heartedly into executing his plans.

The work has been grueling, especially during the war when I went to Washington—because Gregory thought it should be done—and all the traveling, . . . always working—but very little deprived when one considered what the war brought to so many. Then Gregory went overseas and was gone 22 months, and during that time, he wrote me 4 letters, and life was full of uncaptured Steves. He came back in an emotional, somewhat dull, somewhat hypersensitive state, and this year and a half since he got back has been very difficult. I kept fighting to get him back to England, because I felt that a great deal would be resolved which had set in when his mother died, and that perhaps once back in England he could be surer where he wanted to go. Finally we were all set to go this summer and then—another Steve, which was complicating enough, but on top of that, Steve herself, who with her infallible instinct for destruction, saw that she could at least destroy what she could not have. I went to California for ten days and came home to find my life in ruins, whether permanently or not, I still am not quite sure.

The last two months have been a nightmare, worse than those weeks in Sydney in 1933, with Gregory simply breaking every responsible tie, and living out on Long Island, where he could see his present girl, and generally indulge in a deep introspective fit, while I taught his classes, answered his mail, and generally tried to cover his absence, and keep some of his connections going with the world. Everything else happened simultaneously. Larry

Frank, in whose house we have lived since 1942, in a big joint household, had three operations for cancer, my adopted 18-year-old's divorced father and mother, both arrived separately, and had to be dealt with extensively, Cathy's Nannie lost her job very traumatically, and I took her in for a week.

Meanwhile, all plans for the summer kept hanging in the balance, but Gregory still said he was going to England, so I went on with the various commitments I'd made as a way of getting to Europe. But one of those was to teach in a seminar in the American Zone in Austria, and Cathy's military permit kept hanging in the balance. Finally Gregory deserted [sic] he wanted to take his girl to England, so would we please not go to England, so I managed to pick up again an invitation I'd had to be an education officer on a student ship, and we sailed the day before we were to have flown, and then three days before he was to have left, Gregory decided to stay in New York and be psychoanalyzed! Which leaves everything more confused than ever. The moment in which the girl you think you want has agreed to go to England with you and marry you, and your wife has cooperated up to the hilt in helping you arrange that, hardly seems the moment to give the trip up and be psychoanalyzed instead.

So, here I am, going to Europe, by a horribly roundabout route, on a ship that keeps changing its mind—it's an old unconverted troop ship with 600 students aboard—when I had no special desire to go to Europe, when the whole thing was planned from the start and altered along the way, to suit Gregory's plans. I have Cathy with me because as yet she has shown no signs of feeling any strains, and I think I can trust myself to keep her steady, better than I trust other people. The trip from Southampton to Salzburg is going to be something, perhaps the worst trip I have ever taken in all my travelling life.

. . . I am as thoroughly miserable on this boat as I have ever been on any boat, which is saying a good deal, for it seems to me I have almost always been miserable on boats. I can't day dream at all about the future, for I have no idea at all what shape it will take. There is nothing else in my life which matters comparably to what happens to Gregory, or as worth waiting and working for, as even some very attenuated, shrunk partial relationship with him.

His whole future hangs on a very slender thread, because while there are some forty or fifty of the best minds in the country who recognize how good he is, there is no underpinning to his reputation, he has no job, and no institutional place. He has ofcourse money in England, and could return to England, and write another book, which he is all ready to write. But returning to England lies right at the heart of his conflict. And his scientific work has always had as a sort of workaday base, the routine work I did for him, to keep it going, and the field data which I worked up. Our notes are so inextricably woven together that any real break with me, would mean that probably neither of us could use them, but he less than I, because ofcourse I kept them. I mean I wrote them up while he did the photography.

Meanwhile all of my work depends so much on his theoretical leads and formulations, that I simply can't imagine caring about going on working without him working in the same area. He has always only barely been an anthropologist and he might so easily decide not to be one any longer, but to go into some other edge between the natural and the social sciences. He is so far out on the periphery of theoretical developments, that hardly anyone will understand what he is doing, unless I am there, to thread the thing into more usual, more everyday terms.

I feel at once so inside and so outside the whole thing, one moment listening to the beating of my own heart, the next worrying about the whole future of the social sciences and their role in helping to build a viable world. Perhaps it's wrong to build one's public and one's private life together so closely that one jeopardizes the other, but I have never known how not to do that.

And even though it has been a difficult fourteen years, it has also been a wonderful one, and I'd not take any of it back, or choose any other course. Granting that marrying him was a dangerous venture—as you felt when you knew we would meet on the Sepik, and as his friend Noel Porter wrote me 14 years ago, "A Steve will always come along and then I will have to be sorry for you as I was for Bett"[73]—perhaps it has lasted better than anyone would have believed. If he can work out a new marriage and a new life which leads on to better work—I think we can all survive, and Cathy have, perhaps, a good preparation for the kind of marriage she is likely to encounter fifteen years from now. Although I am still deeply depressed, and tend to burst into tears at a moment's notice, I know I can manage any solution which leaves Gregory, whole, and developing as a scientist, and leaves me some sort of relationship to him.

Meanwhile, as usual life is not empty. We will continue to live in the two lower floors of the Franks' house in the village, and I will have Margo [Rintz]—the 18-year-old, who is a sophomore at Barnard—and Cathy with me, and I daresay someone who needs succor always sleeping in Gregory's study. The demands on my time are legion so filling up time is never a problem. All this agony has somehow had much the same effect as falling in love anew, and people keep telling me how pretty and "dewy" I look. Perhaps someday within the next two or three years, I'll get out your way, for the Trusteeship Council, or somehow. I have no feeling that it will be difficult to pick up the pieces through the years. You may already have seen Chin and been puzzled by the news that Steve was here. Poor Chin got rather peculiar shift from us all, but I loved seeing him.[74]

So my dear, here is your long letter, perhaps not quite with the content you'd expected. I am only forty-five, sometimes it seems as if there has been too much of life already, and at others, that there is too much ahead.

My love to Timothy,

Lovingly,

To Claude Guillebaud, an economist and father of the two girls Mead fostered from England during World War II, March 20, 1948. Mead wrote concerning new problems with her second husband, Reo Fortune.

Dear Claude,

I need some help, and I hope it won't be too much of a nuisance to you to give it to me. As you may or may not know, my former husband, Reo Fortune, is at present in Cambridge, and uses the address of the Museum. I don't know what his status is there; there are rumors that he has been appointed as a University Lecturer, but I don't know whether they are true. During the 15 years since I have seen him, although he has devoted a good deal of time conversationally, and what publications he has had, to attacking me, I have had almost no communication from him directly. But this winter he wrote me what seemed like a friendly professional letter, and I mad[e] the mistake of answering it. Since then he has been writing me letter after letter filled with the obsessive content of 15 years ago, in which field work is hopelessly mixed up with himself, Gregory and me. I am seriously worried because I know he has periods of considerable instability and I don't want to contribute to them. But without any picture of what he is doing there, who is backing him, what his prospects are, whether his wife is with him, etc. I can't form a sufficiently good picture to know quite what to write or do. Could you write me a brief summary of how he got there, what his status is, and in general what the estimate of him is, say by Bartlett, without involving me in any way in the inquiry, but simply making it normal curiosity on your part?

I am still hoping to get over to the International Mental Health Congress in August, but it depends on whether H. M/ Government lets me use English money for the trip. I haven't enough American money to do it. I'll write about dates as soon as I know anything, and anyway I promise not to have tantrums of rejection about it, if I have to miss you all again.

Yours always,

To Reo Fortune, her former husband, July 16, 1948, after receiving from him at least eight very bitter letters, one after the other.

Dear Reo,

. . . Do you think it would be any good for us to meet and talk over things, especially Arapesh. You keep writing about the same points and they never seem to get resolved. Perhaps if you could feel how very clearly all the condensations of 1932—of personalities, mood, particular cultures and theories of constitutional type—have been sorted out into their appropriate spheres of discourse, you might feel better about the whole thing. I don't personally think it is very serious if I present one interpretation of Arapesh culture, as seen against my field experience—and your field experience as of 1933, because ofcourse your judgments were included in my understanding of the culture—and you now feel you have other understandings after seeing actual New Guinea warfare among your inland people. We can each present our evidence, the dates at which the work was done, the verbatim materials, and we will have fully discharged our duties to science. Probably every field worker who returns to a field after an intervening field trip, sees new things. But you don't sound—in your letters—as if you felt very happy about it, and perhaps if you saw me and talked with me about it, it would be different.

I make this suggestion with considerable diffidence, because I care so much about your work, and I want the world to have as much of it as possible. I have just finished a long manuscript on sex differences in the seven cultures in which I have worked—minus Omaha—and I am again conscious of how much your delineation of matriliny contributes to the whole theoretical position. But I have never worked in a matrilineal society and can't discuss it first hand. I think my first glimmering understanding of the importance of culture underwriting the paternal role—which has so little support in biologically-based material—came from your discussions about the time you were writing the Cross cousin marriage paper.

. . . Also, because I still have the habit of trying to put my life in order before long journeys, especially by air, I would like to tell you a little about where your things are, which are still in my office. All of your negatives are together in the file where I keep my pre-1939 field materials, marked as yours. All the rest of your notes are in boxes, metal for the important ones, cartons for things like old reprints, etc. labeled as yours and placed on top of a cabinet in my inner office. Bella Weitzner knows they are here, and so does Ruth. So if anything happened to me, they would always be in shape to send you when you want them. I had the best of your 1931–33 negatives copied—as negatives—so that they would in any case be preserved. Did you ever see the set which I had made up in your name for the Haddon collection? I thought it very handsome?

It's almost fifteen years since that lunch in Sydney, the lunch which you said was the right way to make a clean break. It would have been the right way, if you had been a banker or a businessman and I had simply been a wife who could vanish forever from the scene and never publish a line to confuse the issue. But the trouble is that you aren't a banker and I am not that kind of woman, and we had been working partners also, and the work remains, important to you, important to me, and I believe important to history and science. This does not leave us either personally or morally free to handle our own relationship in purely poetic terms, and has ofcourse been a problem for both of us all of these years. Perhaps now, after some of the dust has cleared away, we might get down to some sort of working relationship in which our communications would be related to the material, and to contemporary issues, and not be a reliving of old scenes. It was in that hope that

I answered your first letter this winter. That hope proved to be premature then, but I still cling to it.

Yours,

Mead and Fortune arranged to meet at his office in the museum at Cambridge. It's not clear that they actually met, but this time was the beginning of a truce between them.

To Gregory Bateson, November 26, 1948, in the midst of the crisis of their marriage while he was being analyzed by psychologist Elisabeth Hellersberg. They were separated, and Bateson was living with a friend who was also separated from his wife.

Gregory dear,

. . . I don't know just what you thought I said about tolerating your nightmares, but I was not proposing it, rather I was talking about living through a period of nightmares. Whenever you present me with a specific one, I have done my best to state that I would not tolerate it.

About my wanting to go back to the Past. That is another nightmare—as ofcourse you know it is. I have said that the essence of our relationship was expressed at the Second Goats,[75] and that essence was what I hoped desperately to preserve. And I have also said that it was not dependent upon marriage. I don't want a sociological conformist marriage anymore than you do. I do think that as long as you haven't finally picked out someone else, and as long as you are on the other side of the country, it is perhaps easier for Cathy not to have to deal with a divorce. If you could remember who I am for five minutes running, and so feel comfortable in the knowledge that I would give you a divorce immediately when you wanted one, I would think that was the better course. But I don't think it's absolutely essential. Cathy is a year and a half older now, and wise and loving beyond her years.

I have assumed that to some degree when you finished your analysis you would be a new person, hopefully enough of the person I had loved, but possibly not. I have also assumed that only if that person, which you had become, chose me, as the person whom I then was, would there be any possibility of re-opening the question of our relationship, in other than sociological terms. I have been prepared for the possibility that you might demand repudiations of the past which I couldn't give. I also have always been prepared for you to want more than I can offer at 46, and by that very wanting turn towards younger women, whose youth fits your newfound sense of spontaneity and sureness. I am prepared to accept that circumstance without bitterness of any sort. I am only fighting for your leaving me on the real issues—either between us, or because you quite simply want something else which can only be obtained by breaking this relationship—and not because your image of me has got confused with a raft of unresolved childhood images. I told you last autumn what I felt. I wanted you to have what you want from life, I wanted us to reestablish some genuine sense of each other which would not do violence to all that I have occasionally meant to you, and you have always meant to me. Those are the first things.

After that I look at the rest of life, and I profoundly want it to make sense, and I have no desire on earth to spend it in ways which do not make sense. The best daydream I can muster, and I don't dream it much, is that you would take me to England, and there, in England, we would be able to decide together whether there was anything we still wanted or not. In that daydream we would both know, one way or the other. But I have no desire to go back and wrap around me the grey mists of distortion in which you clothed me, until my very physical appearance reflected the garments which, unconsciously, I wore. You have spent your life offering women impossible alternatives, and being angry at them, whichever they took, angry at those who refused a half loaf, angry at me because I took it, and yet said, perhaps this will become—in time, with luck and love, a whole loaf. No woman who did not to a degree discount sex, initially, would ever have married you. I did not discount it in myself—that you might reproach me for, and justly—I was willing to take the best you had to offer, trusting it would grow.

So you see, I don't know either. I love you better than anyone else in the world. You have more power to move me than anyone else. But whether I want to be reclaimed, I don't know. Whether you can reclaim me, if you want me, I don't know. But we haven't gone through the Hell of the past two years simply to go back to the Past, with you perhaps writing two more pages of manuscript a day.

The logic of human relationships is not as simple as you think. I do not believe one can ever say that because a relationship is voted wholly bad, by someone who had found most of life wholly bad, it is therefore certain to be wholly bad for someone who had found most of life good. Your view of the world is not the only view, nor, even, my darling, your view of yourself. Because you repudiated yourself, I was sad, but it did not mean that I saw what you saw. I can repudiate from the bottom of my soul every moment of suffering you have ever known, and still not repudiate my life with you. To demand that another love what one loves is tyranny enough, but to demand that another hate what one hates, is even worse.

So—

Margaret

The films, with a list, will go off tomorrow.

By 1950, Mead and Bateson had been separated over two years and Bateson was living and working in California. To Geoffrey Gorer, April 8, 1950, reflecting on the status of her relationship with Bateson and the pattern of relationships in her life.

Dear Geoffrey,

. . . Gregory has just been here for ten days. He is better than he has been since the war, or perhaps even since his mother's death. On the whole it was a very good ten days, and although the shape of the future is still cloudy, the possibility of maintaining a relationship within which Cathy can feel she has two parents who are related to each other in some kind of meaningful fashion, seems pretty good. The entry of another woman into the picture might ofcourse change all that, and make the attempt to keep a meaningful picture futile or hopeless, but at the moment that isn't the point. Cathy flourishes, and maintained such a lovely balance the whole time that it was a delight to see. I had to give a picture of leisurely relaxation so unrelated to the normal pace of life that it took several days to get my timing within bounds again.

It's odd to compare the periods in my life when I have been completely contented to fit my every act and thought into a shared life with one person whose chosen pace and scope was less rapid and variegated than my own, and then the periods when there is no such personal setting and my life simply effloresces into such a multitude of contacts and relationships and contents that no one person could even bear to hear about them all. Yet I don't feel one way of life as more "me" than the other, but shifting between them, within the same geographical setting is really complicated. Now I have an enormous residue of unfulfilled tasks which must be dealt with in the midst of Russian *and* our own Easter. . . .

Love,

As she had with Fortune, Margaret kept in contact with Bateson's friends. To Noel Porter and his wife Muss, friends of Bateson's to whom Mead sent food parcels after the war, July 17, 1951. Mead wrote of Bateson's new marriage.

Dear Noel and Muss,

It was good to have your long letter and I sent it on to Gregory. I imagine, however that you have never heard from him. He has remarried, the American daughter of a Bishop, who worked in the clinic where he was doing research, and now has a young son, named John. I waited with bated breath to know what he would be named but John is alright. He had his mother's surname as a middle name, which is American style, so Gregory still holds open some possibility of making peace with the whole question of nationality and residence. He has just published a book, *Communication: The Social Matrix of Psychiatry?* by Jurgen Ruesch and Gregory Bateson, Nortons, 1951.

The principal problem at present is that he won't answer questions about work and field notes and things and then of course feels guilty and then decides he hates me and everyone remotely connected with me, including everybody in England whom I have ever met, like his uncle Edward, whom he hasn't addressed a line to for ten years and whom I called on once. This is all very tiresome, and it's hard to get work done, as I feel responsibility for our joint work even if he [is] willing to forget it. On the other hand, relations with Cathy are very good. She likes her new [step]mother and is looking forward to seeing her new baby brother. They are living in California near Stanford, and his address is Veterans Administration Hospital, Palo Alto. But there isn't much use having it at present while he is in this mood of simplifying life down to immediate stimuli.

Now about the food parcels, if you think they aren't needed too much any more, I would be glad to discontinue them. Life is pretty full of obligations these days and I am somewhat haunted by what would happen if I were ill but didn't die, as in characteristic fashion, most of my savings are in life insurance. Gregory has set up a trust fund for Cathy in pounds which would see her through school, but I could become a problem. So let me know if you think it alright to discontinue them now. People seem to think that the continent is better off for food than England is. . . .

Life continues to be terribly busy, with a variegated mass of projects, books and people, all moving at a ferocious speed which I still think I like and can still catch up with. There is a chance that I may get to England in the summer of 1952 for long enough to get down and see you. One of the drawbacks of modern life is that one has to reckon in hours instead of days, and I've been counting time in London only in hours the last two or three years, always flying, always working against a tight schedule. But I think of you often on your high hill, with the view of all the invasion points of England. . . .

Affectionately,

In 1951, Mead returned to Australia with her daughter, Cathy, to participate in a lecture tour for the New Education Fellowship Jubilee Celebration. To Rhoda Metraux, August 24, 1951, from Sydney, Australia. Mead talked about meeting Fortune there.[76] [h]

Rhoda darling,

. . . I am trying to get a call through to Canberra to tell Raymond Firth[77] that it will be alright if Reo and I are at the same affairs. I'm a little mixed about what I said in my last letter, but I must have told you that Reo was here. He spent all day with me yesterday—a lot of milling and hashing ofcourse—but good talk too—and he has developed a kind of merriment and in some ways greater charm than he had as a young man. Carrie—who is quite entranced with him—says he's like reading Gertrude Stein after a lot of dull prayers. I think it will be good and heal over some of the scars. He's on his way back to pick up the work on one of the island people which government interrupted fifteen years ago. Whether he will ever make any sense of his life, I don't know, but this does I think, give it a better chance. It's all extraordinarily odd, but it seems less strange to be back here because I am—as I always was in Sydney—except for the very brief 1938–39 visit—worrying about what Reo will do next—But I am worrying in a relaxed way,

To Rhoda Metraux, September 9, 1951.

. . . Reo got off last night for New Guinea after spending most of his waking time over the weekend with us. He was Cathy's sitter on Saturday night— a curious reversal from the days when I used to have phantasies that he would come and steal her. I had to leave to give a lecture yesterday afternoon so his last act was [to] carefully [wash] all the dishes for me—before he changed into his New Guinea clothes and went out to catch the plane. His last words were "You must come and stay with us at Cambridge. We have an extra room!" All very good and I think undoubtedly arranged by the angels. . . .

In the late 1950s, Mead and her first husband, Luther Cressman, had begun a cautious rapprochement. Here, Mead did not want to endanger that. To Luther Cressman, September 21, 1963.

Dear Luther,

I learned from Dr. Marks(?)[sic] formerly of your student health department who is here as a resident, that you are going to Alaska, so I have no idea where or when this letter will find you. I am writing primarily to explain about a book for teen agers which is coming out soon, about me. It is not very well written and will not, I hope, have a very wide circulation. I got into it inadvertently, thinking I had agreed to an Encyclopedia article, and found it was a book, published by Encyclopedia Publications, and that a rather miserable little girl was already living off the advance. I originally stipulated that there should be no mention of my private life at all, but it seemed dishonest to write about my field trips as if I had been in the middle of New Guinea all by myself. So the briefest possible mention is made of Reo and Gregory, and then more of course about Cathy. I found that I couldn't put anything about you that could be equally professional and brief so I have put nothing in at all.

As you know, in deference to other people's feelings I have tried to keep my past private life out of the public prints as much as possible. I have thought it would be awkward for others if I did not do so. I have never met your daughter and I don't know how she feels about me. But I didn't want you to think that I had any other reason for saying nothing except that I found that I could not treat the long period of my student days and our marriage with the same professional detachment that I treated my two later anthropological partnerships. It seems as if, if one has either a book or a baby by another person, the relationship becomes one that must be acknowledged. I have so valued the contacts we have had in recent years and Cecilia's[78] kindness to me, and I hope that this book won't lead to any misunderstandings. I have just heard that a book called the *Three Sisters*, by Irving Wallace or Wallis, who wrote the *Chapman Report*, a skit on the Kinsey report, is about three women anthropologists—I don't know yet who, but I am supposed to be thinly disguised. Probably that will be disagreeable. I hope not to you or yours. I am always anxious to know how you are and how you are faring, and I've been so glad you were able to get such good people for your department. With my best regards to Cecilia.

Yours,

To Luther Cressman, June 12–July 5, 1967, from Papua New Guinea, where she was helping Metraux begin her own New Guinea field work. Mead shows how far the relationship between she and Cressman had mended by telling him all the news and her feelings about the American Museum of Natural History and her career there. She was sixty-five years old at the time.

Dear Luther,

. . . I was delighted to get your letter catching me up on the last year or so. I started an answer right away and then got interrupted. This winter has been a pretty tough one, trying to get two expeditions into the field, (Ted Schwartz[79] is going back to Manus this month), teach, finish up back commitments. Marie[80] is running all the office finances, etc. which means absolute safety, plus ofcourse a lot of over-elegant detail. My life is getting fuller of older women, still vigorous in mind, but weak in body, and somewhat cranky. And I suppose it will get steadily fuller of them. Sometimes I worry about modern girls who never learn to get along with women as girls, and then are left stranded by the differential death rate.

Cathy[81] is in the Philippines; her husband has a two year Harvard Ford Foundation grant as a member of a team writing Harvard Business school type case histories (really a form of ethnography) and she is teaching linguistics and anthropology in the Jesuit University there, and this summer planning to [do] field work in a Manila slum—in line with the modern interest the young have in the culturally deprived. Studies of slums are safer for girls than boys in Manila for the boys get drawn into gang wars, which apparently don't [involve girls]. [We moved last year],[82] from the Village, to a large cooperative apartment in Central Park West, just North of the Museum, which means that in bad weather I have virtually an underground passage to the Museum. I don't retire till 1969, and theoretically before that, I have to complete a major hall. However, everything is in the hands of the city, and NY city is in a kind of parlous state. The contract may be let this summer; if so, by June 1968 the hall may be ready for installation. . . .

The long years of creeping around like a mouse as [sic: are] coming to a close. All these years I have financed my own work, found money for my own assistants, and studiously kept from flaunting my work in the faces of my colleagues. Even Harry[83]— who should know better—sometimes reproaches me for not doing more formal scientific work. It's been an odd solution, and a very idiosyncratic one, and one that would only be tenable for a woman, I think. A man, in my position, would have had to take more initiative or found it unbearable. But as a rather odd solution, it has worked. I keep my office as long as I live till in fact, I am deaf, dumb and blind. As I've always insisted on staying up in the tower, there are no competitors for my series of "garrets," as WPA worker[s] called them. You ofcourse had to take some time off to leave your successors free play. I will have no successors in that sense. As for Columbia, that's [a] yearly appointment at a low salary and no benefits which I can keep on at any level I wish. Perhaps in a couple of years I'll give up the evening class which is the hardest thing I do . . . [orig] but it is where I catch the odd maverick who decides to stop designing control systems of polaris missiles and try his hand at anthropology. All of these plans are like

July 5, 1967 Tambunam.

Here I was interrupted and never got back to this letter again, June 12 seems a long time ago. Next week I leave here; the house has been built, servants trained, informants found, the past knit into the present. . . . So I am leaving Rhoda and Bill[84] in a nest of old friends, who, being reincarnation believers, don't at all mind the idea that I am now to be replaced by someone else. . . . I won't be sorry to go. I find that I have less a taste for settling down for months, and spending hours learning people's names, and typing millions of words, than I once had. This has been hard anxious work and I am glad it will soon be behind me.

All the best, and write another good long letter soon. My regards to Cecilia.

Yours,

To Luther Cressman, April 3, 1977. Luther's wife, Cecilia, had died earlier that year.

Dear Luther,

The book has come and it is beautiful, well worth all the time and care that has gone into making it so.[85] But I can understand how sad it is for you not to be able to bring it home to Cecilia. But, as part of what you say about immanence, she surely is part of the book that you recognize as you relive the years that went into the making. Thank you for sending it to me.

I expect to be in this country at the end of April and to leave for Nairobi around the first of May. I have a good many comings and goings scheduled so that if we find we can't fit it in then, I am planning to come out to the west coast, stopping to see Martha.[86] So if I miss you this spring I'll see you in the late summer. Dates for the spring are still a little wobbly so it will be worthwhile telephoning when you finally get to Pennsylvania.

I have just been on the Pacific Coast for a brief two days, with an evening with my brother's family—all step children and none of them really related to me, but they feel as if they were. Two weeks ago I had a reunion of the Ramsey family—the descendants of my grandmother Mead and her six brothers and sisters who left descendants. It seems to me as if I had been going from one large room to another, all filled to the brim with children of various sizes.

It is, I think, something to be thankful for to die quietly with the person one loves most close by.

Please write whenever you feel like it. I will be in Nairobi, Athens and Iran in May and back here by the first of June.

As always,

To Luther Cressman, February 27, 1978. Mead was beginning to show symptoms of her final illness, later diagnosed as pancreatic cancer.

Dear Luther,

It was fun to get the pictures, your silhouette unmistakable after all these years. I sent the one with Louise on to her.[87]

I'm sorry you don't like my saying student husband, but I have always thought of our marriage as the very best kind of student marriage, in which first as an engaged couple, and then as graduate students, we were free to study. After all, the year you had the church in E New York, I never functioned as the minister's wife I had originally planned to be—and the next year after Europe you were leaving the church and starting some new career. So it is my student years and yours that fitted together so well. I'm sorry you don't like it.

I have been quite miserably ill for several weeks while they battered me with tests that haven't been definitive. I think I am at last pulling out of it but I haven't had the energy for anything except what I basically had to do—which was quite a lot.

Keep well.

Affectionately,

All of her former husbands outlived Mead. Reo Fortune died the next year, in 1979. Gregory Bateson lived until 1980. Her first husband, Luther Cressman, lived the longest, dying at the age of ninety-six in 1994.

3

Lovers:
Continuingly Meaningful

———————————————◈———————————————

Mead's philosophy of sexual openness allowed her to be involved with various men and women over her lifetime beyond her relationships with her husbands. This chapter explores this side of Mead's life through her letters to those in the two longest-lasting such relationships, with anthropologist Ruth Benedict and, after her death, with anthropologist Rhoda Metraux. In these letters, Mead explored and reflected on the nature of their relationships together. The letters reveal the ebb and flow of both relationships, the one with Benedict lasting almost twenty-five years until Benedict's death, and the one with Metraux lasting almost thirty years until Mead's death. The letters also reveal and analyze her affairs, such as those with anthropologist Edward Sapir and textiles expert Morris Crawford, as well as her relationships with her three husbands.

The relationship with Benedict has become the best known. Mead first met Benedict when Benedict served as their anthropology class museum assistant in Mead's senior year at Barnard. She described Benedict to her Grandmother Mead in a letter dated October 15, 1922: "The Museum trip assistant is charming—we are all crazy about her—and the rest of the section is so dumb that I talk to her and walk with her. And having read The Men of the Old Stone Age *I shone last time when we gazed at Paleolithic remains." [h]*

Benedict was fifteen years older than Mead, married to chemist Stanley Benedict, and had returned to graduate school in her thirties. Benedict helped Mead through the suicide of Marie Bloomfield, one of her Barnard college friends. When Mead's father was being obstinate about paying for graduate school, Ruth Benedict *gave her the $300 with no strings attached, calling it her "No Red Tape Fellow-ship." It was in the tradition she knew, for she herself had benefited from an anony-mous woman benefactor who made it possible for her to go to Vassar College as an undergraduate. Mead wrote the following thank you letter in 1923 in reply.* [h]

Dear Mrs. Benedict:

Perhaps there is no accepted form for thanking someone who not only opens up all the possibilities of a life work by introducing one to it, but also makes it possible for one to go into that work—perhaps there is no form of thanks because none would be adequate. At least I find it so—dear fairy godmother.

Margaret Mead

P.S. I'm afraid this is not a "no-red-tape" reply.

To Grandmother Mead, August 3, 1925, on her trip west with Benedict. As she was leaving on the first leg of her trip to Samoa, Benedict was going to do fieldwork in New Mexico. By this time, they had become lovers.

Dear Grandma,

Ruth left me last night at Williams and in three hours I shall be in Los Angeles and see Uncle Leland, I hope. We had gorgeous weather all the way across the desert, almost cold, even in the day time . . . We passed through several of the pueblos, so sand colored and flat were the adobe houses that a village was by the window before it was hinted at by a change in the land-scape. We saw many Indians baling alfalfa into square blocks, and we mar-veled that the alfalfa was green. Blue and red and yellow are the colors of this country; green is swallowed up when it does occur. The cattle and horses are lean and lost looking, gaunt and gray on the barren lands. But the trees bloom, Grandma, all of them, pines and junipers and all. Think of a little tree that smells like an evergreen and has daisies on it. And there are also delicate scarlet flowers, growing unexpectedly in the waste land.

Ruth and I got different things out of the Grand Canyon, but we both loved it. She was most impressed by the effort of the river to hide, a torturing need for secrecy which had made it dig its way, century by century, deeper into the face of the earth. And the part I loved the best was the endless possibilities of those miles of pinnacled clay, red and white, and fantastic, ever changing their aspect under a new shadowing cloud.

One minute, I saw a castle, with a great white horse of mythical stature, standing tethered by the gate, and further over, a great Roman wall. A few white clouds would slip over the immediate sky-roof of that part of the Canyon, and the Roman wall had become a friendly street in Thrums, and unexpected cattle were grazing in a forgotten gully. There were many clouds, and in the afternoon a little rain, which did not stop the sun from shining, and later a dark blue cloud which threatened rain in earnest. We had everything except the canyon by moonlight.

Now we are going through a Pass, the name of which the porter mouths hopelessly. It is high mountains and still desert country.

I love you, Grandma dear.

Margaret

To Ruth Benedict, in early August, 1925, on the same trip. After Benedict left the train in New Mexico, Mead wrote the following letter to her, while, as she wrote at the top, "Riding through a most barren place." [h]

Ruth beloved,

And then I get no further for my tears. But I am being very very good. I ate prunes for breakfast as the most hygienic item on the menu. But I had not realized how very detachedly helpless I would be without you. Ofcourse I hadn't wound my watch. I didn't know what time it was this morning and there was a sunrise I didn't want to see.

[Don't] Regret that you hunted out car 302 on the platform. O Ruth seeing you even thru a heavy screen where you couldn't hear me and I couldn't see you was far far too precious to surrender a moment of. And there was something else that last unexpected glimpse gave me. Do you remember when that other train passed in back of you? Against that background, you and I were both moving, both moving fast—and as my train seemed rushing thru space you were always just opposite my window, a lovely windblown figure. And that picture is the one I shall take with me, darling—and it will hold much of comfort.

The bromide lasted until 2:30 A.M.—and I woke to find myself saying over and over, "And Ruth will realize that I always have managed and that I'll be able to now."

Ruth, I was never more earthborn in my life—and yet never more conscious of the strength your love gives me.

You have convinced me of the one thing in life which made living worthwhile. You have no greater gift, darling. And every memory of your face, every cadence of your voice is joy whereon I shall feed hungrily in these coming months.

Margaret

To Ruth Benedict, August 6–7, 1925, a few days later, reflecting on her relationship quandaries. [h]

Ruth beloved,

. . . I gave myself this year, in its irrational terms, to solve three problems: whether I really loved the sweet bread[1] Luther gave me enough to wish to live always with it; what I was going to do about this love for Edward which gnawed incessantly at my heart and would not be silenced, and last, whether I could manage to go on living, to want to go on living if you did not care. I gave myself a year and it only took two weeks. Always when I've finished a task early, I've made holiday of the left over time. Perhaps I can do so here. No work, no worry, just over-time in which to dream of the endless beauty life has given me. You would be glad of such golden holiday for me, would you not, darling?

[End of letter]

To Ruth Benedict, August 14, 1925, still in Honolulu, on her relationship with anthropologist Edward Sapir, who wanted her to divorce her husband, Luther Cressman, and marry him. On the train with Benedict, Mead made the decision to give up her love for Sapir in favor of her husband. She decided it would be better for him to reject her so she emphasized her free-love philosophy, which she knew he could never accept. [h]

Ruth dearest,

. . . Edward's long letter, altho every line has been written in advance in my dreadful anticipation, still is too much to read peacefully. Just reiteration of-course, he will share me with noone, I would love too lightly, I am dominating and egotistical and selfish. One thing I know I can never bring thru this year, and that is my self-confidence and self-respect. They will be riddled into net work before it is over. He can think of so many terrible things to say, and I shall believe them, all of them. And how to live, accounted a parasite on love, and too ill to do decent or consecutive work? I feel dishonest in even attempting the problem for the Research Council. For the first time in my life, I expect defeat, not merely fear it in occasional moments of weakness. Paint the Ariel picture for him if you can—paint any picture he can manage.[2] But his refuge is going to be hate in the end—hate and recrimination. But pride saved, and hate enthroned where love had so short a sojourn, he may not suffer too much. And playing me up as Ariel sets a premium on my personality. That should not be so—I must be despised and scorned.

Oh, I'm sorry I have to write all this. But there is no strength in me. I made out a program of work this morning—8–5 everyday, and not to think about anyone or anything but my work. It is three o'clock and I have come home. You see what the year will amount to.

Does Honolulu need your phantom presence? Oh, my darling—without it, I could not live here at all. Your lips bring blessings—my beloved.

Margaret

While in Samoa, Mead received word that she had gotten the job as assistant curator at the American Museum of Natural History in New York, where she would spend the rest of her career. She viewed the position as a plus for her relationship to Benedict and as a help to her husband, Luther Cressman, neither of which relationship she saw conflicting with the other. She wrote to Ruth Benedict, December 17, 1925, "I'm really jubilant. It's the guarantee of so many things, but mainly that I shall be near you. Also Luther can enjoy his year without worrying about where the money is to come from to start housekeeping with the following year." [h]

To Ruth Benedict, January 7, 1926, in Samoa. In the evening, on a day when the boat came with mail, Mead wrote about the end of her affair with anthropologist Edward Sapir. [h]

Ruth darling,

This is a bad habit I've got into of always heaping the accumulated turmoil of boat day[3] on your shoulders. It's a state of mind which is quite unique and quite awful. I feel as if everyone were talking to me at once at the top of their lungs. Your letters are put to most utilitarian uses. If I'm feeling frightful and frightened of tackling the mail, then I read them first. Otherwise I try to save them for the benediction. And this time they made such a glorious pile—all in separate envelopes. I *like* lots of envelopes.

This was the first mail without a word from Edward. It doesn't get any easier. I don't think I'll ever be able to use Léonie [Adams] and Eda Lou's[4] phrase of having "gotten over it." Settling things and getting over them are quite different. Settling them relieves the awful sense of pressure, and gives space for other things. From your letters I gather Edward has not been as explicit as I counted on his being. The break came, and I thank God for that, purely as a question of attitudes not personal jealousies. That the jealousies flow from the attitude is just history. Anyhow, I saw and see, quite quite clearly that it couldn't be—not for either of our sakes. And I think it's all quite over as far as developments are concerned. I still think if it had been in any way possible I'd have tried it, and sometimes I don't thank God for my clearsightedness. I am so glad, so glad that this doesn't seem to be adding new bitterness to his life. Next to being a blessing is I suppose, being innocuous. Anyway, I have a sense of not having made things clear to you and I want to. I told him I wouldn't say I'd stop writing, tho I have, and that nothing was changed as far as I was concerned. I shall always keep for him the love which he cannot take. Please write me as much news of him as you can.

I felt a clear statement of this had to be made. Your trust in my decision has been my mainstay, darling, otherwise I just couldn't have managed. And all this love which you have poured out to me is very bread and wine to my direct need. Always, always I am coming back to you.

I kiss your hair, sweetheart.

Margaret

To Ruth Benedict, January 11, 1926, from Samoa, reflecting on Sapir and her husband, Luther Cressman, and their own relationship. [h]

Darling,

What you said about divided imagination is most absolutely true. It's healing in itself to set one's mind toward one definite course without continual playing with the idea of an alternate. And I am very happy in the thought of Luther at present.

Curiously enough it was just in the interval before Edward's ultimatum that I came nearest to deciding to marry him. I took no joy in the thought, I felt horribly desperate about it and yet I thought I would probably do it. It was so hard to believe in the stone wall that divides us, altho he always gave me fresh proof. But in that five weeks without mail my imagination got very out of hand.

In one way this solitary existence is particularly revealing—in the way I can twist and change in my attitudes towards people with absolutely no stimulus at all except such as springs from within me. I'll awaken some morning just loving you frightfully much in some quite new way and I may not have sufficiently rubbed the sleep from my eyes to have even looked at your picture. It gives me a strange, almost uncanny feeling of autonomy. And it is true that we have had this loveliness "near" together for I never feel you too far away to whisper to, and your dear hair is always just slipping through my fingers.

Risk my love—Sweetheart, sweetheart, what nonsense you do talk—and will the birds forget to come north in the spring to the land of their desire? When I do good work it is always always for you—That's my wishing. What do you care, really, whether I devise elaborate colour tests for the Samoans? (and such a beautiful test paper of 100 little coloured squares as I stenciled and painted today).[5] But none the less it's all for you. And a day like today when I've worked from dawn to dusk without stopping, I feel very peaceful and it is such joy to go to sleep loving you, loving you—and waken so. I've a hundred details I should be writing about, but if I were there I'd kick all the mss. and proofs under the table and bury my face in your breast—and the thought of you now makes me a little unbearably happy.

Margaret

In a later letter, thinking of Sapir, she wrote Benedict January 28, 1926, "(Perhaps the greatest chasm in understanding is between the polygamous and . . . the serially monogamous. You see, I've never stopped loving anyone whom I really loved greatly. Experiences and imagination alike fail me.)" [h]

To Ruth Benedict, February 18–19, 1926, on her husband Luther Cressman, and on seeing her in Europe at the end of the Samoan trip. Again, her commitment was to both of them. [h]

Ruth darling,

. . . One sentence in your letter puzzled me. (Aren't you weary of my stupid bewilderments). You said how difficult it was to take the bread without the wine day by day and give thanks for it and how that laid one open to all sorts of dangers. You seemed to be talking about Luther—and yet I don't see how you could have been. I don't think in any sense it's only bread to him—I hardly know now whether it's really only bread to me—only it's a wine of such different sort than the other. But as for my course, it's always quite the same. I can't leave Luther on any terms but his perfectionist ones, which means I can never leave him. That's not the counsel of despair, it's just a fact which is true of me *as I am now*. And by the time I get back, I'll have gotten over the inevitable tattered nerves and be a fairly lovable person. . . .

Darling, darling, am I not to see you the first of August? I'd lived so for those weeks, weeks alone with you. It would be heaven enough even if we spent them in Kansas—and to have three weeks with you in Italy. It will be one of those things which we can never make up for—which will never come again. In October you will be coming home to Stanley and your work and the house and a thousand preoccupations. And I will be up to [my] neck in German treatises on Africa. And there won't be any three weeks, or even [one] week. It's the sort of waste that's a crime—like throwing a jewel wantonly into the sea. And I can do nothing but pray that Sherwood[6] made a mistake.

If he did, sweetheart, will you make your plans with Luther? He'll know by now (when you get this) just when I'm leaving Australia, etc. and how much time we'll have, and where he wants to go. He's to plan. I'm too tired to make any plans at all. And if Stanley leaves August first, we will plan to meet then—and Luther will come home too. I hope he'll have money enough to stay to then. I think we will. Then he can tell you his plans and set a place to meet—and then you can plan where we will go for those three weeks. I'll be so blinded by looking at you, I think now it won't matter—but the lovely thing about our love is that it will. We aren't like those lovers of Edward's "now they are sleeping cheek to cheek" etc. who forgot all the things their love had taught them to love—

Precious, precious. I kiss your hair.

Your Margaret

To Ruth Benedict, March 14, 1926, written from Ofu, in Samoa. [h]

Ruth darling,

Last night I had the strangest dream. I was in a laboratory with Dr. Boas and he was talking to me and a group of other people about religion, insisting that life must have a meaning, that man couldn't live without that. Then he made a mass of jelly-like stuff of the most beautiful blue I had ever seen—and he seemed to be asking us all what to do with it. I remember thinking it was very beautiful but wondering helplessly what it was for. People came and went making absurd suggestions. Somehow Dr. Boas tried to carry them out—but always the people went away angry, or disappointed—and finally after we'd been up all night they had all disappeared and there were just the two of us. He looked at me and said, appealingly "Touch it." I took some of the astonishingly blue beauty in my hand, and felt with a great thrill that it was living matter. I said "Why it's life—and that's enough"—and he looked so pleased that I had found the answer—and said yes "It's life and that is wonder enough."

So you see I'm alright again, if dreams mean anything. Edward's defection—he counts it as such—has given me back my confidence in my philosophy of life which his love tried so hard to destroy. It is being very close to despair when all one's dearest sanctions are threatened—brought to a test which sees them bringing only unhappiness to all concerned. But now I feel that I was right—that Luther and I are right—to trust in the possibilities of life itself instead of coercing them to a pattern. I feel immensely freed and sustained, the dark months of doubt washed away, and that I can look you gladly in the eyes as you take me in your arms. My beloved! My beautiful one. I thank God you do not try to fence me off, but trust me to take life as it comes and make something of it. With that trust of yours I can do anything—and come out with something precious saved.

Sweet, I kiss your hands.

Margaret

To Ruth Benedict, August 26, 1926, Connaught House, 8 Montagu St., London, written after a stormy meeting with Benedict in Europe. On the ship to Europe, Mead had fallen in love with New Zealander Reo Fortune. (See Chapter 2, Mead to Benedict, July 15, 1926, from Paris—Benedict was traveling in Europe then with her husband, Stanley.) [h]

Ruth dearest,

I am very happy and an enormous number of cobwebs seem to have been blown away in Paris. I was so miserable that last day, I came nearer doubting than ever before the essentially impregnable character of our affection for each other. And now I feel at peace with the whole world. You may think it is tempting the gods to say so, but I take all this as high guarantee of what I've always temperamentally doubted—the permanence of passion—and the mere turn of your head, a chance inflection of your voice have just as much power to make the day over now as they did four years ago. And so just as you give me zest for growing older rather than dread, so also you give me a faith I never thought to win in the lastingness of passion.

I love you, Ruth.

Margaret

To Ruth Benedict, June 12, 1928, on her way to Mexico to get a divorce from Cressman, reflecting on her relationship with Benedict. [h]

Darling, I reached Tucson this morning to be greeted by the pleasant news that there had been a wreck, or a burnt bridge, or both, which had changed the train schedule so that there would be no train to Mexico until tomorrow night. And I waited a whole day in Philadelphia so as not to get here a day early! It's hot as hell. I forgot how it felt to be forever seated in a running brook.

However, I discovered a nice cheap place to eat where they have a "merchant's lunch" for 40 cents—a boon after that exorbitant train. I've a comfortable enough room and I've finished and typed my *Cosmopolitan* article. I'm sending it to you along with a letter to Mr. Morrow who volunteered to "look it over."[7] If you don't think it too awful, will you send it on to him. Or if the changes you want to make are of the sort he'd not see anyway, and can be made without upsetting the whole article, will you send on to him. Anything which offends you, ruthlessly remove.

As I was writing the part about regarding sex as play, it came to me how completely we don't regard it as such. We'd always lapse into that sudden deathly seriousness which David Garnett described so vividly in *Go She Must.* I tried to imagine any other way, but I couldn't.

. . . Do you know for all my love of competing, I've never competed with you. I've emulated you, I've taken comfort in your least successful efforts which were later crowned by glorious success, but I've never pitted myself against you. Maybe competition and passion are mutually exclusive, as Ray insists. Anyway it came to be [sic: me] yesterday that I must never compete with him—and I'm afraid I'll be tempted to.

I've never been ever tempted to compete [with] you—perhaps too because it would be a competition of *being* not of mere *doing* and that would be unthinkable. Come Fair or no Fair, I don't try to whip you. I'm not that foolish. I'm reputed always to like my own possessions and what I am, but I wouldn't risk one of your qualities in my attempted possession. That's a mixed sentence, but you'll get what I mean—Oh, sweetheart I'm lonely for your arms.

Your Margaret

Margaret was traveling by train toward the Pacific for her marriage and first field trip with New Zealander Reo Fortune. Right before she left, she experienced a serious attraction to another man, Morris Crawford, an expert on ancient and modern fabrics who consulted at the American Museum of Natural History. In several letters on the train, she reflected on this. To Ruth Benedict, September 3, 1928, on her way to St. Louis, Missouri. Mead was still brooding over the situation.

Darling,

. . . I am not very clear about my own psychology. I don't know whether my gradual depression over those last letters of Ray's [Reo Fortune] when I reread them after coming back from Westport and a growing feeling of a possible long postponement of our marriage has reinforced my worrying over this other thing or whether it is the other way around. If I hadn't sent that letter to Ray when I did I'd be hopelessly suspicious. As it is all sorts of wild hypotheses drift through my head. Perhaps only one person can make a sufficiently fundamental impression on me to hold me to unswerving fidelity. Perhaps the capacity and attention which I have left for other people beside you is somewhere a little off center and incapable of rising to such heights. The psychoanalysts could fix that up to suit themselves but still I think that it might be explained in terms of a basic orientation of the personality, the only orientation which that personality was capable of. And maybe what I give any man is less than half.

This whole thing is much harder for me to understand than anything which has happened yet. Schematizing my life, there has been you and you steadfastly since you came into it. Nothing has ever threatened that fact. Looking at the heterosexual side of it, I chose Luther and remained perfectly steadfastly interested in him without swerving or temptation for seven years. Then out of a definite unfulfilled need and complicated by symbolic association with you I chose Edward [Sapir]. I haven't any illusions about his having chosen me. I wanted him and without the moves which I made he would never have realized my existence. And when that was over as definitely and finally as anything could well be, I chose Ray. Again I did the choosing, it lay in my hands whether that relationship should ripen or not. It was my will which wore down all my resistances, not any move on his part, only the things in him which I recognized and loved. And then for two years again I was quite peaceful and unregardful of the landscape. All of that is clear and acceptable. I have always chosen when I should pay attention and when not, whether I wished a new point of focus in my life or not.

My old fears about whether I was made for monogamy or not were based quite simply on a knowledge that I could love more than one person, not upon a fear that meteors were going to come trailing across my sky. My feeble attempts to go on with my marriage once I had rejected it don't count in my sense of having willed what I wanted. But I didn't will this. I have a sense of very definitely not willing it, of having felt no place for any other important relationship in my life, and of having quite clearly done what I could to avoid it. I didn't even have a sense that it was unwise to see Morris again. The first qualms I'd felt about generalized excitement passed away with that first dinner drive. I was safe and amused I felt. He took my denial of love making with good grace; he was the last person in the world to really fall in love with me and so I didn't have to guard his feelings any more carefully than my own.

But when I get through saying all those things which are quite true, where am I? It's no go saying that he is a unique person, who would inevitably attract me, and then one doesn't resent the inevitable however much one flouts it. It may also be explained as inevitable that I fell in love with Edward and with Ray. And there is such a thing as overdoing inevitability.

I suppose it might be said that if our relationship and any relationship to a man are as separate and incomparable as they seem, operating on different sets of wheels, then it might be laid to pure biological urge which had been frustrated for too long. That's a sweet and abominable picture and not one that I can possibly make anything of. If it's true, I'd have to fix over my scheme of life radically, but I wouldn't fix it over so that I was sure my husband would never be away for a month. Better no husband at all than a price like that. Anyway, I don't think that is the point. But it wouldn't help any if it were. (Damn these written monologues. I am hopelessly, impossibly spoiled. I can't tell one idea from another till I read an answering comment in your eyes.)

It would make a fascinating study to work out just in what respects two people could gradually come to depend upon a common mind, selecting one function from one mind and one from the other, counting one person's experience to explain one set of points, drawing on the other's memory to clear up others, etc. We come awfully near to doing that in everything from science to love. I wonder if you'll feel as mentally amputated as I do. I have just one definite urge and that is to write to you, write to you, write to you.

. . . The great pieces of space, the steadily falling hours of time which are passing without being woven closely in the net of our common knowledge, terrify me. It's as if in a long, woven strip suddenly blank spaces were to appear where before all had been rainbowed and patterned. Something has happened to the weft, it runs brown and gray, gray and brown through my hurrying fingers. I weave desperately fast, but under my window pass fields gold and lovely with flowers which you will never see and my elbow is sore and irritated from a bad cut which you didn't know I'd gotten by falling down on the Museum steps. Brown and gray and only every twenty or thirty threads can I slip in a colourful one and regain one note in the pattern which winds woven and beautiful all about me, woven by our four hands in the last six years.

. . . Is it just a trick that I don't feel this as a trick of sex? I don't feel it nearly as much so as I did with Ray where his looks and youth counted so highly in the initial reckoning. While I am raking all my sins over the coals, I suppose there is one point that I should add and that is that I am enormously intrigued by someone who knows more than I do in every respect. I realize very clearly how definitely I've played up to myself the picture of the time when Ray will know more than I, how eagerly I have snatched at all the points where his judgment has been better, his knowledge fuller. I loathe being teacher. I loathe having the most to give in a relationship.

Ruth, I suppose I'll get used to being [away] from you or in Castorp fashion "get used to not getting used" to it.[8] But right now it's all one awful misery. I have no genuine clarity about any other course in the universe except my love for you and I go ten thousand miles from that. Oh, darling, I'm so lonely.

Margaret

To Ruth Benedict, September 4, 1928, from the Kansas City Station, ruminating on her relationship with Fortune and with Benedict. [*h*]

Darling,

. . . I have a feeling that I shall gradually get this new perplexity written out of my system. I have a feeling that it's an intellectual difficulty I'm in more than an emotional one—that it's more a documentation of possibilities than any real conflict right now. But that may be due to time. I'd not like to have spent another month seeing Morris—and yet I've no proper justification for such a convent attitude towards life. I suppose that's one of the reasons I look for that demand in someone else—because I can never really make it myself. Isn't that a pretty picture. Suppose I had a tendency to drink too much, eat too much, read too much, undertake too many different jobs, smoke too much, stay awake too much, etc. and had to go about picking out strong personalities, stubborn personalities with particular biases on each of these points, get attached to them and so regiment my life. That's my idea of strictly nothing. And yet to choose someone who will make a demand of limitations which I seem quite unable to make myself, is just about equivalent.

You have never asked me to do anything I wouldn't myself elect, never demanded or even wanted one inch of formal compliance, and you've gotten more from me and of me than any other living soul has or is likely to, I'm thinking. If with an endowment of warmth and response I carry also the possibilities of promiscuity on the one hand and too many complicated involved relationships which will hurt innocent people on the other, do I have to go outside myself for a check and a balance? The only check I would give this of my own will would be first a further trial and second, to end it if it proved incompatible within myself with other things which I wanted more. But ofcourse I could do nothing of the sort—but remark, "I've promised to marry a man whose whole set is different from yours and mine, who could never understand it if I let you kiss me, or even if I wanted you to kiss me, and I mean to try to satisfy his demands." What is wrong with this, Ruth? Or am I phrasing it wrong? But it looks awfully like the same old point— that I can say "Whither thou goest I will go and there will I be buried. Thy people shall be my people"—but I can't add "and thy God my God!", in any shape or form or size. And maybe that's the ineradicable root, the one homosexual twist there is no getting by.

Sweetheart, you didn't realize you were going to get such a dose did you? I'm glad of this trip—of its great spaces in which to think and to write to you. Maybe by dint of it all I'll get straightened out in my own mind. I want to get it straight *before* I see Ray, undoubtedly when I see him I'll throw every consideration to the wind again, but I've no use for that type of decision.

Darling, you will never know what a priceless and so undeserved gift you have given me in giving me a perfect love no least inch of which I need ever repudiate—Oh—I love you, my beautiful. I kiss your eyes.

Margaret

To Ruth Benedict, September 5, 1928, on the train and almost to Gallup, New Mexico, still thinking about their relationship together. [h]

Darling,

. . . I've slept mostly today trying to get rid of this cold and not to look at the country which I saw first from your arms.

Mostly, I think I'm a fool to marry anyone. I'll probably just make a man and myself unhappy. Right now most of my daydreams are concerned with not getting married at all. I wonder if wanting to marry isn't just another identification with you, and a false one. For I couldn't have taken you away from Stanley and you could take me away from Ray—there's no blinking that. So it's not the same thing. Oh, dear, I oughtn't to be writing this. It's not a reproach, heaven knows, and I'll probably feel differently tomorrow. But everything seems principally important in its relation to you. When I get my mail at San Francisco I'll be first interested in what you've got to say about Morris and second in what Morris has to say himself. Oh well likely enough I can make the identification completely and not fret. But just at present I am so terribly conscious of the instability of every emotion I have which is not centered on you. Beside the strength and permanence and all enduring feeling which I have for you, everything else is shifting sand. Do you mind terribly when I say these things? You mustn't mind—ever—anything in the most perfect gift God has given me. The center of my life is a beautiful walled place, if the edges are a little weedy and ragged—well, it's the center which counts—My sweetheart, my beautiful, my lovely one.

Your Margaret

Mead and Fortune married and went to do field work among the Manus of the Admiralty Islands in the Pacific. To Ruth Benedict, November 7, 1928, on a ship near Samarai, Mead reflected on her two loves. [h]

. . . Shipboard has its complications. Reo vacillates between wanting to have nothing to do with people and wishing he had—and naturally once having taken up an attitude of aloofness, he can't do much about it at this stage. But on the whole it's been a good trip.

I suppose in time I'll accept the perfect balancing of my two loves with complete understanding. There are so many strange parallels. Just as I used to see Reo's face with sudden clarity when I was walking home from the Kingscote two years ago, so now I see yours—Reassertion of the other sphere. But I'm untroubled by any mad desires for a single allegiance. I suppose that will come later, as it came this winter—and have to be fought through in silence this time instead of with your perfect understanding, my sweetheart. But for now, I am blessedly happy, touched with a happiness which I have done nothing to deserve.

I love you, darling.

Margaret

To Ruth Benedict, May 8–11, 1929, from Manus, on managing their relationship when Mead and her second husband, Reo Fortune, returned to New York. [*h*]

Darling,

. . .

May 9

Heaven only knows what next year is going to bring forth. Reo has a trick of looking at things with a kind of ruthless realism and simplicity which is frightening. It's not going to be a matter of tact and skillfully arranged lack of conflict, etc. It's not going to be any of the things that it is with Stanley.[9] He's going to see quite clearly, count his losses and his gains and minimize the former not in the least. As he phrases it, he doesn't want to hurt me, and as I say my relationship to you is so infinitely precious that any disturbance of it would hurt me, very well he'll give me up. But there is to be no blinking about what it is all about and no pretending that he isn't losing. Heaven knows just how it will work out. It's the kind of problem which you and I have never had to face. Your peculiar living arrangements were too special a blessing to be repeated again. And ofcourse he's correct when he says that I've never tried it. I haven't really. I've never tried living with two people with both of whom I was deeply in love. But you have[10]—that's one thing which gives me courage. It's going to be the hardest nut we've had to crack yet. I know Reo is partly influenced by the fact that I've so little strength here that to imagine me managing two intense love affairs seems impossible, and also by a dread of New York's bigness and loneliness. He'll persist at present in thinking himself to be abandoned if I'm out for an evening, but I'm sure that's a temporary point of view.

I was going to tear this letter up remembering what difficulties letters have caused, but I think it will be all right. You'd probably be constructing edifices which were worse than the truth. So if I can only put the truth down straight. He's not concerned over spiritual issues. He has no thought of grudging you your place in my soul, no [thought] of grudging physical love to us. He's simply and realistically concerned with how it will work. He's haunted by no pictures of our year together last year. That is, I don't think he's jealous. And that's the principal point because it being irrational, we would have no remedy against the havoc it might work.

Instead, we have only to deal with a man who has definite ideas of the way he'd like his life run, which is practically what we have to deal with in Stanley. The fact remains ofcourse that his idea of arranging his life may be much more inconvenient. Still it's the kind of problem which doesn't need to make for rancour or misery necessarily, as jealousy and wounded pride would.

He's genuinely concerned over the difficulties which may ensue over brushing so closely against another emotional set in me, and apprehensive of ignorantly trespassing too closely and being unwelcome. That's a problem which need not arise, I think. But to obviate it he wants to know the time that I spend with you. It makes a quite different ritual of relationship than the ritual of silence necessary with Stanley. But I think I can manage it better. Any other course would mean out and out lying—for it will be a long time before his life and mine run on separate enough wheels to make mere avoidance a possible means of obviating friction. You'll know I can handle explicitness better than silence.

I think that it is only imperfect and inherently selfish relationships which need threaten each other. It's the strong but selfish parent-child relationship which threatens marriage, not the strong but generous one. And aside from the mere grasping tendency of one relationship over against another—it's the light which one relationship can throw upon another and less perfect one which is most likely to be destructive. It seems to me that that would be an almost inevitable result in trying to manage two relationships with 2 members of the same sex i.e., 1 woman + 2 men.

May 10

There would be such a tendency for one relationship to reveal the weaknesses of the other person—difference would have to be put down to difference in personality not to differences in sex. But bet. [between] a relationship with a man & a relationship with a woman, the differences, varying rewards and penalties can all be ascribed to sex.

And my relationships to both you and Reo are so perfect that I feel as if neither would threaten the other. The beginning of this year while Reo's relationship to me was in many ways difficult and perplexing, I was continually fighting comparisons which might have been very destructive, whether I had chosen to shut my eyes to them or to recognize them. I don't believe I'd ever be willing to say of my marriage "This is an inferior personal relationship which I accept because I live in a society which automatically associates its principal institutionalized rewards with marriage. I want children. I want cooperative field work. I want the kind of home which only marriage will permit." I wouldn't have been willing to say that—nor would I have been willing to take Louise Bogan's solution—when she's serene—of pretending about my marriage. It was in some such mood as that that I wrote you that querulous letter saying "Why did you give me a standard which is bound to find all the world wanting?"

May 11

Now that we have gotten past the first roughness and stumblings—I can give my relationship to Reo the same quality of assent which I give to yours and mine. The differences are as outstanding as ever but they do not stand as criticism but rather as documentation on the difference between loving a man and loving a woman. They seem to me more generic differences than differences inherent in your two personalities, altho ofcourse those differences are striking enough. . . .

Mead and Benedict struggled with Fortune's reaction to their relationship. According to Mead herself, they did suppress much that they'd had together, in deference to Fortune's feelings, for the two years Mead and Fortune stayed in the United States while Fortune finished his Ph.D. at Columbia University. (See

Mead to Benedict, March 9–14, 1933, in this chapter and February 14 and April 9, 1933, in Chapter 2.) Sometime in the fall of 1930, or the spring of 1931, Benedict left her husband Stanley to live briefly with a free spirit named Thomas Mount, who left advertising intending to win fame as a novelist. By summer 1931, that relationship was over and so was her marriage. That summer she fell in love with a young woman she knew casually from visits to California to see her sister's family, and that fall they began to live together in New York.

Before Benedict returned in 1931, from leading a summer field school for anthropology students in New Mexico, Mead and Fortune took off for their second field trip together, this time to the mainland of New Guinea. Mead wrote the following letter October 22, 1931, from Sydney, Australia, on receiving Benedict's letters about her new relationship with Natalie Raymond and commenting on Tom Mount. [h]

My beloved,

Your letters came yesterday. I was so desperately relieved to see your handwriting. I don't think I'd have minded if they said you'd eloped with Edward [Sapir] which is about the worst thing I could think of. But I don't think that it's only against a background of tremendous relief that I'm just quite simply glad about Nat. I more or less expected it from your letters of the summer— and I can't find anything but happiness that you are to have someone living with you who will make the world real for you. I think it's better than Tom. He always seemed a broken reed for any enduring day by day happiness to me—and then you have pity for women's failings where you have instead high standards for men. At least that is the way I read it. So I can think of your having greater security with a woman. I know Nat just enough and not too much, either. It would be tantalizing not to have seen her, and I think it's just as well, not to have known her too well. I just don't believe that I know what jealousy is—in the ordinary sense. I've no feeling of insecurity at all, nor doubt, only a tremendous thankfulness that you are to be companioned in love, instead of alone in an apartment cell. I could bless Nat.

Tom's marriage was, as you say, to be expected . . . I'm glad it didn't come against a foreground of lonely winter for you, though.

When I got the letters I felt like dancing and singing for joy—I've felt as if I were hanging over a bottomless chasm these last ten days. I wrote you all the addresses and the damn letter must have gone astray. If you were ill, Nat would write me word wouldn't she? I'd figured it all out that no one would let me know and you might be terribly, terribly ill. There won't be a proper quota of letters on this boat for I just couldn't write. There was no use treating you to one long wail. And now I feel as if a ten ton weight had been lifted off the back of my neck.

I've got to go downtown now and see if I can vamp Burns Philp[11] out of charging us freight—I'm going to take them some pictures for their magazine.

Darling, darling, I'm so glad you're happy.

Margaret

But an excerpt from a letter to Ruth Benedict, December 7, 1931, showed her to be uneasy.

. . . Darling, you do seem so far away. I dream of you very often but you are always angry with me; angry because I am going away, or have not come back soon enough, or something. I've had exactly three letters, two written on succeeding days and the lost one which far predated them, since I left America. It's the worst luck we've ever had with mails, these last few months, and it's hard to preserve the illusion of being in communication with you at all; the angrier dreams seem so much realler, and God knows I can't write to them. Heaven knows whether I will get any letters before I go inland, as the *Marani* will leave its mail for us at Wiwiak and Mr. Macdonald won't be back from the bush for some time. The old boat is always most here. Oh, Ruth, my beautiful, I love you so.

I love you. I love you.

Margaret

To Ruth Benedict, February 20, 1932, from New Guinea, reflecting on Benedict's new relationship to Natalie Raymond. [h]

Ruth dear,

. . . One phrase in your last letter makes me wonder. You said Nat had been worried about the difficult adjustment you counted on my making, knowing that she wouldn't be able to make it. That must cut both ways. If she couldn't make it—as me—can she make it to me—as herself? Isn't my existence a stumbling block? Ofcourse it may be just the being in possession point which she means. She has been able to have you all to herself at the beginning of her love, while I had years of adjusting myself to a non-resident position. If it's true she scorns herself for being unable to make my kind of adjustment, you could decide on that circumstance—or perhaps you have. I remember when I first fell in love with you how I used to ponder on how wonderfully proud Stanley must be to have you as his. But of-course I've never had much opportunity from the circumstances of my life as much as of yours, to develop much possessiveness towards you. All of which is at the moment fortunate. . . .

An excerpt from a letter to Ruth Benedict, May 14–15, 1932, from Alitoa, New Guinea, reflecting on Mead's relationship to both Benedict and Fortune.

Darling,

. . . When Reo is here I am completely simple, integrated all focused on a point. I work through one day after the other, plan meals and enjoy them, sit and talk with Reo in the evening, sleep, all with a kind of lyric simplicity and with an almost complete unconsciousness of self. That is what has made writing letters to you so difficult. In my relationship to you I have always been anything but simple, every other aspect of my life, from a reaction of a book, or a passing street scene, to the most complicated of relationships to other people, has been brought into service, to variegate and intensify my re-lationship to you. Possibly that is because we have never had the rhythm of everyday existence to bind us together, but even that summer on 124th Street, I was not conscious of much difference.[12]

Around you like a luminous and differently patterned cloud has always hung your relationships to other people, many of them obscure to me, any of them likely to debouche a whole flood of partially understood emotion upon our peculiar scene. And for myself, there have always been the swiftly passing current of emotions aroused by other situations, and brought to you for comfort, clarification, or just pure literary interest, or as footnotes to statements about my own moods or emotions. Your comments, the breadth of your understanding has been a white light into which I wished to haul everything which was vaguely seen, partly comprehended, more merely differently coloured. I think perhaps also this is an old pattern brought down from the days when I was terror-stricken about those "fifteen years" which [you] had ahead of me, and which were bound to leave me a contentless hulk by the wayside of conversation sooner or later.

Or it may be because you are a woman, and having a sympathetic interest in other people which Reo lacks. Anyway, there it is. While with him, my relationship tends to turn in upon itself, to shut out the passing scene, and the complicated response. Ofcourse when there are other people for whom I care anything at all about, I retain a fair degree of responsiveness to them. But here in the field where responses to humanity and methods of ethnology are so hopelessly intermingled, I seem to lose all personality, as you have known me, and become as simple as a plant, quite happy, serene, industrious and enthusiastic, but all one colour, with nothing to report. . . .

But the worst aspect of the situation—the effect Reo has on me when we are alone—is the effect on my letters to you. I feel such simplicity as I am reduced to as a sort of treachery to you. Instead of the usual complicated response which I am accustomed to giving you, filled with half tones and overtones, and asides, all of which serve to throw the peculiar character of our love into strong relief, from this one-tone simplicity I just say "I love you" in one way instead of in a hundred different phrasings, and I feel as if I were failing you. Anyway, this is more explicit than I have been able to be before, and I can always trust you to understand anything half way clear. . . .

F. Phillips sent me a copy of the Blue Ribbon Samoa.[13] Reo is really impressed with my having gotten into a popular edition, instead of disapproving which is a help. I have been rereading it at meals since he left [on a supply trip] and I find there is very little in it that I regret—the journalese of the first part of the introduction, I do. And all over again, I have decided that Edward's [Sapir] accusations of cheap and sensational are unfounded. What I don't understand is why the general public ever reads it at all. . . .

In a letter to Ruth Benedict, July 12–13, 1932 [first page misdated June], Mead has just told her she and Fortune may stay in New Guinea for some extra time and laments the loss of time in her relationship with Benedict.

. . . But three years away from you; do you remember my mad plan of taking ten years to go round the world, that we used to use as a nightmare afterwards? And this will be almost as bad. Three years just gone, lost forever, out of life, three years which could have been so full and so precious. I'd decided not to say anything, but I couldn't risk Priscilla's telling you, and I had to tell her because she was counting on our return in their plans. For you to get any news of me from other people seems outrageous. Those lost letters from Sydney explain why you kept saying you knew nothing of my field plans. I should have realised then that the letters about Barter[14] might have been lost. If anything I furnish you with too much detail about plans which never come off, just to feel that you know what is going on.

I don't feel that we will be farther apart because of the long absence; I've no sense that we won't take things up just where we laid them down, but it's the loss, the waste, the time that cannot be lived together because it has been consumed apart, with only letters every thousand years to try to bridge the distance, that's the part that hurts. If there were a few thousands of years more of life instead of only thirty or forty, I'd still grudge the week I didn't see you just the same. It's a year since I saw you last, almost thirteen months, no it's over thirteen months, for you left just before your birthday, and there's got to be more of it, more of it. . . .

In a letter to Ruth Benedict, part of a long letter of twelve pages written over several days in August 1932, when Mead and Fortune were preparing to leave Alitoa, Mead talked again of their relationship in terms of Benedict's new relationship with Natalie Raymond. [h]

August 12
. . . I don't want to pry and if you would rather leave all your comments at "Nat and I are very happy," you know it is alright. But I do wonder sometimes whether you feel any possibility of permanency in this, or whether you expect it to vanish away—if not in feeling at least in presence. You talk about Nat as dependent and adolescent and that [this portion of letter ends here]

Aug. 25, Karawop

and then Reo came back and we got out in three days—and I've not had a minute since—and today came mail—your letter of May 24th with the answers to the questions I was starting to ask. Your letter sounded so sure and serene—it made me glad to know you are not armouring yourself with threats of insecurity, but can look forward to years together. I know your confidence was stated with provisos but still for you, it's [a] confident statement. It gives me a much clearer picture. I couldn't tell, you know, how much it all seemed transitory to you. You've said so little—maybe because you weren't sure either until you wrote. I feel you safe again—as I've not really felt since you left Stanley. And I don't worry as much about the elasticity of our field funds. You know I feel quite differently about "coming home because you are lonely" and "coming home because we want to see each other." The last has far more urgency but no anxiety attached. I know that you know that if I have to stay here longer it's just because it's in the nature of things—and we accept them. But if you were all alone, I'd feel there was something wrong somewhere—in my marriage perhaps—that somehow I had betrayed you impersonally, but nevertheless completely, by not being a permanent companion for all your hours. When we are each happily companioned, then the blessed balance reasserts itself. Thus Nat not only makes me happier about you but happier about us—you and I— makes our relation more perfect—. She bears only gifts.

. . . By the way Ruth, what is Nat's last name. I don't think I have ever known and it seems strange not to.

Good night, sweetheart. I love you, love you.

Margaret

To Ruth Benedict, November 3–4, 1932, while working with the Mundugumor people of New Guinea; part of longer letter, on feeling that they had come to a point of change in their relationship. [h]

November 4th—noon—

I'm still not back at work—I hope I'm not just lazy but I've little impulse beyond lying on my back. I decided in one of the intervals of lying on my back that we had come to the end of an epoch in our relationship. Then I had a long tussle trying to decide whether to say so or not—but I decided that if I thought so and said nothing, then we would should [sic] have reached something worse than the end of an epoch. The cause of all these decisions was that you did not answer my long letters written while Reo was away—you just said how nice it was to get them. So I thought that while our affection for each other hadn't changed, our intimacy had. For while I was conscious of the periods when I wrote extrovertly and with no essential comment about myself, you weren't even conscious of the fact that you were writing extrovertly. But then I reread your last five letters—closer together than your letters have been for a long time—and they are so warm and breathing that I think I have perhaps mistaken a change in us for a change in our relationship.

Perhaps we have both passed into a non-introspective stage—got through our spiritual adolescence for a time anyway. And I suppose it's luck that we got through it together—and now we can be very brisk about anthropology and careers. And of course it's very true that after one has said "I am happy" this is never the comment which remains after one has said "I am unhappy." Possibly that is the reason people like Papa Franz never have intimate friends, for if Papa Franz has been unhappy he must have always been either unwilling or unable to make his feeling articulate. Perhaps too that is the reason that friends made in youth seem the nearest and most intimate. For happy people—I prefer the term tho Marie [Eichelberger] would say for "adjusted" people—companionship and mutual interest take the place of their souls as a basis for discussion. I wonder if you find you are less interested in "queer" people now than you were. I know I've found myself less interested in the unhappy—not less sympathetic, but less interested—and interest is a better support to listening to another's woes than sympathy or love. The most we can hope for is that this warm companionship in happiness will bloom steadily so that if we are flung back into the mires of unhappiness ever again—we'll both be there together. Oh—sweetheart, what a blessing you are—happy or sad—extrovert or introvert.

[End of letter]

To Ruth Benedict, February 23–28, 1933, from among the Tchambuli (now Chambri) in New Guinea, in the middle of the tension between her sec-

ond husband, Reo Fortune, and her future third husband, Gregory Bateson,
also doing field work nearby. Cracks were also beginning to show up in the rela-
tionship between Benedict and Raymond, as Raymond withdrew sexually from
Benedict. [h]

Ruth darling,

Mail at last—from Nov. 5 to December 23—the longest letters I've had
from you this trip—6 of them. I suppose that's what the point [was] you
made about feeling no block about letters this time. I think it's just the situation here which made me seize on the less happy aspects—but I ended up
with your two sentences in my head, the one about people always being
sure of your love as they went away from you, and the one that you could
trust me to get to Europe before it breaks—or as you said literally "we reach
the breaking point." The first point made me know that you've lost all that
heavenly sense of serenity in Nat about which you wrote last summer and
into which my letters were written—the ones which you were answering
this time. I keep wondering if I overemphasized it and so made you realize
that it wasn't there anymore. But I'd been so thankful that you had it, and
now to know it's gone makes me frightened for you again. Oh, I know
you'll manage—only I'd thought you weren't going to have to, for a few
years anyway.

. . . From that—or on top of that—came the *New Republic* poem—and
because this last two months has brought everything to the surface of consciousness—I began to realize just how thoroughly I'd put poetry and everything connected with it out of my life in order to adjust to the paring down of
your-and-my relationship to suit Reo's demands. Reo is now saying he wants
me to go back to you—completely—and that he should never have played
the role he played—but I don't trust it all—It's probably just that he likes you
better than Gregory—and trusts your powers of self-abnegation more—

I've worked out I think why Reo holds me so strongly against everyone
else—it's because of his violent, almost mad sense of the reality of his own
feelings. I've decided that what I really am is an extrovert without a sense of
reality—and that's not such a contradiction in terms as it looks. I believe
completely in other people and other people's feelings—but not in my own.
Therefore that long period of lack of belief in you which only ended with

the trip west when I went to Samoa. Your own uncertainty was too great to give reality to my sense of our relationship—at the same time my sense of you was so vivid as to be unbearable without some such assurance. Reo is the introvert whose every sensation has a violently real, almost eidetic existence of its own. He abolishes my own sense of my unreality as no one else can—I believe in his pain, and, it convinces me. It's a bad point that pain should always be the convincing point. But I think it's so with you—a happy relationship still leaves your sense of unreality—when it is jeopardized your pain becomes real enough to you to convince the other person. And then there is always the chance that the situation which aroused your pain has already gotten too far underway. But Reo because his essential lack of at-home-ness in the world and his enormous vulnerability to hurt from the slightest point—is always documenting the strength of his feeling with pain—and he continuingly convinces me. But it's rotten that it should be so. What you would find as perfect is someone with a complete and exaggerated sense of reality like Reo's—who could sweep you away in the current of their emotional conviction and have enough left over to convince themselves. You talk about my giving reality to things, but that was just where I fell down. I couldn't give reality to your feeling—instead I responded to your own doubts of them.

Now that I see parts of it all reenacted with Gregory, I understand it all better. His ivory tower is more inaccessible than yours because passion is no drawbridge to it—or at least I think not. Anyway, it's a point I can't find out about or use. And I find myself caught in the same maze of unreality which I used to find with you. I'm not the sort of person to storm an ivory tower, my aggressiveness dissolves before it. I think that is what used to happen to you and me—and maybe what happens to Nat now. You only believe it's not an ivory tower yourself when the attack is withdrawn for a moment. . . .

It's rather awful to be so self conscious but there it is. Anyhow it's God's mercy to be able to write it all out to you—who always understand everything. I can't talk to Gregory much—it's not fair when there is so little time. And ofcourse I can't talk this way to Reo. I suppose you feel the same way about Nat—you have no one to whom you can talk. . . .

Feb. 28

This is good bye for this mail, darling. Don't worry about me—it will all work out somehow in the end. I have a certain conscience about sending you such worrying stuff—but I'd feel horribly cut off if I couldn't write you about it—

Oh, my darling—when, when will I come home to you. I wish I knew. I love you, Ruth, I love you.

Margaret

To Ruth Benedict, March 9–14, 1933, still in the middle of the tension between her attraction for Bateson and her feeling for her husband, Reo Fortune, and analyzing her relationship to Benedict. [h]

Mar. 14

. . . Gregory's fortnight in Aibom is over—and I think it's been a good thing, and I've got some sense of proportion at last. I feel on the whole pretty annoyed with myself. My feeling for him has been composed of three things—a 16th year old delight with a large amount of childlike play in it— this in response to the strong element of this in him—a strong maternal feeling for him, probably more than he deserves, just as one can overdo a maternal yearning over Léonie, and an attempt to weave my excitement about him into a religiously romantic picture of the stained glass window variety. This last is strictly inappropriate and I have a sense of its being all got up, and being slightly inappropriate—the same feeling I had about Edward—that it's a reflex of my feeling for you—gone off at a wrong tangent of identification with you. For there are ways ofcourse in which Gregory is to me what Tom was to you. He brought the same kind of light to my eye. Talk of patterns and repeats in life! There is just a whole section of me which belongs to you as inevitably as life itself. Let that set—made up as it is of poetry, and consciousness of myself—and a sense of the Heaven to be seen through the stained glass window—be upset—and I try to attach it all to someone else—and end up by feeling a fool in ever having thought I could find it except in your face.

The sense of estrangement from you this time is not Nat, nor any sense that you don't care—but the result of these two years of conflict in New York when

I wasn't strong enough to stand Reo's hurt when I left him for you. I've as little power to confine my love to one person as I have to stand making him miserable. It's a rotten combination really. However, I think the practical results of this whole thing are really going to be good. He felt the essential inappropriateness of my romanticism about Gregory—and thinks that he'd see me more whole if all the barriers were removed to its expression where it belongs. The change in his feeling is ofcourse the vital thing. If I can just escape being torn in two by a sense of his hurt when I go to you. That's the good thing to get out of my present sense of distance from you. It does not depend upon what terms of expression our love is cast in—that will I presume be determined by Nat, and we've proved it's not an insurmountable difficulty anyway. . . .

Anyway, perhaps I've succeeded in removing the thing which made me feel you couldn't trust me—"to come to Europe"—that is Reo's feeling about my love for you and my complete weakness in the face of it. Also, he's realized that the fighting I've done over Gregory was a fight in the light of all my sense of deprivation about you—and it's gotten through at last. . . .

To Ruth Benedict, June 27, 1933, coming to the end of her marriage with Fortune, reflecting on where her life went wrong. [h]

. . . And I've a very definite picture of trying to finish off the Museum obligations as rapidly as possible—getting the hall installed and the collection written up. I shall write the Arapesh monograph first too and have that off. Reo is absolutely refusing to discuss the future from any angle—and I'm pretty definite that it's a question of seeing him happy and placed and then ending it.

I feel now that when I violated every dictate of my own temperament in Paris and went away with him and left you—I started on a course which had nothing in the world to do with me really—and never really came back to myself until that two weeks I had alone in Alitoa. That ofcourse was only an interlude—then I pressed myself more furiously than ever back into the mold I've tried to fit—and then at Ambunti at Christmas I came back to myself for good. Now that you will be able to say what you really think, it will be very interesting to know what you have thought of these last five years. I keep wondering and wondering and inventing possible phrasings.

Altho there is really a little too much turmoil and uncertainty at the moment for perfect happiness, I am really far nearer to it, than I have been for years. I think, you know, that everything started to go wrong that summer in Paris when I let you go away and went off with Reo. That was all crooked and everything has been odd since.

Darling, I wish I knew how worried I've made you—and am still making you. I'm alright, really, and more simply nearer to you than I've been for years.

I love you.

Margaret

To Ruth Benedict, September 13, 1933 [h]

Ruth darling, we land tomorrow—and I'll be on the same continent with you. Think of that's being a point to note! David [Hannah Kahn] radioed the boat today that she's meeting the boat. I'd been worrying about making connections with her, so that was a relief. If I have to go through Portland to see Luther, I'll probably just take David with me. It's a comfort to know she's there.

I'm quite plump which is going to help in making everyone think life is smooth and usual. I've been drinking quarts of milk for so long now that it's told. Noone will possibly think I'm unhappy or worried when I'm fat, will they?

It seems to me now that for the last five years you and I have just been marking time, holding on to what we had—it was so perfect that nothing could really destroy it—and trusting to that perfection to bring us safely to some far shore. I wonder if now you will tell me how you have felt about it all. I've only been living with a part of myself these five years and that must have been even plainer to you than it was to me. Oh, my darling, my darling, I'm coming back to you—back to you.

Margaret

But on her return to New York, she found her relationship to Benedict significantly changed, because Benedict was not willing to resume physical intimacy, as Mead was. In a letter to her future husband, Gregory Bateson, from New York City, October 3, 1933, Mead talked of Benedict in relationship to her new idea

of "the squares," a way she, Bateson, and Fortune tried to analyze both human
temperament and cultural ethos by seeing them as polarities characterized with
the names of compass points: northerners, southerners, easterners, and westerners.
Mead saw herself and Bateson as southerners.

If only the world would stop spinning long enough for me to catch. But alas for "pious planning." The great news of the moment and the reason why I have felt so desperately unreal the last two weeks is at last revealed. Ruth is a westerner. All my original placing of her as a tetraploid[15] had meaning, and it clears up so much. These days since I have been back have been frightful, I have never felt so confused in talking to Ruth and about everything in general. But now it is all clear. And I am really very glad for it gives me a first class westerner to make me respect them, and we needed that badly. I have a strong notion you are going to find Noel [Porter] is a westerner. The symbol I used of you as crystal and Ruth as stained glass had a very definite point and when I tried to treat her on the premises on which I talk to you, everything became so confused and muddy. I haven't told her yet, but I think it can be done, especially as I am practically sure that Léonie [Adams] is also a westerner. You fought that you know, but that is because we were underrating westerners so badly. Once get good enough ones and that point goes.

It clarifies all the things in my relationship to Ruth which have been unclear. She is no longer in love with me, or with anyone, and that is just as well. I sometimes doubt how much genius for actual polygamy I have, as over against a little less documented love ofcourse. It is simply amazing how confusing it is to wrongly identify anyone. We've got to get some criteria.

Last week I was all for writing you that the shoulder point was no good, for Ruth has square shoulder[s], but the soft southernish lines in other ways. So does Pelham, and I was placing them both as southern. But I am back again to feeling the physical criteria as sound, and by them and by all the data I can organise at present, Léonie and Ruth are both western. It was curious to what a degree your attitude and mine was unintelligible to Ruth. And she has been making a valiant effort to understand, and ofcourse being miserably confused herself also. I feel as excited as I usually do by a new square point, and temporarily too interested to be as miserable as I have been. And anyway, one can live as long as there is clarity and one feels one's square position strongly enough. The minute one loses that everything starts slipping again. What amazes me is how anyone has any sense of reality at all without it. I rather doubt if they do. I think we pass over into a new level of integration which is incomparable, but it's no rock of ages until we get to be a little better diagnosticians. . . .

Mead continued to talk about the change in her relationship to Benedict in letters to Bateson, as in this one, December 6, 1933.

. . . Ruth is really a very darling westerner and never a bit of a Turk. I've been having fair fits all fall, and I suppose their basis was due to the withdrawal of the physical excitement which used to accompany our relationship. It's the excitement one misses, not it's fulfillment. Ruth is as placid as a field of butter cups, it's just something that has gone, probably forever, from her life. And as I don't know physical frustration in any of its usual forms, I didn't quite know what my fits were about, but I think they were a progressive doubt of her affection for me because I missed the old forms of affirmation. It's a great point to be the same age. It's better if two lovers can lose that particular nerve together, rather than serially. Even so, a southerner is infinitely better off than a westerner who has to face, not merely a temporary maladjustment and feeling of unreality, but actual physical frustration as well. I get more and more happy about being a southerner all the time. . . .

In January, Mead and Benedict spent time together and talked as before. Mead wrote Bateson in a letter, January 11–12, 1934, that "Ruth and I devoted four solid hours to the possible politics of the future last night." By February 14, 1934, Mead wrote to Bateson concerning Benedict, "We are very happy and easy at present and enjoying each other."

To Gregory Bateson, March 19, 1934, in reply to a remark of his.

My darling,

. . . One probably should be in love with someone who is around in the Spring, a cry of "absence absence in the heart" is more manageable in winter. For myself I feel it most unlikely that any new contact would awaken responsiveness in me. My responsiveness to Katharine [Rothenberger] was ofcourse fourteen years old and merely a reaffirmation of an old sharp poignancy, rather like it might be to meet in heaven, those whom one had loved and lost, long ago on earth. And although all love making and so all physical integration depending upon responsiveness is out with Ruth, still my delight in her, and ability to weave my feeling for her into this Spring as I have, more vividly, into so many other Springs, probably makes me less lonely than you are. I do have someone to bring a daffodil to on a fine bright day. Only mostly, I feel the Spring as part of my feeling for you, and it re-sharpens an already unbearable longing. . . .

In a letter to Bateson, June 18, 1934, Mead wrote of Benedict, "Ruth is more like her old self this spring than she has been for years and we have been so happy and peaceful and every moment has been so meaningful."

In 1936, Mead left for Bali to marry Bateson and do field work there with him. Another rich set of letters to Benedict came out of this 1936–39 field trip.

To Ruth Benedict, August 16, 1936. Mead reflected on why people thought she was angry when she was arguing, while she saw herself as fully engaged.

Ruth darling,

. . . In working over this [printer's] proof, I have had another incident like the famous seminar at Papa Franz' when I tried to find out from Bunny[16] how planning was done in Zuni, and you scolded me afterwards for being belligerent, and I wrote to Papa Franz afterwards and apologized and laid it all to fever. But it wasn't fever, I'll bet. It's two things; it's the discrepancy between my voice and my intentions about which you have commented so often but which I never really understood before thoroughly; and second, it's the effect produced by someone who can really have a fit about some impersonal intellectual point upon people who very seldom have fits about anything. I can see how several of your and my discussions which have grown heated it seemed to me by a kind of black magic fell under the same head. I get interested in the point and my interest is so emotionally toned as to *sound* angry, and the listeners come to believe I am angry.

I got this worked out because I was trying to get Gregory to see that in a particular paragraph he was laying himself open to criticism; I was not angry, and even when he reacted to my vigour as if I were, I still didn't get angry. I was simply Hell bent to get the paragraph changed. This is not an alibi. I agree that people should learn to speak the emotional language of those about them so that they won't be misunderstood, especially in terms of anger. But I know now why I have remembered so hard all these years, your tone when you said: "you'd needn't have been so belligerent," for I quite honestly had not felt in the least belligerent. But it was *Bunny* I was talking to, and everybody was against me, that is, judged that I had been belligerent, and I did get good and sick afterwards, and I half believed that I must have had fever, for only some explanation like that could have produced the discrepancy between the way I had felt, just thoroughly interested in how a Zuni household did arrive at a decision, and the way you and Papa Franz reacted.

I remember Papa Franz said I over-formulized the problem, but that was an attempt to shut me up using a theoretical stick. Because I didn't. Gregory has been working here on the way in which this dance club comes to a decision—each person adding little more than a head shake, or a throat clearing, but it is *still* possible to locate leadership . . . [orig].

However, I suppose now that I really understand just why people think I am bellicose and violent and belligerent in intellectual discussions, I can stop being. There used to be a reflection of the point at home. I remember Mother used to complain that Grandma always thought people were quarreling when they discussed anything. Certainly part of my discussion voice is just family imitation. The other half is probably quite disproportionate interest in the points. . . .

To Ruth Benedict, September 5, 1937, from Bali, about issues of trust and loyalty in their relationship.

Ruth darling,

. . . I have been thinking about that pause when you didn't write because it seemed unnatural not to tell me Reo's news, and trying to systematize the theories of conduct involved. As far as I know there have been two other similar occasions, when Eda Lou was ill and borrowed your apartment, and when Marie hasn't wanted me told something.

And yet there have been hundreds of instances when we have told each other things which would have infuriated the one told of, always I thought on the assumption that as long as we trusted the other never to act on the information in such a way as to show we knew. The time you told Eleanor Phillips about Louise's mother, I thought the only difficulty was that Eleanor is untrustworthy, and acted outrightly on the information, whereas you could have told Marie a thousand similar things with safety.[17]

When it has been the affairs of someone close to you, you haven't hesitated to tell me, but when it is someone close to me, you do. If I couldn't be trusted, if I were likely to write Reo, or do something which would show I knew, it seems to me it would be different. But if my communication with you is to be held up and distorted because of Reo's fits of guilt over marrying Eileen and being pleased about it, until somebody sends me a clipping the size of a house announcing his marriage, that seems a bit heavy tribute to pay to the Goddess of Fidelity. You wouldn't have gone in on the "millionaire who endowed his New Guinea trip" scheme, if you hadn't been sure you could trust me never to tell him the truth. Anyway, it seems confused, and I wish you would write me your rationalization of the position.

I think I have always accepted the full implications of using a selective and intelligent confiding instead of formal fidelity. I remember being furious with Léonie for telling Leah, that she, Léonie, had told me all Leah's affairs, when finally Leah decided she wanted to confide in me. Leah ofcourse was upset, and it was stupid and unnecessary on Léonie's part. I should have listened to the whole story over again, and never shown I'd heard a word of it, and Léonie should have known I would.[18]

If it were a question where the information was so drastic that the behaviour of the person confided in would have been inevitably altered, as perhaps with Marie—that seems to me different. It is not not having the news of Reo that I am worrying about, but the distortion of relations with you. Why should I have to lose two weeks of communication with you, to pay for Reo's animosity. When he went thru Sydney he swore Guthrie not to tell me, but said she could tell Gregory! Further, if I was to be allowed to spend every cent I had on him, secretly and in spite of his will that I should give him nothing, why shouldn't I be allowed to know that he is going to be happily married. You've told me everything distressing about him in the last five years which I could help or remedy, why should he be allowed to bind you with a promise that only keeps me ignorant and worried, when he is happy.

Oh, dear, I am cross with him. He evidently infected Mishkin[19] with the idea that I should be told nothing of his plans too, for Mishkin wrote me that he was on fine terms with the world.

If he were motivated for instance [by] a desire that I should not know he was in need, because I couldn't afford to help him, and yet he knew I would, then I think there would be some sense in the request because it would be motivated by a real reference to what the knowledge would do to me. And you might easily respect it on those term[s]. In the same way, when we agree not to worry Marie with some bit of bad news, we act in reference to a reality situation. But Reo's theory is that I will take some low and malignant attitude towards his remarriage, that I hate him, and will pursue him with my hate if I can, and that is a figment of nonsense. Well, it just shows how tiresome he is, that it has taken up all this paper when I might have been writing about pleasanter things.

But two months later, in a letter to Benedict, November 5, 1937, from Batoeon (now Batuan) in Bali, Mead reaffirmed her trust in Benedict and her appreciation for all of the things Benedict did for her, saying, "You are the only person whose judgment I can trust you know; everybody else is willing but they do such funny things."

To Ruth Benedict, September 6, 1938, from Iatmul, in New Guinea, where Mead and Bateson went to gather comparative data using the same fieldwork techniques they pioneered in Bali. Mead was fascinated with dreams and recorded and analyzed them. It was one more way in which she reflected on her relationships with others. She continued to dream of Benedict even after the older woman's death, especially while she was traveling.

Ruth darling,

. . . Night before last I dreamt of you in an odd new fashioned hat, a kind I'd never seen. It's funny how persistent that dream pattern is, whenever I am unhappy about you and think you are going away from me, you appear in an odd hat, all harking back to the day you came in your plumed picture hat to teach the Museum class and I was so terribly disappointed. This hat was very odd indeed, its oddness was all based on its being a new style in hats which I had never seen. Everyone—you, Jane, Louise Bogan,[20] were beautifully dressed and I felt very odd and grubby in field clothes and finally, at the end of the dream, it suddenly occurred to me that I could go and buy a new dress at once, one that would look like the others. It was, in that sense a variation on the old examination dream in which I discover at the end of the dream that I passed those exams years ago.

    . . . I love you, love you, darling,

To Ruth Benedict, September 16, 1938, on the lack of mail from Benedict.

Ruth darling,

    I have a new dream now, that the letters which come, with their differentiating aspects, on blue paper, typed in blue type, etc. are really disguises that look as if they were from other people but then when I get to the second page they turn out to be from you. Only ofcourse they aren't—ever. We have our big six weekly mail which gathers up everything that has been scattered about in the interval, missed air mails, etc. with all the newspapers, parcel post notices, etc. this week. Marie's letter written some week or so after you got back from Guatemala was there, so I know you are alright and looking well and had a good time. It looks as if all the weary hours I'd put into writing duty letters to Marie at times when I didn't want to were coming home to roost at last, for without her letters commenting on having seen you, on your plans—to go to Guatemala, now to take a sabbatical next year—I should be so enshrouded in unreality and sick from active anxiety that I doubt whether I should be able to work at all. Which is after all what we are here in this hot, crowded, uncomfortable public way of living for. . . .

    I don't suppose that my letters sound very natural. I vary between wanting to wail, and bewail, or say with properly gritted teeth, it's alright if you are bored with writing to me and I'll try not to be so verbose, to tending to react normally to the fact that there are so many many kinds of things that I want to tell you, and that not even the fact that you have only written 4 times in 6 months can properly damp. And as long as I am in doubt about the cause of it all, I tend to be optimistic, with ofcourse attacks of despair.

    If you only have written 4 letters I think it argues for a kind of forgetfulness of time, and a kind of lack of urgency about our whole relationship which hasn't even caught your attention. That means I haven't done anything to seriously displease you, or you would be more conscious, would decide what course you were going to take, would not I think, let matters drift. It also argues that nothing else very drastic has happened in your life because whether you meant to tell me or not, you would be careful not to seem other than usual in that case. To face the probability of your not even knowing what has happened ofcourse is not very easy for me, but it means at least that you have had no unhappiness, no sense of loss, no disgust with humanity which permits such boredom to smother the soul—all of which you might have been having. And having just a dulled negative aura is perhaps better than having an actively bad one. I don't suppose I'll know whether that is so until I see you, but in the interval I can pretend.

. . . I suppose the two things have something to do with each other, my conviction that the type of field work that I have put so much into developing has been all waste motion and my growing feeling that I bore you, that I have had to stay away too long. Maybe the mechanics of psychic energy will work so that I will feel in my unconscious that if I invent methods of field work which are better, which will produce material which you can use, which your students can use, then I will be reinstated. If that should happen, I'd have a fine drive for the remaining ten weeks here. But in my heart I know well enough that while I may begin over again as an anthropologist, there is no such beginning over again with you if "our cup is broken."[21]

And the better part of wisdom is to stop this kind of writing from which no good can come. Marie says you are looking beautiful. I hope you are happy.

[End of letter]

To Ruth Benedict, October 6, 1938, from New Guinea. Mead revealed her relief at finally getting a letter.

Darling, your lovely long August 9 letter from the farm has just come and I feel as if the heavens had opened after a long drouth. I haven't been sure from one day to the next what I thought, some days I've thought that at the bottom of my heart everything was really alright, that I thought so that is, but others, the world has just been black and I have been sure that horrible unconscious forces were at work in you, even if you didn't know it. I never got the letter from Guatemala, so you see what a gap there has been. And the article didn't help any, it rather hindered because I couldn't understand what you were trying to do exactly and I was in a sufficiently morbid state so that the fact that in an article which dealt with points that I'd worked on so hard you never once referred to a thing I'd done, it seemed that you thought my work was utterly worthless too, and that cast me further into the depths. Out of it grew the decision to try to revise all my methods or at least development [of] alternatives and ways of making them clearer, and that was probably a gain. Today I started to read *Swann's Way* because I had such a vivid picture of your delight that time you read it over a weekend, the summer we spent on 124th street, and it gave me a sort of sense of being with you again, because I could be sure you had read just those words, all those words, in just that order.

. . . Darling, you don't know how happy your letter made me; I've mainly been trying to keep from writing for weeks for fear I'd say the wrong things and as it is you'll have an accumulation of wails to go through, but even so, you'll know I'm alright by the time you get the [letters], and you'll know I loved you when I wrote them. Thank heaven I wrote that early wail to which this letter of yours was an answer. But when you don't write poetry about your "country" and I can't see your face, and the only outward and visible sign—in a way I know it is an outward and visible—is a renewed ease on your part in handling the small mundane affairs of the world—well, I need just a word or two occasionally to be sure. It only needs that when it's specific enough. But it was heaven to [have] answers to actual points in letters too. I haven't time to discuss the point about the social organization and the men's characters. I will in the next letter. Oh, Ruth, I'm so happy.

[End of letter]

In a letter, August 28, 1938, Benedict again reassured Mead of her love and her trust in "the permanence of our companionship."

In 1939, on the way home from their field trip to Bali and New Guinea, Mead and Bateson found out she was pregnant. Mead wrote to Benedict, August 24, 1939, that they had named her the guardian of their unborn child in their will. Bateson decided to return to England since the war had started and would not be there for the baby's birth. Mead assumed that Ruth would be with her for the birth, but Benedict had left on a sabbatical to California before receiving the news.

Mead sent Benedict a letter on October 10, 1939, that opened: "Ruth darling, well, after a couple of hours of futile tears and a sleepless night, I have got myself into the mood to call it 'bad luck' with at least fair grace. They could always just keep the baby in the hospital until you came East, if anything happens to me. Perhaps if it's a husky baby, and I myself am reasonably husky, I could come up to the farm in the spring before going back to work."

Benedict sent her a letter, which explained that she thought the baby was going to be born in early January, by which time she would have been back in New York. It turned out to be impossible for her to go to New York for the baby's birth at any rate, for she fell ill with pleurisy in November, and the baby was born in early December. She knitted a pair of baby booties instead, her first in fifteen years, and sent them to Mead. During this sabbatical Benedict also met and fell in love with psychologist Ruth Valentine, a relationship as important as the earlier one with Natalie Raymond.

World War II took them both to Washington—Benedict to work for the Office of War Information and, through Benedict's intervention, Mead was named executive director of the Committee on Food Habits of the National Research Council. But through summers, lecture trips, and other times apart, letters still flowed, such as the one of August 24, 1945, from New Hampshire, where the combined Bateson-Frank household spent its summers, which Mead ended by saying, "and always I love you and realize what a desert life might have been without you." [h]

Benedict supported Mead through her separation from Bateson and enlisted her to work for Research in Contemporary Cultures, the project that came to be known as the study of national character. But she herself did not have long to live. A sudden heart attack put her in the hospital, and she died there September 17, 1948, at the age of sixty-one.

After Benedict's death, Mead found another special woman, Rhoda Metraux, who originally had been her assistant on the Food Habits Committee during World War II and then moved on to work for the OSS for the rest of the war. Rhoda was married to anthropologist Alfred Metraux and was twelve years younger than Mead, reversing the terms of her relationship to Benedict, who had been fifteen years older than she. During the 1940s, Metraux was in graduate school working toward a Ph.D. in anthropology for herself, which she finished in 1951. Metraux and her husband, Alfred, were living apart by 1950 as he resided in Europe to carry out his job with the United Nations. Mead and Metraux collaborated in the Research in Contemporary Cultures project and by 1955 were sharing a house at Waverly Place in New York City, although each had her own space, with their children, Mary Catherine Bateson and Daniel Metraux. Mead wrote to Metraux in a letter on July 23, 1964, "other people are fun—but no one makes the world continuingly meaningful." The following letters reveal some of the dynamics of their relationship together.

To Rhoda Metraux, December 22, 1948, three months after Benedict's death, September 17, 1948. Metraux had taken her infant son, Daniel, and gone to Haiti with her husband, Alfred Metraux, where he was going to do field work.[22] [h]

Rhoda darling,

Last night was the really bad night—coming home for the first time since you left—at 10:30—and you not there to visit. Saturday night I lay awake following your plane through the dark—as if somehow the intensity of my thought might turn into the necessary angels.

———

Your cable just came—and I feel better. Cathy's comment,[23] "but all love isn't safe!" Bless you for sending it.

———

Here there is an enormous air of Christmas about the place—door bells ringing, flowers arriving, Cathy opening cards and finding bills and solicitations among them—but I could turn back the clock to Thanksgiving.

Poor Jim![24] All his plans are in rags and he is going to San Francisco with his parents—and perhaps become a Dutch son after all—and I don't feel sure enough that isn't the answer.

———

Lisbeth's picture-Christmas card to me appears to be the Immaculate Conception! Meanwhile Martha has a wonderful theory—that EH[25] is the Frankensteinian embodiment of all the envy which has been directed towards me but which I have failed to reciprocate. Which seems like a fascinating idea—especially from a Freudian.

———

I discovered that I had buoyed myself on Collingwood's *Principles of Art* this summer, but I hadn't reabsorbed that fact. I think I shall give it to Allen[26] for Christmas—and just see—

———

These are all bits—as if you were moving about the room with the baby in your arms and I was giving you the day's news—specifically they are nostalgia for just that—and much more.

[Stops here]

To Rhoda Metraux in Haiti, January 2, 1949, from New York. Mead was in the middle of working out a new relationship with her husband, Gregory Bateson, during their separation. They would divorce in 1950.[27] [h]

Rhoda darling,

This is the first time I have written the figure 1949. My sweet, my sweet, it was so good to get your beautiful long letter. It took an unconscionable time to get here—Christmas rush, probably only—and I'm very deeply grateful to have had the cable. So at least the area of my imagination could be a little limited.

On the whole things go well here. We have had a great deal of talk—and very little sleep—and for the first time in a long time are talking *with* each other instead of *at* each other. All eventualities are as unclear as ever, but the present is—on the whole—good, and to be kept close to the heart.

Why—or what is the mechanics of de-evaluating backwards—or only defining by results—(as the Russians do)—rather than saying This moment is itself absolute—in its own right? I have so little feeling for the other position.

There has been quite a bit of Gregory's saying how differently he thinks he felt about some past moment—quite differently from what I thought—but in some strange way this doesn't take the past away from me—it creates another moment—a different one, in addition.

I suppose I feel that the same moment could be seen a hundred times—and all have value. Is this anything about the past-present-future point in your letter? It must be so different for some people. I remember how shocked Paul Radin was that Edward [Sapir] had wanted to give me Florence's ring—and I couldn't understand why.[28] It's got something to do with my Martha part—of having no imagination—and so bounding nothing with a fixed image. Your face—close to my eyes—and so a different loveliness—does not bind or erase the other images I have of it—it remains—to place beside the others—seen across the room, half-glimpsed—or curiously wrought from a few lines in the Botticelli Venus above my bed—

But no violence is done to you, because I have seen a part of your face—in the Venus—nor to the Venus—who before her special qualities of beauty—as clearly seen as ever—I don't like the word "additive" exactly—and yet in the sense that this is no *reduction*, no *subtraction*, no divisions it is true. But for some people the frames are so tight that if your face were to be glimpsed in the Venus—then the Venus is—in a way—when done, to become you. I have been dreaming a little about the three women's faces paintings—on my wall—the little strange Holbein Girl, the Venus, and the St. Ann—and how much I have seen in them through the years—and yet, they stay themselves . . . [orig].

In a way even trying to write is senseless because the whole matter of selection is so impossible. I could talk now, for a while—without stopping—or choosing which came first because all would be relevant. We could talk about Tulia's[29] way with Daniel—which makes me happy to think of—and go back and forth between Mr. T. S. Eliot—whom Edith Cobb had decided suffers from never having slept with the other half of himself—and your face, and my Marthaness—and perhaps—even a little comment on the bright coldness of New Year's Day—only there it would be soft—fertility wine. There are no grounds for selection of what I say to you—except that a letter came, sealed up—made into a unit—in which the weighting of a phrase attains unreliable proportional value. You will have to have a little table entitled "Table of appropriate weightings and loving corrections"—and for a two-page letter—read—"a 20-page letter would have been fitting here"—for once and one way of saying "I love you"—a thousand better and different ones.

On New Year's Eve Gregory decided to stay another week—as the analysis was going so well. This means a violent reorganization of duties which had already been posted off—three weeks—and I'll have no time for a proper long letter for another week. But oh, my darling—there is a little singing at the back of my mind, which is your name.

Margaret

To Rhoda Metraux, January 8, 1949,[30] *to Haiti from Perry Street, New York.*[31] [*h*]

Rhoda darling,

. . . Life goes on here—pretty promisingly. I have bits of aches and pains and little outbursts of extreme irritation—all of which show that there *is* a strain—but it is not unbearable, and blessedly—you are not chosen as a battleground, so that much of my dread on that score is now really gone.[32]

I don't like that fever you had at all—although I suppose it could have been "newcomers fever." Still—be careful, my darling, please.

I think we're making a fair amount of progress towards my being allowed to use my own insights—partly because as Gregory gets more articulate, what he says makes particular sense, whereas what "they" say only makes general sense.

There is something very repetitious about the themes I seem to pick to write you about—in that they all cluster around the idea of "no substitutes" each moment, each person, each event lives in its own right. For now what I want to say is that writing and being written to, is a great pleasure—as long as one doesn't say "how poor it is in contrast" which I'm afraid I did in my last letter. Letters are not a substitute for anything, but they are lovely to have. Perry Street is still a lonely place to come back to eat ten o'clock at night. I can't get used to you not being here—used—in the sense that my feelings become blurred and the sharp sense of deprivation is gone.

. . . You know, when you ask I [sic: a] question, like how did I know to give Emily Brontë to you, and did I know how you love *Wuthering Heights*, I just don't know the answer. Did I know that concretely, or not? I just would know that you would care about *Wuthering Heights*. It's as if I knew all the general premises so well and had known them so long, that the details seem *given* and so I forget whether or not they have ever been actually concretely communicated. I've lived with a kind of very intense concentrated knowledge of you for so long, my darling, peeling off the special gossamer bit that fits the particular occasion, seems so simple—so "natural"—in the sense that it couldn't be any other way.

When Jane[33] comes down get her [to] take some pictures of you, of your room, with you in it—of the view from your windows, of you and Daniel in the garden, so that I have some pictures of what it is all like.

. . . Cathy looks simply exquisite in the little bonnet—and I wear your scarf all the time. It is really too fragile for every day, but the color and texture which you chose—and which everyone says "looks like Rhoda"— seems to have been meant to touch me, and comfort me. It's color that one can *touch*; —or is there only a remembrance in my fingertips.

My lovely sweet—you are here in my heart and just out of focus if I turn my head—and felt in the tips of my fingers—and there to read this letter—

I love you,

Margaret

To Rhoda Metraux from Mead's sister Priscilla's house in Beverly Hills, California, January 30, 1949. This was during the time when Mead's mother was in the hospital after having had a stroke.[34] [h]

Rhoda darling,

. . . Out here watching my tiny incredibly energetic namesake,[35] whom my sister and brother regard as just like me—while my aunt who was here too kept claiming that she was not in the least like what I had really been as a child—I got another dose of realisation of what the long battle has been to be myself, and a battle which will never really end till I die because it is joined with the whole organized unconscious of the age. The only question is whether I can learn to look less like a botcher.

It's lonely to have had no word from you for a week—and it's so terribly far away. I'll be back in New York when you get this—I fly back on Mon-

day night—and there will doubtless [be] a letter waiting for me—and life
will knit together again over the strange gaps introduced by a week without
word—and flight across a continent. I love you, my sweet.

Margaret

*To Rhoda Metraux, St. Valentine's Day, February 14, 1949, from Oxford,
Ohio, to Haiti.* [*h*]

[No salutation]

Both your flowers came before I left, darling, probably in disobedience,
but so fortunately. On Friday I had a couple of hours at home writing and
stopping to enjoy the sunlight through the windows, and then the long box
of spring flowers—and the little cochise hat note—so familiar and precious
a form—came. Pink tulips and long long stemmed daffodils and that new
sweet scented yellow flower arranged like many branched candlesticks on its
stems—and blue Iris. I arranged them in three vases—there were so
many—one slender one to be reflected in the mirror over the fireplace—a
squat gay bunch for my back window in the bedroom where the sun could
catch it Saturday morning, and one for the home upstairs where Cathy is
fed and made welcome when I must be away. And thank you my sweet.

It's really very odd how little guilt over special symmetries I have. I suppose partly because I never chose asymmetrical relationships on purpose—
out of any masochism or desire for self-sacrifice—but simply chose people
who couldn't pay attention, and then accepted their not paying attention as
part of the world. Now when you write me long letters to my short ones, or
poems where I have only prose—I simply feel touched and delighted—a
little strange, but with a pleasant kind of surprise. I felt the other night—
after I had arranged your flowers, and [saddened and][36] in a chill dusk, just
cool enough to make one know one didn't have a hat on—to buy the children Valentines on 8th Street—as I walked, remembering the flowers and
the poem, and the way people seemed happier about my looks, as if somehow Ruth had bequeathed me a little of her beauty—by giving some sort of
care into your hands—for me. That's all rather complicated, but you will
understand. I have such a firm belief that nothing in the world is too

complicated—nothing that I can think up—but that you will but understand and add to it.

. . . For the flowers, for the poem, for your love, my darling, thank you—

Margaret

To Rhoda Metraux, March 9, 1949. Mead was in Indianapolis and Metraux was still in Haiti.[37] *[h]*

Rhoda darling,

Here I am traveling again before another letter is written—while your picture—with Daniel in your arms—smiles at me—only I'm not there—on the mantel piece in the back room at home.

Last Friday I sent you some arbutus—I wonder if it ever arrived. I stood woefully in front of the florist's window looking at it, and you—to whom it should be given—so far away. And then, I thought, why not. If only they didn't decide it harbored some new kind of pest and forbid it. And I asked so fast that the shop clerk had it all packed before I'd thought about the need of writing a card. But really there was no need—You would have known, without a customs declaration. . . .

Probably my letters will be sparse and peculiar until after Gregory comes and goes. I find I am in a way more anxious than at Christmas. . . .

But however scattery my letters—I love you, my dear—

Margaret

P.S. Better write to the Museum, with a big "Personal," on the envelope.

To Rhoda Metraux, April 9, 1949, from Perry Street, New York to Haiti. Here, Mead saw herself, Metraux, and Bateson in a triangle; what happened depended on Bateson's future decisions.[38] *[h]*

Rhoda darling,

. . . I've just reread your last two letters. There's a terrible temptation just to say wait till she comes home—to say anything about anything. Words face to face are better than words a week apart on paper. I don't quite understand your paragraph about my really pursuing what you essentially denied and I want you to know I don't in case somehow something else I say gets misinterpreted. *But*—don't worry about trying to tell me till you come. . . .

You know I don't really feel that you have been away, but I feel that Daniel has—

There is one other thing that would be clearer perhaps face to face, but somehow I feel it's important for you to order your expectancies while you pack Daniel's sox and vests—neatly—to come to New York. This plateau on which you and I live—will have to go on being a plateau until the other pieces of the picture fall into place. That is one conviction which Gregory's last visit brought me.

However much we may deny it—a triangle with an unresolved side is so unstable that the other sides are necessarily less stable (clear) also. It might of-course cease to be a triangle altogether—but till it does—and that I know now will never be of my choosing—there it is. As long as any possibility of choice lay in my hands—things looked less clear. Do you see, my darling? God knows, I wish it weren't a triangle—but it is—not in action—but in feeling. Please heaven, this won't make you feel you don't want to come. We have so much. (Something of Léonie's about "God's lamb" keeps coming into my mind here.) I'll look it up—in a minute—after I've kissed your eyes as if you were here—and I couldn't go abruptly from your side to a bookcase—to look for a quotation. (It's in "Heaven's Paradox" p. 18 of *Those Not Elect*.) Did I see us as still rather young, immature—and so able to live on the thin celestial fare? I'm not sure.

I loved the angel poem—and as I read it—I realized—that in some strange way I am not Martha anymore—the name now is a name you give me, but the old meaning, the "Martha who would be Mary, shall carry a troubled heart" is gone. Perhaps because I am really not Martha to you in my old sense. Anyway, it's gone—slipped off like a well-worn garment—the kind one doesn't give away perhaps, but keeps tucked away for years after—affectionately—in a bureau drawer—after all it once kept one warm, though in strange fashion. I somehow feel that I will never again break my heart be-cause I am unable to be someone's unreal dream—what have you done—my sweetest—perhaps you know—Oh, it will be so lovely to be able to see you again—often—Do beg Murray[39] to search hard in the village—so I can drop in on you at odd hours again.

Goodnight my sweet

Margaret

To Rhoda Metraux, April 20, 1949, midnight, from Perry Street.[40] [*h*]

Rhoda darling,

Your letter brought you so close, my sweet, and now I walk about the Village—loving the streets again—because soon they will be streets that can lead to you. Murray has been consulting me about apartments and I plunked her for the nearest—being very careful ofcourse about such questions as where a baby carriage could be put or how much privacy would be allowed— But hoping that it would be very close. In a way ofcourse, I am furious that you can't just come here and stay—which is physically possible—but not otherwise I am afraid. It's that sort of inevitability in *la condition humain* which I resent—and—like your promised amnesty—it is not in terms of any person—but a fury at the nature of things. But aside from this one piece of rage that you must be perhaps two, perhaps four blocks away—I count my-self as so blessed—in that you are coming—and so soon.

Cathy is building all sorts of imaginative pictures of Daniel—being es-sentially uninformed as to how a baby of his age can behave. And I am find-ing all the things that I can say to you and to no one else springing up from my heart—and trembling on my lips. It will be so wonderful to have your answer while I can still remember exactly what I said, or see in your eyes which part of my sentence you are responding to.

The spring is lovely and still young—there are lilies on the barrens—and the small boys are playing with words sounds on Perry Street. And in 10 days you should be here—my own, my sweet.

I love you.

Margaret

To Rhoda Metraux, October 13, 1951, from Perth, Australia, where Mead and her daughter were visiting while she spoke as part of a lecture tour for the

New Education Fellowship Jubilee Celebration. Mead also used the opportunity to renew contacts and make queries as she scouted for her next field trip. Here she talked of her relationship with men.[41] *[h]*

Rhoda darling, a whole [batch] of letters from you—all good—and I feel so relieved and peaceful about [you]—and Daniel—such lovely bits about Daniel—and I'm so pleased nursery school is going so easily. . . .

Life has now assembled itself into a pattern which has been familiar through the years—the intensive exploration of a new role in a new place, focused in feeling on someone—it always is a man, I think, who is also deeply involved in the same general set of operations. It was Bob in Washington, Erwin in England, Clemens in Salzburg, Alex in Dallas. Odd—they are all so similar—a just not falling in love—in any direct sense. It's like shipboard—except that it has all the intensity of work instead of the irresponsibility of play. I am now completely enjoying what I'm doing, feeling gay and purring and relaxed, and counting how few more days there are. Charles is an extraordinarily sensitive creature—Australian born of English parents, in love with democracy—which be it noted—so were all the others. The other thing that isn't like shipboard is ofcourse that it lasts—for life. I suppose I ought to add in Monte Phillips too and the time I did the life of Mrs. Parkinson.[42] Times out of life which are nevertheless in life—acutely.

This is a lovely little city. The people here like living here. I have an enormous room, with a chaise lounge and linens—large chests with long mirrors and a balcony. Some utterly beneficent and unknown gods put Charles and me here—for this eight days, and the other two—which I regard as just so much waste of precious time—half way across town. So we have breakfast together every day and dinner most nights and go and come to meetings together and talk and talk—about Australian culture and the nuances of the various sessions. Charles comes to all my lectures which [makes it] a terrific challenge not to repeat but a delight too because he follows the clues—how a remark to him at breakfast and then an illustration used ten hours later in a lecture fit together. So, my dear, I have my sunshine and my guardian angel in all—as you wished. And it did seem so unlikely. The trip through Melbourne was a variety of nightmares—mostly Carrie's.[43] Tell E. Cobb[44] that

I'm getting all the details of the "Little Black figures" story together and she was mostly right about the phoneyness.

My sweet—I am wearing your earrings again—and I sit up in bed and drink my tea, in the pink [wintry][45] jacket you gave me—and look at myself in the longest pica glass and smile—for you—I love you.

Margaret

To Rhoda Metraux, from Honolulu, June 1, 1953, at the beginning of Mead's return field trip to the Manus people of Pere, New Guinea. Metraux was also preparing for her own first field trip by herself, to the island of Montserrat in the Caribbean. They had decided that they would try to go into different field areas at the same time so they would have parallel experiences that would lead in discussion to new levels of thought.[46] [h]

Darling,

I'm sitting in the airport, facing a tropical sunrise and bathing in the air that feels and smells so differently—the whole of my life folded back on itself—beginning with *three* hours with Gladys in the Chicago airport. But that too was probably arranged. She retires in five years and that gives us something definite to work for.[47] But it's strange too because Gladys is an old woman, a vigorous energetic old woman—but still an old woman. I'm in the airport now because I didn't want Kenneth[48]—whose hair is grey—to get up at 5 in the morning. But five o'clock in the morning means nothing to me yet—more than it did twenty-five years ago.

I telephoned Val: Ruth Tolman had a coronary three weeks ago.[49] She is going to live—but Val's life with no one ongoing in it is just a maze of shadows and burials. It's so wonderful that you are so much younger—It's as much a blessing as I used to feel Ruth's being fifteen years older was a threat—I feel that there are so many years ahead—for life and love and work and new theoretical advances. For the first time I go through Honolulu without a carking worry or a memory bringing immediate tears. Angels still seem to "ring our islands round" and I love you—our field years will match each others'—and later years also. I love you.

Margaret

To Rhoda Metraux, January 10, 1954, visiting the Barlows in England on her way home from her 1953 return field trip to the Manus of Pere Village, New Guinea. Mead was fifty-two years old.[50] [*h*]

Rhoda darling,

It's nine o'clock of a cold foggy winter morning. The fire has just been kindled in the "morning room," the outlines of the garden seem all but dissolved in wet as one looks out the window. Coming back here always means ofcourse coming back to thinking about Gregory—this time it's to accounts of all Nora's struggles when he finally decided to send for all his scattered possessions—after not answering a single letter for four years. So the Blabers are gone from the walls of the dining room, and there are no things here any more just memories. But I've not cried yet. I haven't cried—except when I said goodby to Pere—and to J.K.[51]—for six months—the longest six months without tears in—I think—my whole life. And I haven't cried simply because there has been nothing to make me unhappy.[52]

Your letter describing Tao's[53] visit came—I'm so glad it's working so well—and glad for her. It will be her first really meaningful anthropological work. It's funny to think that I've been trying to get her to do some for almost thirty years.

Geoffrey's mother broke her thigh a week before I arrived, but stood the operation very well and is now home again. One of the things that I desperately hope won't happen is that I will be here when she dies. Altho we both know perfectly well that the hour has passed when her death would have changed our relationship, still I'd like to spare us—and especially him—having it dramatized by my being here.[54] Geoffrey himself is placed—hair getting gray and stringy—and pretty plump. He looks a little unbuttoned, is finishing his English book which is dull to the point of torture to read, but he's got some good ideas out of it. He's definite about no more field work—unless it were Denmark—and is playing with doing ancient Athens—which is a possibility I've been cultivating for years. He has the necessary training to do it.

The WHO conference is going beautifully. I wasn't cleared because it is necessary to have *new* fingerprints—but I am very much there—as a guest—and I had my ticket anyway. Then yesterday I went to a meeting of the Soc.

Anthrop.—very odd to see all at once, so many people one has known as individuals and ofcourse it was a great shock to find Reo so grey and grizzled and somewhat ravaged—I've got so used to living with the pictures of him in all his young beauty. One of the real problems is that everyone seems so old. But the WHO Seminar—because they are such extraordinary people aren't so disappointing—physically—and they are magnificent intellectually—things are just falling into place, like the tremendous impetus at the end of a jigsaw puzzle.

Grey Walter has made a model of "imprinting" and a model for the different kinds of learning. Lorenz has made a film of a baby—as he would young birds—I'll type it all out as soon as I get back to a typewriter. Is Tao using Beule's Gestalts? I think we ought to have a series of those if possible—as they are proving very useful for Manus and they are simple to do. Also—as quickly as possible, she should add to the battery anything that seems essential, which we don't have.

I wonder if you are going to like the "me" which I now am. Do you place the smiling picture that was taken at the same time as the sad one which I used to have standing on the linen bureau—in an off-the-shoulder dress with a tiny oval locket. For that, people say, is how I look—this is ofcourse someone that even Geoffrey has never seen. I've lost 28 lbs.—so the regression to a younger person is overdetermined. Ofcourse it may all wear off very quickly when I get back to the present. I want to keep the lightness if I can—it's very pleasant and I've got used to it. At first my body seemed sleazy—as if I'd exchanged good quality heavy wool for rayon—but it is odd—to be physically—someone you never knew. Now if you had my obsession about the lost years—which I had about Ruth—this would be a real break, but you haven't, have you? The contrast between the WHO group who have talked and talked about how different I look—with cautions against getting any younger or I won't be allowed in—and the anthropologists who didn't know I'd changed at all—to whom I am simply an ego, thrust timeless and bodiless—was very odd. Anyway, it's all one of those pleasant little bonuses from Fate out of which I'm going to get a lot of fun—especially as I thought it very likely I might, given enough tropical illness—come back looking like an old woman. Johnny Whiting is a little old man—white haired—Nora, on the other hand, is apple cheeked and young again.

It is felt, in London, that a decent study of the "structure" of the West Indian family is very badly needed. Leila[55] says you sent back *940* pages of notes—I think you've outdistanced all of us. Claire Jacobson is going to be my secretary to tide me over the first weeks back—till I get someone. Leila has to go to help her daughter have a baby. Be happy, darling,

Margaret

To Rhoda Metraux, January 29, 1954, from Perry Street. Mead was home in New York, but Metraux was still in Montserrat in the Caribbean doing field work.[56] [h]

Rhoda darling,

My original image was right—it is like blinkered vision—not looking down streets—which should lead to your house, but don't because you aren't there. I don't so much feel unreal as surprised—like putting one's foot out to find a step that isn't there. I am dreadfully afraid I'll repeat things too—perhaps I ought to type always so that I'll know what I said.[57]

. . . Chuck[58] came down today to ask me into the Department on virtually any terms I'd like—but starting next year as Visiting Professor in General Studies—but within the Department doing Ph.D. theses etc. I chose [a] 1/2 time Professorship—teaching 2 evenings, Introductory Anthropology and a graduate course—"Cultural Character" one semester and "Methods" the other, with 1/2 my salary paid directly to a research assistant! It isn't entirely settled but looks alright tho no one can understand why I want it that way— because they can't imagine paying one's own secretary. But Chuck was sweet and it's pleasant to have all the old hostility of the department gone. He's asking Bunny to give a course too. And I'll get 1/4 time off from the Museum next year—(Harry[59] agrees one shouldn't spend Museum time "preparing" for one's lectures. God save the mark)—which will give me a week off a month! So it's all very felicitous. Oh and I'm being asked to be a Senior Fellow at the new Ford Center. I'd better really examine myself—there must be something wrong—all this respectability. Our big batch of films got back today, safely.

The house purrs pleasantly—this is the sound of the gas grate—Cathy turning the pages of the *Canterbury Tales*, 2 cats gently frolicking on the floor—the whole room filled with memories of you—a room where you might come in the door—any minute—There is so much to do that in no time at all it will be spring—then summer—and you will come—

Margaret

To Geoffrey Gorer, December 8, 1960, from Boulder, Colorado. Mead had hurt her ankle and was still using one crutch.

Dear Geoffrey,

. . . This will be about my own private world. This last year has been pretty frightful, but at least it has given me a chance to test out and think about my own attitudes towards any diminution of freedom due to age or illness, once and for all. I still [have] no answer to the question of having a servant of my own, because no servant of mine will ever meet Rhoda's requirements in the house, even if after Tulia retires we could manage it. As a result there seems to be no answer to how I am to obtain even a modicum of private life, for Rhoda keeps taking over jobs that Tulia should do so that every detail of life is inextricably mixed. I play with the idea of getting myself a hidey hole somewhere that no one will know about, but it would cost so much money and time that really is needed for other things. Ofcourse life will be easier when I am completely mobile, but at present we live in the same house, work in the same office, ride in the same taxis, and it is simply too great a burden on one relationship—the same point I'm making about American marriage, although for other reasons.

Rhoda still is extraordinarily fragile, with shifts in bodily states that no one can explain satisfactorily. It's not yet the menpause [sic] they are sure. It may be the extra kidney but they don't know what to do about it. She has fits of extreme sense of pressure in her head, of dizziness and inability to think. She is very anxious to go and take a look at Germany and I think I have that worked out for next summer for her, with this Outdoor Commission money. Maybe that will help—but I only say that in the sense that anything *might* help. . . .

To Rhoda Metraux, October 30, 1965, from Pere Village, Manus, on a return field visit. As time went on, Mead began making visits to her old field sites to follow up, rather than making full-fledged field trips, with all that they entailed. This typed air letter is marked "Personal letter #5."[60]

Rhoda darling, my writing is [so] very bad that I keep thinking it is better to type—without carbons—these personal letters. If you [would] rather have them handwritten, please say so. They are ofcourse less likely to be read by the casual prowler if in my script.

This is in answer to your long letter about shyness, and Paul's[61] troubles. Don't you think the point is that your shyness lies very close to a kind of familial core constellation, and that when something touches any member of that constellation, the shyness as your special expression of your early childhood experience is evoked. Marie's examination trauma must be the same sort of thing. And my fear of hurting other people just by being. Anyway you *should* write these things out—even more if they can't be talked about.

. . . I am having a lovely time doing the linguistics of Manus imagery for your project.[62] I hesitate to write you any yet, because as I understand you want to go back to Montserrat to review and clear your own conceptions up and you might not want this yet. It may also be that what I think is contributing to your problem isn't. You have gotten angry once or twice when I have seemed to be laying premature and clumsy hands on something that is, I know, extremely delicate and fleetingly caught, yet. But I do want to get things as clear in my mind as possible about the Manus end of the material, so we will have that to report, and I think I have a useful set of clues here. . . .

After a certain amount of fussing and feeling and fretting I have settled down very happily, go to sleep contented and wake up bright-eyed and bushy-tailed. I find I like the Donald Allen anthology very much and I am reading very slowly, trying to learn each new poet. I've got such an impressive lot of work done, now that all my potential guilt is assuaged, and I can work at my own pace. My typing is bad, my fingers stick to the keys a lot and I tend to skip letters and words, partly from typing PE[63] shorthand. It is indicative of nothing more.

. . . This letter is now far enough along so that if there is a sudden chance to send mail, I can get it off quickly. The American survey team on the next island may want to rent our canoe to take them in, and then I'll have to work fast.

[handwritten] They've come—

All my love, my sweet
Margaret Mead

To Rhoda Metraux, February 22, 1967, from Manila, in the Philippines.
Mead was visiting her daughter, Cathy, and her husband, Barkev Kassarjian,
who were living there.[64]

Rhoda darling,

This will be a sort of winding up the first lap of the trip letter. I am installed in Cathy's very attractive house, unpacked by her little maid, politically briefed by Barkev, but I haven't seen Cathy yet.

California all went very well. Everyone is in an uproar over the political situation; they have refused to handle Yehudi's[65] big grant so he will be leaving;
San Francisco State expects to lose 99 faculty just from the cuts, and be unable to take in 2,500 students. Aunt Fanny was very frail but still herself. The
conference on the pill was on the whole good fun although ofcourse I was
misrepresented in a headline which said I said give the pill to all high school
girls. But the two forms of marriage got a good deal of play too. And did you
see that I have been named as one of America's 11 sacred cows, who are invulnerable to attack—in an LA paper, and then Russell in the *Times* last Sunday
named a set of profane cows, fair game for everyone, but matched no one
with me in any way. Then the CIA student stories broke, given a great play in
California, none in Hawaii. I have put all my weight behind emphasizing the
need to separate free money that wasn't whittled to pieces by Congress, from
the specific spying activities. I suggested to *Newsweek* that they make a list of
what different countries do for their Olympic teams.

Hawaii; I was greeted by Ted with news Lola was in the hospital, they finally decided to take out the ambiguous lump in her lung; she is recovering
well, but it was a rather dreary little household, Adan whining and miserable,
Debbie housekeeping without enthusiasm, Ted, very low keyed, and planning to do far more writing than he will get done.[66] He had his appointment
confirmed at LA while I was there. Big party with untold numbers of anthropologists. Kenneth brought me a lei he had made from his own garden. He—
and Marguerite—will be in Oregon for the next semester—I said you might
go through Portland and he brightened.[67] All very nostalgic, with Kenneth
on one side, and Larry Snyder on the other, and someone bringing me Peter
Buck's copy of my *Social Organization of Manu'a*—with all his notes in it—to

autograph.[68] A very tiring day yesterday, but a good one. I spent the morning with Gregory who is in wonderful form, just spouting ideas in all directions. The conference he is going to at the Wenner Gren castle is on primitive art, then to England to a conference on Existentialism, no less, so he will at last go to England [and] see everybody. Paperback *Naven* is selling well. Lois[69] has had an operation for a disc, and is still in pain. She is working as a psychiatric social work[er] advising Hawaiian University students and studying for her Ph.D. in Psychology. Gregory has been dredging up Iatmul and is planning to make you a tape. All absolutely excellent Ph [sic], and I had two breakfasts with Mark and Regina and Mark is blissfully happy, running half the programs in the hospital with a lovely house and garden.

So, darling, that's it for now. I can only hope Jimmy [Mysbergh] is looking after you and you aren't getting too tired.

M.

To Rhoda Metraux, July 12, 1969, from Athens, Greece. Mead's life was getting more crowded, more frantic, more overscheduled, and her letters were turning into notes. [h]

Rhoda darling,

We don't know yet whether the Apollo flight went up! Not till I go down stairs—but last night no one seemed to know.

Aside from losing all my luggage—left in Zurich they think and I hope—otherwise it's all gone to Turkey—all has gone well. But I had just 35 minutes before the stores closed to get anything to wear. I had on my [wintry][70] suit and walking shoes! However, with that beautiful American beauty colored scarf—I almost looked civilized at the party.

Zurich—week was very very good and I am rested just from being able to attend to one thing.

Next summer—or early fall—when the text book is finished—we must plan to go away—without any manuscripts of any kind—somewhere where you can attend only to the sea and the beat of your own heart—and mine.

All my love,
Margaret

To Rhoda Metraux, January 3, 1974, when Mead was seventy-two years old and Metraux sixty.

Darling,

It seems pretty silly to write three copies of a love letter but you sometimes don't find letters no matter where I put them.

I have to go in half an hour. I don't understand anything at all, why these endless depths—after what seemed such a good height like last night.

I don't know where you are. It's frightfully worrying. It would be very very good if you would telephone me tonight so that I could know that you are alright. . . .

Oh, I love you so, and I feel so desperate.

M.

To Rhoda Metraux, March 6, 1974, from Honolulu. [h]

My darling,

I've just pressed the violets and 2 little violet leaves within the unread pages of a new book. Be better, my love. Life does make sense—you make it make sense for me—I love you,

Margaret

To Rhoda Metraux, April 23, 1976, at a particularly stormy point in their relationship. Here, Mead detailed what she saw as some of the problems that had to be addressed if they were to stay together.

Rhoda darling,

This won't do. It won't. We have come to a point where I simply don't know which is true, the things you say to me when you are angry, and/or depressed or the things you say at other times when you are the delightful person whom I love. . . .

I believe that people who live together, and I mean this in the profoundest possible way, should be allowed to take some responsibility for each other, to tell the other that they are too tired, or too thin or too fat, or need a vacation, that this is part of life. I also believe that there ought to [be] some reciprocity between people who share a dwelling! Endogamy is terribly exacting and as you know I tried to avoid it, after I had lived through two marriages that couldn't stand it. I specified, very carefully when we went to Waverly Place, that if we lived together, we shouldn't also work together. This was an attempt to protect our relationship. It failed. Gradually, and not at my instance, everything we did became completely entwined, *Redbook*, the Museum, the office, the House. This represents a degree of dependence which, in some moods, you hate and resent and berate me for, and I don't know what to do. . . . I have fought your insistence on anonymity in editing my work rather than our doing cooperative work. I am and always have been happy to do joint work with you, with both of us taking credit for what is done. . . .

I devoutly believe that there is no difficulty between two people for which both are not responsible. But at the same time, I have always believed that depressions like yours and Ruth's and Gregory's were not created by other people and not curable by other people. That all I could do is to hang on, stick it out, be there, no matter what I was called in the heat of agitation. . . .

I love you more than anyone else in the world. I have been perfectly willing to dedicate my life to you as long as we two should live but I can not do it unless I feel that you are better off with me than without me. . . .

You feel that my picture of you is totally distorted; perhaps your picture of me has nothing to do with what I feel that I am doing, and feeling. There is no way of knowing as long as you refuse to talk it over.

What do you suggest?

I love you but that doesn't seem to do much good.

    Margaret

To Rhoda Metraux, June 9, 1977, from Kenya, when they seemed to have worked out their difficulties.[71] [*h*]

Darling,

Everything goes well. The new blouses are just what I needed—and everything is visible and findable in the little plastic bags. As always when you pack for me, I feel surrounded by your love.

I hope you've had your glasses fitted. The three days with wrong glasses are still taking their toll in fatigue—So if you haven't, please do.

I feel very well repaid for coming here. I've been able to correct for a lot of things that otherwise would have gone uncorrected—for example a conference on desertification with no mention of human settlements—which is supposed to be part of UNEP—an agreement with UNESCO to keep the environmentalists out of a conference on environmental education—all tiresome—but necessary.

It's pleasantly cool here after a month of heavy rain that has "restored the water table" but washed out most of the road.

Lovingly,

Margaret

In the last year of Mead's life, Mead and Metraux were estranged from each other. Their relationship had always contained heated arguments, which now included what kind of treatments Mead should have and Metraux's objections to the faith healer, Carmen de Barraza, whom Mead employed. Mead took a summer trip to California, and when she returned in the fall, she took an apartment with her sister Elizabeth in a building across the street from the apartment at the Beresford in New York, where Mead and Metraux had lived since 1966, but on the same floor. Mead, seriously ill with pancreatic cancer, decided she needed to focus her energy on positive healing. She felt she could not live with Rhoda for awhile, as their relationship was a major source of stress at this time, and she needed all her energy to try to heal her body. She also wanted to help her sister, Elizabeth, who had been ill and living alone.

She did not see this move as permanent. She wrote to her English friend Geoffrey Gorer on September 10, 1978, "I have told Rhoda I won't move back into 211 until she sends Tulia home and we can make plans for the next year—when she can sell 211 without as much tax. Meanwhile I have a myriad places to stay now that Elizabeth's apartment is there to save the picture for the outside world." She did not stop loving Metraux, as a message in a little card Mead sent her August 26, 1978, from California shows. Metraux was at her remodeled one-room schoolhouse in Vermont that summer while Mead was traveling in the West. [h]

Rhoda darling,

This is the only small stationary I have, and I don't feel like typing much yet—

I am much much better. Barbara [Roll] is inventive and persistent and is getting between 1200 & 1500 calories into me a day. There are still setbacks—which the new doctor, Goodrich[72]—says are like a phantom limb. But they hurt just as much.

Barkev and Vanni[73] arrive tomorrow & will go later to visit Barkev's brother & the Beirut cousin who lived with them last year—now settled near San Rafael.[74]

At present I expect to come back by the end of the month.

I hope you are having a good time.

I love you,

Margaret

4

Friends: A Genius
for Friendship

<hr />

Margaret Mead made new friends throughout her life, while nurturing existing relationships over time. She valued each relationship for its own qualities and made friends across all kinds of backgrounds and ages. Mead, further, made friends of the spouses and other relatives of her friends—and with her friends' friends. She put her own friends in contact with one another, offered advice, mediated disputes, and sometimes angered friends who found her meddlesome.

Mead's family moved a great deal during her childhood, though their homes were clustered in several distinct areas: Delaware and Bucks Counties in Pennsylvania; the city of Philadelphia; and Hammonton, New Jersey. She made numerous friends in her childhood travels and kept some of them into adulthood. The largest cluster of her lifelong friends, however, came from her college years. From her first year of college there was Katharine Rothenberger, and in the remaining years there were her close friendships with a variety of fellow students, including Marie Eichelberger, who remained a constant presence until the end of Mead's life, Eleanor Steele, and with several generations of "Ash Can Cats," those living in and circling around her apartment at Barnard. The latter included Léonie Adams, who became a famous poet; Louise Rosenblatt, who became a prominent literary theorist; Eleanor Pelham Kortheuer Stapelfeldt; Leah Josephson Hanna; Hannah Kahn ("David"); Deborah Kaplan; Viola Corrigan; and Eleanor Phillips. Some of her other New York friends, such as Jane Belo and Jeannette Mirsky, eventually worked in anthropology, and others helped Mead with her work, assisting in her office and with publishing and

other tasks. (The next chapter includes letters to her friends who were anthro-
pology students and colleagues.)

Mead often put herself in a caretaker role relative to her friends, sometimes
choosing people as friends because they seemed to need her to care for them. In
other cases, Mead allowed people to express themselves by nurturing her. Marie
Eichelberger devoted years of her life to caring for details of Mead's daily exis-
tence and, later, became "Aunt Marie," a prominent caretaker for Mead's only
child, Mary Catherine Bateson. Some of Mead's other friends were involved in
rearing Cathy, including Sara Ullman and her painter husband Allen, whose
daughter, Martha, was a year and a half older than Cathy. Mary Frank, the
third wife of Mead's colleague and friend Larry Frank, also played a significant
role in Cathy's upbringing. For most of the first fifteen years of Cathy's life,
Mead shared a household with the Franks—at 72 Perry Street in Greenwich
Village most of the year and in New Hampshire, in the summers.

Within her circles of friends, Mead played a variety of roles, as she nurtured
and was nurtured. She sought kinship and succor in the company of her friends.
As Mead traveled, she kept in contact with friends, old and new, through corres-
pondence, maintaining a connection over time and space.

While she could not choose her family of origin, the myriad friendships she
cultivated and sustained provided her with another kind of family as she moved
out into the world.

In her youth and college years, Mead looked for a special "best friend" in each
new school setting she found herself in. The following are three examples of this
search at both DePauw and Barnard, where she saw unique qualities in and made
a strong effort to get to know Katharine Rothenberger, Léonie Adams, and Lee
Newton. The first two became her friends for life; the last fell away over the years.

To her mother, January 13, 1920, from DePauw University. Katharine
Rothenberger had come to DePauw as a junior transfer student at the same time
Mead entered as a freshman. Later in life she would become a history teacher. [h]

Dearest Mother,

. . . I've had a wonderful time this weekend really getting to know
Katharine. After all one's college year is worth while if one gets nothing out
of it but a friend like Katharine. She is lovely, and I know you would love

her. She comes from cultivated people, who have lived in New Orleans a good deal. She has a splendid original mind, plays a violin wonderfully, is lots of fun, and a real inspiration and stimulus. I was very fond of Ruth, but I would never have sought her out, if Fate had not thrown us together.[1] But Katharine is the sort of girl one would go any distance to know.

She has no intimate friends here, because she is even more reserved than I am. A good bit more so, in fact, because she doesn't gloss over her reserve with an outer layer of apparent openness and familiarity the way I do. I know there are lots of people who think that I am not reserved, but simply don't really feel deeply about any thing. People who meet me casually I mean. But Katharine just pads over all her inner thought and fires with ice and snow and sleet, and one has to be very brave, and possessed of a good beating heart to be brave and persistent enough to cross those deserts of ice and snow.

I don't mean that she isn't lovely and gracious to everyone. She is, but that makes it all the harder to know her. She is a patrician from the top of her red-gold head to the toe of her little pointed shoe. You would love her. A French count with chateaux in France and much property in Louisiana and a Croix de Guerre and so forth, was visiting her over Sunday. He is perfectly adorable and madly in love with Katharine, but unfortunately she does not reciprocate his affection. . . .

The next year was Mead's first year at Barnard College in New York, where she was delighted with the diversity of students. As she wrote to her Grandmother Mead, September 20, 1920, "Barnard is lovely. I'm so glad that I came. There are girls here from all over the country. In this little apartment where we are temporarily lodged, there are girls from Vermont, South Carolina, Alabama, Connecticut, British Columbia and Wisconsin. For religions we have a Baptist, a Methodist, a United Brethrenist, a Roman Catholic and a Jew. My roommate (for good) is French, but I haven't found her yet." [h]

At Barnard she met Léonie Adams, a year ahead of her in school, who would later become consultant in poetry to the Library of Congress (now poet laureate). Mead had also started creating her group of Barnard friends as a family, with she and Léonie as parents, others as children, and later, grandchildren.

To Grandma Martha Ramsay Mead, February 6, 1921.[2] [h]

Dear Grandma,

. . . We have two new Freshmen in our apartment, known as "The Children."[3] We are trying to bring them up carefully and set them a good example.

It was so nice to see Dadda on Tuesday. We both enjoyed *The Prince and the Pauper* so much.

I am finding someone to partially take Katharine's place in Léonie Adams, one of my roommates. She is very brilliant and very lovely, and she is someone to take care of, for she is as delicate as she is lovely. And the fact that she doesn't care about me makes it all the better. It isn't a drain on one's time to care about someone quite quietly from a distance. The little folks down at St. Luke's take so much affection out of me—they need it so, coming from overcrowded homes where they are kicked to bed and kicked up. But I get hugged and patted 'most to pieces first by the little ones while I wait for the others. . . .[4]

Lovingly,
Margaret

Mead's Barnard family of friends grew and Mead was the acknowledged leader. As she wrote to her mother, October 11, 1922, after Léonie Adam's graduation, "Our room is very attractive, and is the rendezvous of half the college. Léonie, Leah, Bunny Bell, Eleanor Steele, Vi, Marie Bloomfield, etc. ad infinitum.[5] I used to think they all came to see Léonie, but it seems they must come to see me." [h]

After Léonie Adams graduated in May 1922, although she still saw her often, Mead looked for another "best friend" on campus and found her in Leone Newton, with whom she had a romantic friendship in her senior year, with Lee taking the name Peter and Margaret, Euphemia, in a series of passionate letters of which only those to Mead exist.

To her mother, November 19, 1922. [h]

Dear Mother,

. . . Saturday, Lee Newton and I went down to hear Scott Nearing[6]. . . I am having a very good time—my chief difficulty being to get all my friends in and not offend any of them. Léonie is up here several nights a week. I see practically as much of her as ever. Lee is taking Katharine's place more and more. She's about the only person in my class, tho ofcourse there are lots of casual acquaintances.

. . . We will get down in time for dinner I think on Wednesday night—probably just Pelham and I then Luther and Eleanor will come later.[7]

Lovingly,

To Grandmother Mead, February 6, 1923. Margaret mentions Marie Eichelberger for the first time. Eichelberger became a loyal friend throughout Mead's life, tending to her clothing and appearance, managing financial matters, helping her raise her daughter, getting involved in the Institute for Intercultural Studies, and, most importantly, keeping Mead's secrets. [h]

Grandma dear,

. . . There is one little freshman whom I am particularly interested in. She is about twenty-five, tho she looks twenty, and has been ill for years, so that she is just entering college now.[8] She came to college "prepared to change every idea she had," and she's very interesting to watch. About two weeks ago she asked me to go out to dinner with her, because "she might not be here next semester,"—and informed me that she was sure she was going to flunk everything. I spent the evening putting the fear of God into her. After my experience with Ross Greer I'm good at that and so far she has made 9 hours of A and four of B. Next Thursday night she is taking me to see *Peer Gynt*. It's being given in this country for the first time since 1909—and the wife of Ibsen's grandson is taking the part of Anitra. That will probably interest Elizabeth—she has danced to "Anitra's Dance" by Grieg I imagine. . . .

To her mother, February 11, 1923. Mead was enjoying Barnard College life and her friends, when, in her senior year, a defining event occurred. In this letter, Margaret discussed the death of her college friend Marie Bloomfield, the young woman who had been her ally in getting to know Ruth Benedict on the trolley. [h]

Dear Mother,

I presume that you have not seen the newspaper accounts of Marie Bloomfield's death or you would have written. She committed suicide last Wednesday, and the days since then have been very full and very dreadful. With all her brilliant mind and fine aesthetic sense she could never be convinced that she was not totally inadequate and doomed to failure. She thought it took courage to die, and she wanted to prove that here at least was one thing she could do. Poor little lonely thing! I was the best friend she had in college and I never loved her enough. She was just one of the group of younger girls and often I did not have time for her. This last weekend however—I went down and brought her home from the hospital. She left the *Little Book of Modern Verse*—which I gave her for Christmas all marked up—showing quite clearly her purpose. She was so inextricably bound up with our lives, that it is very hard to go on without her.

So that explains why I have not written. I wrote everybody Tuesday night. Sallie is here this weekend and I am trying to give her a good time—she knows nothing of Marie's death—as I didn't want to spoil her so short holiday.[9]

Can you let me know from just what part of Italy your Italians come so that I can look up the ethnographical background?[10] Please do that as soon as possible as I want to get started. Did the library books come?

Lovingly,
Margaret

To her Grandmother Mead, February 22, 1923,[11] offering a different view of Marie Bloomfield and her own renewed commitment to friendship. [h]

Grandma dear,

. . . Your little note about Marie was so sweet. It is so hard for people who did not know her to understand—but you succeeded so well. I loved her very dearly, and I miss her frightfully, more and more every day. I just went to look for something in my bureau drawer, and it was neatly put away in a box where she had put it when she last straightened up my bureau drawers. Everybody is always so much lovelier to me than I deserve.

. . . People have been so lovely to me lately and the room is just full of books and flowers and candy. I just let them. I had been rather avoiding friendships or at least new ones because they took time. But I'm thru with that—Marie's death was too terrible a lesson. . . .

Lovingly
Margaret

To her Barnard College friend Leah Josephson Hanna, October 30, 1925, from Samoa. Hanna was one of the commuter students who became a part of Mead's Barnard College group. In a draft of her autobiography, Blackberry Winter, *Mead said of her, "she provided us all with warmth and a never failing sophisticated and sympathetic ear."*[12] *The name Ash Can Cats had been given to a group of Margaret's Barnard friends by a favorite professor, drama teacher Minor Latham.* [h]

Leah dear,

The gods made a mistake. You should have been born a Samoan. I can just picture you here passing the days gracefully, pleasantly, lying on a mat looking out of an open Samoan house at the endless white spray of the surf. I wish I had you here—you would be preeminently at your best as a companion in this climate where hasty movements are an offense against all good manners. Much time is given to singing monotonously sweet songs and there are no dishes to wash. There, aren't you mad with me now. S.P. or rather it was P.S. we dubbed you was it not. Still, I quite seriously mean that you'd be a perfect companion for these parts. In Spain all things are reputed to be relegated to "*Mañana*," "tomorrow," but here they are postponed to "*de isi aao*" "someday" and most likely never done at all. You are up and go to sleep wherever and whenever you like, and stay up all night if there is a moon. You only cook twice a week—if you're a Samoan—and leaves are the dishes. It's all most excellent economy. I'd like to transport all the weary ash can cats here for a joyous *Malaga*[13] to my pet village where I am the honorary *taupou* and a gorgeous guest house would be at our disposal.

Saying "weary ash can cats" makes me wonder desperately how you are. No word from you for three months and no news of you either—except the occasional recurrence of your name in other people's letters which teaches me you are still alive or were six weeks ago. It's a rather ghastly thought to know that my last news was of six weeks ago. About this near mail days (it's two days off) I get frightfully nervous and when the letters do come I'm afraid to open them. And you said you'd write. Why don't you, damn you!

Nov. 3—No letter on the boat. I officially announce, I'm done writing till I hear from you.

<u>M. M.</u>

To Eleanor Steele, November 17, 1925, from the village of Ta'u, Samoa.[14] *Eleanor was the friend from Barnard who left to go to the University of Colorado. In 1924 she was diagnosed with tuberculosis. In a December 26, 1933, letter to Gregory Bateson after Eleanor's death, Mead described her as "one of the people I loved best in the world." [h]*

Eleanor dearest,

Why and how and where are you? I get increasingly worried—the last letter described an episode of a weekend long ago in a remote N.J. town— and promised your address as soon as you had one. If you haven't one by now, I'm afeard you are become indeed dissolute and addicted to transient motor cars or gypsy van. You're really very bad for you know perfectly well that I worry a lot when I don't hear where you are and more than ever now. Please, dear, a word, I demand of you at once. I hope you haven't quarreled with Lee[15] so that she will not forward this.

I'm progressing. I can do everything except swear in Samoan, and that only because there are no swear words in this easy going tongue.[16] It is a land of gestures, even the dogs and chickens know you don't mean it and come in again at once and cats have tears in their eyes if you speak crossly to them and myriads of *tamaiti* (children) follow me in disrespectful but gay fashion. A small girl is minutely examining all my books and pictures and saying *fa'afetai* [thank you] in a still small voice. I'm 68 miles from the station—by boat—there is one other white woman on the island, but I'm discouragingly safe and protected even tho I conceal carefully the fact that I'm married.

Write, damn you.

Margaret

To poet and friend Eda Lou Walton, Palm Sunday, March 28, 1926, from Samoa. Mead had come to know Eda Lou through their mutual interest in poetry. Walton had her Ph.D. from the University of California at Berkeley, and was teaching English literature at New York University. Her apartment in Greenwich Village became a gathering place for writers and would-be writers. She was also attracted to Luther Cressman at one point. [h]

Dear Eda Lou,

Because I felt that such a long letter couldn't be answered by a wee note either, I got no acknowledgment of your note and book off by the last mail. The book is lovely, from the cover clear thru (I took special delight in finding again the poems which I had hunted out of old magazines when I first met you—but that was just a private bit of sentiment.) You've really done it amazingly well. I like the proportions of the book, the variety, the arrangement. It's a very real gift to be able to so skillfully reinterpret an alien people—and with such an argument to confute you, I challenge you to give me one of those pessimistic lectures on your complete literary worthlessness which you used to exult in. You were a dear to send the book to me here and I thank you so much. I had tonsillitis very badly when it came and my eyes hurt, but not enough to keep me from reading it at once.[17]

I leave these blessed isles six weeks from tomorrow. If there were no white people on them they would be blessed indeed. I am perfectly happy [to] be a Samoan princess. I could stay—if there were no definite ties to call me back—and marry a Samoan and live in a back woods village where I would never see a white person again—and be happy. But I would go slowly mad if I had to stand even a year more of the miserable white people down here—all of whom have "missed too many boats."

I have met one Samoan, too, whom I think would make a perfect lover, for a bit—However I remembered the white population and that I represented the Bishop Museum, the Research Council and Columbia University and refrained. But he does have such a very accurate notion of the place of casual amours. One could be quite sure that ten years later he would tell the story, as he told me a dozen—like these—"And so we loved each other very much. But three weeks later I eloped with another girl and she wrote me and told me she was thru with me—and I say, I am sorry, but what could I do—"

I'm engaged in polishing up my work, filling in gaps, getting lists of materials, measurements of artifacts, etc. All of which is dull. But I've excellent incentive to get it all written up and out of the way on the long voyage. . . .

Forgive all this. Ruth says what one needs in field work is a "substitute for conversation." Letters have been that for me. . . .

Despite your own somewhat encouraging reports of your state of mind, I'm not awfully relieved about you. There was so little of you when I left, and I still remember the dream in which all of you had disappeared except your eyes. To Rome!

Affectionately,

Margaret

To Eleanor Steele, May 5, 1926, from Samoa. [*h*]

Eleanor, my love,

I wept with joy over your letter. Marie has educated me to the point where the mere mention of T.B. scares me into a state of chills and fever where all judgment deserts me. Lee [Newton], with her customary picturesque brutality, was declaiming darkly how you ought to make up your mind whether you wanted to live or die and stick to it, etc. I was scared stiff. And I wasn't feeling jubilant enough myself to write the sort of gay exhortation which the situation demanded. I'm terribly glad you're better—I can't tell you how glad. But you better take that 10% of air at a safe distance from New York for some time to come. Only do come near enough so I can see you occasionally.

I'm out visiting my beloved Samoan "family" in my favorite Samoan village, done forever with canned "dogs" and "peasoups" and "chicken tamales"; done with the constant friction of living with several unworthy representatives of the proletariat, done with living on sufferance in other people's houses and roasting therein. Here they love me and I love them, and "disembodied mind" tho I be, I wept for joy when I came back to them. I see a steady vista of tears, mostly joyful, but not all, lying ahead of me. After all, my naturally affectionate nature is used to more enduring food than kisses of ten-year-olds. That's all I've had for nine months. I return to a simplified world. My introverted lover [Edward Sapir] has taken unto himself a new mistress which is as well, as I'd never have been able to muster a continuous store of steadfast intelligence sufficient to pilot my triangle over several years of rocks. Still—

By the way, Lee knows nothing of that little episode, so don't inadvertently enlighten her. It will be nice to see Lee again and have her dramatically clutch my hand and call me "woman!" Does she call you "woman!"?

There have been a wealth of experiences this year with just the elements for you to appreciate. Mr. Holt is so "pure" minded he didn't want to read Thunder-on-the-Left,[18] and I used to sit five feet away from him discussing incredibly infinite details of sex with his Samoan assistant. We talked in Samoan, in a pleasantly impersonal tone. . . .

My dear, your ambulating epistle delighted my heart, please don't stop writing again. . . .

All my love,
Margaret

To Gregory Bateson, November 20, 1933, after coming home to New York from New Guinea, about her old friends.

Darling,

. . . I've completed the ten days of Marie's vacation. She has gone back to Pittsburgh, rested, peaceful, blissful over ten perfect days. I was so tired by the end of it that I welcomed cramps as an alibi, but it was a job well done. Ruth was undoubtedly right that it was the only thing to do, to leave her in complete darkness about everything, but it does put an unreal film over the world to do it. . . .

Saturday I had to go shopping with Marie and I discovered what was the basis of my almost physical illness which always accompanies shopping. I'd put it down to some childhood trauma, but it's not that, it's a genuine dislike of choice. We dragged through the crockery department, because Marie wanted to get me a teapot. There were dozens of teapots, all ugly and awful and I began feeling sicker and sicker. Then suddenly I saw a darling little southern, squat vinegar cruet, and bought it. And I enjoyed that. I like seeing something which catches my attention and responding to it, but to have an enormous array of things presented from which I must choose just bewilders and repels me, while it gives most people a sense of power. You should hear Marie describe the antique shop at Bergdorf Goodman's— that's the famous shop from which she got me my tiny vanity case so I could have something purchased within those sacred portals—and how people wait while respectful clerks bring priceless sixteenth century tapestries and spread them before their critical and wealthy eyes.

My whole idea is to walk along the streets, glancing casually in windows, and suddenly seeing something which is right and lovely, and then buying it if I can, but anyway enjoying having seen it. We had one bit of luck, a lovely copper shop where a man makes all his own mad designs and I found a lovely array of animals and birds, all cut from shining copper and built into ash trays, one for each ash can cat.

We have an ash can cat party just before Christmas, and now I have a lovely wise owl for Léonie, and a heavy bronzy peacock for Louise, and a light cold peacock for Eleanor Phillips, and a beautiful fey mottled fish for David, a fine very mannered horse for Deborah, and a penguin for Pelham, and a delightful silly pukpuk for Leah. And in an old jewelry shop where I had meant to buy nothing I found a tiny round crystal pendant with three brown leaves set in it which is just right for Ruth Holt, and a deep blue cold one, with a delicate tracery for my Mother, and lapis earrings in one shop and a pendant to match for Ruth. This was all luck, for I hadn't intended to do any Christmas shopping, and there were the things, unlooked for, just scattered about and looking like the people I got them for. It's such fun to find things which just have their essential fitness for someone written all over them. But I don't feel that as choice at all. They fit so well that there is no question of choice, it's just recognition. . . .

After Marie left last night, Léonie came up. Her Irish N.W. enner [sic: Northwesterner] had been having a mild temper tantrum and Léonie has found that the thing to do is just to walk out. She seems like a wise little bird now, so fundamentally uncaught in her deeper emotion and so res-olutely determined to have a peaceful nest. But she admits Bill as much more manageable than Louise ever could have been, and then she is so much less caught. I admired and respected her determination to have no nonsense. Bill had been rude, he had been cross and hungry and he had misunderstood an innocent light-hearted laugh of hers. Very well, out she marched and left him to cool his heels. But while I see its wisdom so clearly, I don't want that kind of wisdom, I want no nest which has to be purchased at such a price, tended by such persistent careful self-respect and refusal to be unfairly moved. I think I'd rather live all by myself, forever, than pay that price. . . .

To Gregory Bateson, December 5, 11, and 16, 1933, from New York, on the death of her friend, Eleanor Steele, from tuberculosis. Margaret had invited Eleanor to stay with her and taken care of her through the final months of her illness. In her last days, Margaret got her into a hospital.

Dec. 5

Darling,

. . . Right after I finished my last sentence the hospital called to say that Eleanor had been placed on the *danger list*; that was Saturday morning and I have been there practically ever since. She died last night at ten. I feel a great sense of relief that she, who has led such a harried, insecure, and often so unhappy life, should be out of it. The alternative would have been another attack of tb; she has gotten well twice and she hadn't the reserves to get well again; she would have been sent away to some sanitorium to spend meaningless years, oppressed by the sense that she was a burden upon other people. . . . Last night the doctors said an infusion might lengthen her life a few hours and we ordered it, but she died before they could give it to her.

It was a strange group, Gordie, her husband, the Turk Methodist one whom she left Howard for and then went back to Howard from; Hugh, her lover, whose presence was completely ambiguous as he had a wife, who had developed an illness as soon as he wanted to come and see Eleanor in the hospital; a little girl named Sarah Bachrach, very southern and a new friend of Eleanor's whom I've never known, and her lover, a young man who had been raised on Voltaire in Paris.

I had never seen anyone die before, but there was no surprise in it. Watching the physical degeneration which went before was dreadful, and to see her wasting away like a badly built house at high water, while she kept her high spirit and her loving response to everyone, gave me a strange sense of unreality and yet pride. It's God's mercy I was here, because how the tangled threads of Gordie and Hugh and Howard and her uncles would ever have been managed otherwise, I don't know.

And she worried and worried about me, was I getting enough to eat, was I resting enough; she even worried for fear the nurse would work too hard. And Gregory, she never asked for Howard, the man for whom she has left everything, sacrificed love and allegiance so finally, that although she worried about him, and even tried to get out of bed to go and get a newspaper to see how his speeches were going, yet she didn't want to see him. Once I asked her if she wanted him to come back and she said, no she wanted him to stay out west forever. And if there was ever an allegiance which I had felt as unbreakable and almost of a different order from ordinary marriages, it was her feeling for him, and in many ways, his feeling for her. The diagonal seems doomed to one kind of tragedy or another. . . .

December 11, 1933

. . . I am at a low ebb, darling, but it's just physical, in the sense that I am only now beginning to feel the results of lack of sleep and worry of that two weeks. I keep dreaming and dreaming about Eleanor and about all the complications which seemed likely to arise and which were only avoided by great good luck. They are just anxiety dreams coupled with snap shot pictures of her as she looked just before she died and it's just an overstrain phenomena, that's all, but it keeps me at a low ebb. And that makes my letters poor. . . .

December 16, 1933

. . . Somehow I think that it is not Eleanor's death which has proved so upsetting but merely being at a focus point of all these disparate people's emotion and having to listen to their so different versions of reality for hours on end. I'd quite gladly throw all the Turks in the Hudson for Howard, outrageous as he is. But with the exception of Howard, I never knew any of the others except through Eleanor and now they have all stepped out from the dim stage of another person's relationships and are just all over the place. It may end by Christmas. . . .

Mead gradually got over Eleanor's death, but gained lifelong friends in Eleanor's friends, Sara Bachrach, and her then lover and later husband, Allen Ullman.

On her field trips to New Guinea, beginning in 1928, Mead made friends among the Europeans and Australians who had either been posted to New

Guinea or had come to make their fortunes. One of these friends was an Aus-
tralian, Judge F. B. Phillips, whom they called Monte. She had met him on that
first field trip with Fortune, when they studied the Manus. Phillips had sug-
gested that she write the life story of Mrs. Phoebe Parkinson then, while Fortune
returned to do follow-up work among the Dobu.

To Judge F. B. Phillips from Bali, November 15, 1937, as she and Bateson
were planning to return to New Guinea.

Dear Monte: We enjoyed your account of the earthquake or is its proper name the Eruption, so much, and put it away among our treasures. Subsequently I have had newspaper accounts from all over the world, all ofcourse definitely inferior to your vivid account. I was so glad to know you were safe, and glad for Rabaul's sake, that you were in charge, as luckily for Rabaul, you always seem to be whenever there is an emergency. But I hope it didn't add too many grey hairs to your head. If you were an American Indian you could count coup on them "This is the grey hair I got in the strike"; "This for the Eruption when it began"; "This when that ominous little puff appeared"; "This when the *MacDhui* caught fire"—and I daresay you have a few more now, since your last letter. It was extraordinary how much more closely news of the Rabaul disaster touched us than news, say, of European or Asiatic wars. We both felt as if it were our own town which had been threatened and sat, going over and over what might have happened, in the ways [sic: days] before we got any news. . . .

I do wish you could have visited Bali while we were here, but at least we can tell you all about it, and perhaps fire your imagination for next leave. And by the way, what would you like from Bali. No use being shy, because we shall certainly bring you some things, and they might as well be the sort you'd like. . . .

Of your American friends; Ruth is acting head of the Department at Columbia still, Jenny is going to South America, or rather Central America (Guatemala), Bill and his family are in the S.West, and he is slated, so rumour has it, to be head of a research laboratory there. Eddie Barsky, Jenny's doctor brother-in-law, whom you probably hardly remember, he was so quiet and apparently colourless, led the first American Hospital unit to Spain and has returned covered with distinction to raise more units.[19] Did Reo's fiancee from

New Zealand go through Rabaul on her way to China to join him? Now I am wondering if all the disturbances in China have interfered with their plans.

Is there any sort of a small map which shows the new air bases in New Guinea? We try to work out what, for instance the air mail routes are, but we haven't enough data. Just after we decide that New Guinea must be riddled with airdromes, your letter comes speaking of avoiding being marooned at Wau for a month, which sounds as if there were no air communication. It is all up on the table at the moment as Eleanor Oliver is here and her husband, from Wewak [New Guinea], has been trying to get in touch with her, with an enormous amount of resulting confusion; I have had to make two trips the length of Bali, and telephone all the wayside post offices where cables have been marooned, until I almost hope that New Guinea hasn't got too civilized.[20] All the best for Christmas and the New Year,

    from

    [Margaret]

To Marie Eichelberger, May 15, 1938, from Margaret's field trip to the Iatmul in Papua New Guinea with her third husband, Gregory Bateson. This letter reveals perfectly the bases of their friendship. Eichelberger was concerned with Mead's clothing and appearance, with her health, and with providing Mead with material things she needed. Eichelberger had taken over Marie Bloomfield's job of organizing Mead's life. In a letter to Bateson, April 23, 1934, she had written after a visit by Marie, "She has spent the entire weekend doing things for me and I have accepted it all with no sense of guilt; it's her principal form of expression in life and it seems nothing but cruelty to deny it to her, on a set of moral premises which have nothing to do with her."

Eichelberger also served as nexus of news on old friends and to old friends, since Mead trusted her to pass her bulletins along to them. Being able to love Mead, and to show that love in her care for her material possessions, appearance, and friends, and later her daughter, Cathy, was for her a fulfillment.

Marie dear,

    . . . I am enclosing a long bulletin which will give you the main facts. All except the fact that my clothes are going to pieces much too rapidly. Do you think you could get me a new dozen of those kickernicks?[21] Ruth would give you a check for Wanamakers for them, and mine are beginning to split right down the back. I get so hot and they stick to me and then . . . [orig]. If you write regular mail, put a handkerchief in. They are always welcome. I use one every minute here to wipe the perspiration off my upper lip. It's terrifically hot, but a pleasant healthy feeling baking heat. . . .

A couple of ordinary cotton pajamas tops with long sleeves would be a big help. And a new sanitary belt of the kind you make. I have some store ones, but I still cling to the old one, and it's a strange grey colour. I think my dresses will hold. If necessary I can just wear the short-sleeved pajamas as extras. The dresses are getting old, a few are tearing under the arms, the snaps are coming off, but I think they will hold. Ofcourse they have to be washed so much oftener here, one can only wear a dress a day at the most.

An occasional magazine is welcome, about twice as welcome if you send only one or two. Packets of the same kind of magazine intimidate us. . . .

I like the idea of your doing a journal and I for one will always love reading it, specially if you type it. We now type the carbons of our letters on punched paper and put them in loose leaf notebooks. It's much easier. You are quite right, I think, in considering a journal a form which fits your temperament. Go ahead and do it.

When I was sick I had malaria tropica, or sub-tertian. It has no periodicity like my old form (tertian), is much more dangerous because one gets no rest in between attacks, but is also much easier to cure. I have taken the full course of atebrin and should have none left in me. As far as we know neither of us has any malaria in our systems at present. Right after I got over that, we both got flu, and also I was lame for a long time from the intramuscular injection they had to give me. But we are both well now.

Will you pass this bulletin on to Pelham, for the cats.[22] Unless I do that, I can't send Carrie[23] one, and she knows so much about life here, and does so much shopping and liason work for us in Sydney, that she really ought to have one. She's a grand friend, not once in these five years since I've seen her, has she fallen down on a commission of any sort . . . [orig]. I had, in this last mail, my first letter from Léonie [Adams] since I left America, very sweet and all herself.

I am so thankful to know that you are in New York and away from that dull baffling trying position.[24] Be happy there. If you are furnishing, there may be some of my stuff around. Don't expect another letter for six weeks; at present we can only be sure of a six weekly mail.

Lovingly,

To college friend Léonie Adams, from New Guinea, May 16, 1938, missing her friends. Letters were not only a source of information for Mead when she was far from home but a way of reaffirming her bond with those she loved.

Léonie dear,

A mail with a letter from you in it, is like having a real birthday for the people who are born on February 29th. Just about the time you must have written it, I was considering writing you a plea for a little word of you but I decided that there was no reason that you should be plagued into writing if you didn't feel like it, as I know quite well that you can think about people without letters. But it was lovely to see your writing again, and to have such a good report of the parts of you that matter most. And I hope that the economic frame of your existence will work out satisfactorily soon.

I get very little news of anyone and most of that indirectly. I haven't heard from Pelham since her baby was born, from Leah for over a year, even Louise writes rather seldom, and David and Eleanor and Deborah not at all. I know that Ruth or Marie or Elizabeth will write me of births, death, marriages and publications, and I have to be content with that. I suppose it means that we are all growing up. I have heard no word of Louise B.[25] but I cut out the picture that was published when her book came out and pondered over the curious truth that it could be the same face as the picture taken when her first book came out, which was published in another paper. Ruth sent me the book, but it must have got[ten] lost on the way. Elizabeth writes of raptures and alarms, her letters illustrated with a mass of water colour sketches which are steadily getting better and while [sic] tell more than the words. I was delighted that you found her possessing what once you and your mother thought she would never acquire, I think your mother called it "sense." If only the fear of anti-semitism didn't hang over the world, I would be very happy about both of them—Elizabeth and Priscilla.[26] The worst news that I have had since I went away was Ethel's death.[27]

We have had a very good two years of unclouded peace and growth . . . We read a good deal, especially aloud when we are tired, from the more peaceful authors, Trollope and Jane Austen and Lady Russel. The news of the world seems very far away, and rather like a tale of horrors too bad to be real. We never get anything but the starkest statements of the most calamitous events, Gregory's mother hardly ever deals with smaller than Germany or This Age, or Democracy and my mother never deals with larger ones than the conditions of relief or education in Pennsylvania politics. The two together make an odd pot-pourri and one that doesn't increase one's sense of the reality of the scene. However I have very little feeling that the most rabid half-convinced communist could prick me into a sense of guilt that it would be more useful to mankind to be organizing the world for the dictators than to go on doing research that may some day give us some clue as to how to avoid them.

I must stop this and start typing up notes. I began it in a dawn so grey that I could barely see the keys, but the dawn has now become a dark wet morning, a lamp is lit and there is now no excuse for shirking a notebook full of pencil scrawls that will rapidly become unintelligible. Gregory has to go and record the cutting up of a pig.

Love to Bill.[28] If you feel like writing again or like sending me any Works, it will make me very happy.

Lovingly,

To Leah Josephson Hanna, June 22, 1938, from Tambunum Village, New Guinea, reflecting on their friends.

Lee[29] dear, It was good to get your letter, and the pictures of the children are lovely and duly ornament my new "room wire" where the native children come to draw, and to model me birds and fish and masks in brittle clay. I hear so very little from the crowd these days, so that every scrap of news is precious. Léonie however, did write a letter, which I [sic] arrived right after I got here, and I agree that she sounded much more peaceful and also as if she meant to work. It was in her usual packed cryptic style, but it was very reassuring. Aside from a letter from Wissler saying our publication program at the Museum was at a standstill, and a salary reduction slip, I have had no proper news of the recession or of how it is effecting [sic: affecting] the world.

And I haven't heard from Pelham for a long time, and I can't help worrying about what is going to happen if there should be a war with Germany, or even the increasing hostility to Germany. . . . Meanwhile, we are free from all the worst rumours, until after they have become fait accompli or been rendered abortive. It gives one a sense of living with a great sweep of historical perspective, rather than of any contemporaneous participating in the world. The lives of my friends seem real and current, what the baby weighs, or whether the book was accepted, or whether the flowers are blooming in my mother-in-law's garden. But the world, it might be going on at a different period entirely, especially as I usually know nothing of the personal reference of any event for months afterwards. The Cincinnati flood killed Ethel; the Rabaul eruption made my friend Judge Phillips a CBE, but these all seem ex post facto, as the fact that your father came to America brings you there, part of a long sequence in history, the course of which is already jellied beyond meddling.

This is an evil country, although it is beautiful and more exciting in many ways than Bali. But there is something in it that does not like the white man. Gregory has had fever almost since we arrived, and has some obscure disease—we don't know whether of blood or skin, or whether it's a fungus, so that every cut and scratch tended to spread and spread in a series of encircling blisters. He has to go about draped in bandages, as he got the thing when he was putting the finishing touches to the mosquito rooms, canoe with its outboard engine, etc., and necessarily got a lot of small cuts and bruises. So far I am reasonably strong, but a little headachy and irritable. The mosquitos leave one with a prickling skin; I don't show the bites, but instead retain an after image of the sting which recurs at intervals with all the original intensity so that I am always slapping at emptiness.

. . . As long as I don't ask for any extra field funds the Museum is satisfied to let me stay here; as long as the game and the water-spinach and the yams last, we don't have to buy much food, and practically all our expenses were initial overhead. We have a superlatively comfortable camp, with a fan turned by hot air, and a canoe with an engine to use as a taxi along the river front, and electric torches that sit on their haunches and are as satisfactory as electric lights in places like a storeroom or a bathroom. I have a bathroom with a cement floor—you don't know WHAT that means in comfort; I even feel moved occasionally to powder my nose in it . . . [orig]—This was begun in the ten minutes lapse of time before breakfast and is being finished while I wait for hot water to dress Gregory's stubborn blisters. I am so glad you have your apartment. Love to the children, and Nat and Denah and everyone.

[End of letter]

To Marie Eichelberger, September 16, 1938, still among the Iatmul in New Guinea. With the vagaries of mail delivery, letters could take months to reach Mead in the field. Redundancy in correspondence was necessary to make sure information got through. Some letters were never received and others would arrive out of order, further complicating the attempts to communicate at a distance. Paying extra to cable a message was sometimes necessary to convey information more quickly than was possible by boat, and in this period, by the new flying boats that were part of an emerging transcontinental air fleet.

Marie Darling, five letters from you, spreading from the regular mails with handkerchiefs in them mailed in June to the last air mail written just after Ruth got back from Guatemala. And bless you for them. I think I'd be half beside myself with worry if I didn't have your documentation at each mail that Ruth was apparently all right, that nothing catastrophic had happened to her. There was no letter at all from her; her last letter was written on the ship on the way to Guatemala and she hadn't written for six weeks before that. I presume that one letter, perhaps two, have been lost—perhaps in the China Clipper that went down.[30] But she has slipped into the habit of writing so seldom, of trusting a six weekly page and half condensed bulletin to cover the ground, that one lost letter can make a big hole. So all the detail you gave was a great help in somehow adding to my rather precarious belief that she is still alive and happy and well.

The pajamas came too—thank goodness, for I had only 1 bed jacket coat, and was wearing one of Gregory's as a substitute and his have started to rip so that I was going to have to give it back. Now I am well equipped. You will probably have discovered by now that I meant it was safe to write air mail *from* America *up to* October 1st, but don't blame yourself for my imperfect English. The kickernicks I have will last alright. Send them to Bali, addressed Care Afscheepzaak, Den Pasar, Bali, and mark "hold for arrival in January." These are going alright, but the two the rats chewed are holding up better than I expected. They were brand new and the rat holes haven't spread so although they are a sight they are alright. I imagine I shall wear kickernicks in summer in civilization so you needn't worry about the expense. Ruth will give you a check for the amount. When they last over two years under these conditions, that isn't an exorbitant price to pay for them. It was fun to get the handkerchiefs with your perfume on them.

I didn't think the kickernicks worth a cable, but I would have thought your appointment news was worth one. I keep going over in my head your good luck and bad luck patterns, and trying to see where this is likely to fit. Because although most of your good luck has been in the form of reparations from a rapacious life—(I've been reading Melanie Klein hence the vocabulary) there has been some good luck, and it ought to be about time for it. I am praying hard.

This mail brought us a lot of new books which was a great help in adding to our zest for living. I have already read Anna Freud's *The Ego and its Defense Mechanisms* and like it very much. Have you read it? It's one of the clearest statements of the intelligible parts of psycho-analysis that I've seen. But the English school with their spontaneous generation of conscience and immaculate conception of the idea of goodness, have me completely baffled. I am very anxious to get to England and talk to some of them, try to get at some of the material on which their statements are based.

I have finally decided that I have been on a wrong track for ten years in my field methods in that I have exploited my special gifts instead of my general gifts, and so developed techniques which are so special that no one else even tries to use them. For they haven't tried you know, not in any way that matters. So now I am going to try to work out ways in which more simple methods will get the results which so far I have got from idiosyncratic methods, by using both sets of methods at once, and then I can compare them, e.g. I will compare an introspective life history account with actual observations on many babies' real history, and see how they check, what allowances one would have to make etc. It's a decision which I have made with some chagrin; it seems so stupid to have spent so much time on a sort of piece of solo virtuosity which can only end in sterility. However, it's not too late to retrieve it. The use of photography is ofcourse another such method accessible with people of a lower memory span, a narrower attention span and less energy and interest in people than I have.

. . . I am alright, still no malaria and quite strong, and neither very thin nor very fat. I suppose I weigh something around 118, and am quite comfortable. Gregory has these occasional attacks of malaria but he recovers from them beautifully with a real lift so that I am not seriously worried about him. We have fits of being very anxious to get out of here, but this isn't one of them, as I have got this new notion about methods which I want to work out.

Love to you, my dear, and oh I am hoping so very hard that the job will be alright. HOPING. HOPING.

Lovingly,

To college friend Louise Rosenblatt Ratner, October 4–6, 1938, from among the Iatmul of New Guinea. Louise was a "grandchild" in the original college family, who received her doctorate in comparative literature from the Sorbonne, France, and became a college professor well known for her theories on reading.

Louise dear, it is months and months since I have heard from you. It is true that the mails are erratic here, the *Hawaiian Clipper* was lost, etc. etc. but I know I have no letter from you unanswered and that I answered the last over five months ago, before we came to New Guinea. A correspondence which runs on a six monthly basis doesn't amount to much, does it? One might as well put it on the yearly basis at once, when it becomes something quite different, and one says to oneself "Next autumn when I write to her, I must remember to tell her . . ." There always seems a chance that something that I say will annoy you, and there is also the fact that you dislike writing from a morass of difficulties and so when I don't hear from you I am naturally worried.

Time is very askew here, mails come only six weeks on the average, the big ones, and one tends to think from one mail to the next without allowing for the enormous gap in between. But the fact remains that I not only owe no letters that I have owed for more than the last month—between mails, although I am typing eight to ten hours a day here on the job, and writing with a pencil the rest of the time, so that writing letters is only an intellectual relaxation. It is that ofcourse. I am inclined to think that a first training for field workers should be practice in letter writing as a way of taking them out of themselves and also crystallizing their ideas.

I could weep over some of my ethnological correspondents. Bob Redfield writes, in answer to a letter I wrote him about a culture contact problem: "I am so glad you are doing more work among the Iatmul; in a seminar lately when we were discussing *Naven*, we agreed that there were several things more we would like to know about the culture. Sincerely yours." And Cora [DuBois] writes 2 long handwritten pages saying how odd her kinship system is, how surprised Lowie will be, how the terms don't fit as they should, etc. etc. never mentioning WHAT the system is.

People seem to have lost the power of communicating ideas in letters and descended to a sort of impressionism about their feelings about their feelings [sic] about the world. This isn't directed at you because you don't even do that; you have confined yourself to a recital of the most relevant news about yourself and Sidney which is ofcourse what I want to hear first. But I have no idea what you [are] thinking about, and I am sure you *are* thinking. I don't even know whether you are drifting further towards communism or further towards the decision that all totalitarianism is poisonous to the series of values which we care most about. I find when I play at the game of three wishes that my wishes have become very impersonal, 1. that war may be averted, 2. that we may be able to develop anthropology into a vital subject which can contribute to shaping social institutions—where five years ago I would still have had to put my own personal affairs—even if phrased as "may I do good work"—first. . . .[31]

Oct. 5, 1938
. . . I am having great fun here with the children's spontaneous play with toys, something I've never had native children do before. . . .

Oct. 6, 1938

. . . I have just read *The Story of an African Farm* which I missed at the proper time. Has it ever struck you that there is something very strongly in common between Olive Schreiner and Emily Dickinson, both working on their own and almost without a tradition, and both concerned with Eternity, Truth, and God, as seen through the simplest every day imagery. If you compare the last page of the *African Farm* with Emily Dickinson it seems to me the likeness is very striking. It also seems to me that their imagery was perhaps moving towards a genuinely feminine set of symbols, half formulated and rough, but nevertheless not the masculine set of symbols with which poets like Elizabeth Barrett Browning works. One might look at women poets as those who use imagery in such a way that their sex is completely lost sight of, Léonie for instance, those whose imagery is definitely derivative from male imagery, EBB, and people like Emily Dickinson.

There may be nothing in it; I am not sure that the most gifted woman has much chance if she drinks at all from the stream of poetic tradition, and that is ofcourse just what the gifted woman will do. Have you read Virginia Woolf's *Three Guineas*—I've only seen full reviews, and it seems to me that she is always in a fog if she tries to work above the symbolic level, or maybe I shouldn't say above, but if she tries to deal in rational rather than affectively valid sequences. On the face of it, her highhanded repudiation of the man-made world is the veriest nonsense, the very language in which she does it was all invented by men and women's contribution to it has merely been to produce more men to keep it alive. But underneath the superficial nonsense, it may be that she is trying to say something about the possibility of a different set of symbols which will be credible to women. It's thoroughly unsatisfactory this speculating from reviews instead of from the real thing, it must seem to you very second hand, but it is all we have to go on.

We've very few books, everybody stopped sending them at once for no known reason. So we have to get what food for thought we can out of the reviews. . . .

Late at night. We've had a mail, and a pinnace and the news of how near we were to war and the typhoon and the Hollywood flood, and by the same canoe came some of my old Mundugumor informants I'd sent for and I feel half dizzy with the impact of it all. . . .

Our love to you both, See you about April,

To Judge F. B. Phillips, November 8, 1941, from New York City.

Dear Monte,

Bless your consistent and unflagging friendship which writes even when I don't. I've carried your last letter about with me in a little packet marked "answer as soon as possible" for almost a year, and never got it done because I wanted time to make it long. And there is no time. Meanwhile comes your letter. I can guess all the triumph in having got in a second time and not only "in" but "overseas," and share your happiness about that.[32]

And I shared also all the joy and pride of your first letter and also took your strictures on America's attitudes on the war properly to *head*—NOTE I said *head* not *heart*—for the thing we do with such articulate data is to use it to think with. It's an odd role to have in a war, just thinking, watching, thinking, and writing memoranda. Sometimes at ten minutes notice one may have to produce the psychological rationale for a policy that may effect the world. What effect will such and such an announcement have on American morale, on the east, or the west, or the middle west, or our relations to South America. There are times when one feels that one is doing nothing, and times when one's role seems almost too important for proper sanity.

However, Gregory and I do make a very good liaison team, especially for translating between British and American thinking, judging what will work in one country and fail in the other, and all the rest of it. A lot of time goes into maintaining the sort of organization that can use non-citizens to their fullest—which ofcourse government can't do at present. I do a great deal of speaking, to all sorts of organizations, all over the country, which means a lot of traveling, but gives me a lot of fresh material to think with all the time. Life has been a little more complicated this autumn because I have had a phlebitis in both legs and can't walk at all; it means taxis and wheel chairs and aeroplanes rather than trains whenever possible, all of which is clumsy and expensive.

. . . Write when you can and I'll try to do better, and a Merry Christmas from us all, and bless you.

Yours ever,

To her college friend Eleanor Pelham Kortheuer Stapelfeldt, August 20, 1948, at a difficult time in both of their lives, looking back. Pelham had been one of the "children" in their college family and Mead described her as one "who had an extraordinary gift for humorous and sensitive insights." [33]

Pelham dear,

Your letter took a very long time to reach me, and it is just that sense of flux in your affairs—and mine—which makes it easier to answer a letter than to write a second letter into silence—even though the answer will also fall like a flung pebble into an unseen pool behind a barrier mountain. I haven't written any book, I have done only those things which I had to do, lectured and taught here, kept my eye on the structure of this venture, spent an intensive week in Paris, working with a UNESCO workshop and working over Geoffrey's manuscript of a book on the US, doing a Broadcast script on "the English character," into which I note with pleasure, no single note of bitterness or alienation seems to have crept. But as for the book, how can I plan to write when the advisability of publication hangs so in the air. So I have written only a few thousand words, and reorganized the outline, and done some thinking. [34]

Your vacation will be over now, and I wonder how you fared. For me at least the most healing thing is good close contacts with human beings, mostly old friends, I do not have the energy to make completely new ones, but with old ones, with whom one can wander in memory over days before one's present over-obsessing commitments. For that, ofcourse, Europe was especially good, as it has nothing to do with Gregory at all. Sighting Plymouth in a silver mist was hard, and the two days in Southampton—when the boat changed its mind in Mid-Atlantic and took us back to England instead of to France—were also trying.

But even London, where I will have five days before flying back, has nothing to do, directly, with any memories of Gregory. It was Reo and I who went to London and rode on buses, and went out into the suburbs to see the most moving performance of *The Idiot*. It was with Ruth that I sat in the garden of St. Julien le Pauvre and looked over at Notre Dame, it was with Louise and Luther that I saw France. As a result, ofcourse, my life has no time structure at all. I get into an empty carriage and ride through the streets of Salzburg, with no knowledge of where the plock plock of the horses' hooves will take me, to the barouche, to an evening in Dublin after eating oysters, to a day in Kandy in 1926, to the time that I gave the horse's reins to Deb, in Bucks County in 1921.

Erwin Schuller, who was my best friend in London in 1943—an Austrian—is coming here with his Canadian wife with whom I worked in Washington during the war—for my last two days. What I seem to need most is close, aware human relationships, which somehow reinstate my sense of myself, as no longer living "in the season of the narrow heart." And Matthiesen[35] reads the "Wasteland" aloud, and students read the poetry of their country, in eight different languages, on the terrace, against the lake and the mountains, in the dusk, their eyes straining a little towards the dimming words, and Cathy shudders with delight over a passage from "Venus and Adonis."

My dear, I wish I knew how it was with you. I will be back on September 4th, the weather permitting, starting this plane trip with the feeling that this would not be [a] bad time to die, which is a feeling I have when I feel myself, rather than a being battered by someone else's emotions. But tomorrow's mail may change all that and introduce some new uncertainty into the world. We three must meet again, soon, soon.

Lovingly,

Mead's life reached a turning point in 1947–48. The war, which had occupied much of her work time in that decade, was over. Her marriage to Bateson—and with it, their collaborative work—approached its end. (They divorced in 1950.) And, on September 17, 1948, Ruth Benedict died. Benedict, Mead would say, had been the only person to have read everything she had ever written, a distinction that was sacred to Mead and would not—could not—be duplicated.

Despite her grief and her losses, Mead forged on. She took over running the Research in Contemporary Cultures project Benedict had been managing. She

Margaret's brother, Richard
Mead, as a young man.
Mead Collection.

Marie Eichelberger in 1916.
Mead Collection.

"Ash Can Cats," left to right:
Margaret Mead, Léonie Adams,
Deborah Kaplan, Pelham Kortheuer,
and Viola Corrigan. MEAD COLLECTION.

Ruth Benedict at
Lake Winnipesaukee,
New Hampshire.
MEAD COLLECTION.

Franz Boas in
Ruth Benedict's
living room.
MEAD COLLECTION.

Margaret Mead and Luther Cressman
at their wedding in September of 1923.
MEAD COLLECTION.

Margaret Mead and Reo Fortune, on their first field trip together, in Pere Village, Manus, 1928. Courtesy of the Institute for Intercultural Studies.

Self-portrait by Gregory Bateson, ca. 1933–35. Photo credit: Gregory Bateson. Courtesy of the Institute for Intercultural Studies.

Gregory's mother, Beatrice Bateson, from her visit to Bali with Lady Nora Barlow, ca. Dec. 1936. PHOTO CREDIT: GREGORY BATESON. COURTESY OF THE INSTITUTE FOR INTERCULTURAL STUDIES.

Mary Frank [left] with Claudia Guillebaud in the 1940s. MEAD COLLECTION.

[left to right] Margaret Mead, Catherine Bateson, and Philomena Guillebaud, ca. 1944. MEAD COLLECTION.

Larry Frank and his son Colin.
MEAD COLLECTION.

Catherine Bateson with her grandparents, Edward and Emily Mead. PHOTO CREDIT: LEO CALVIN ROSTEN. COURTESY OF THE ROSTEN FAMILY LLC.

The Mead family at Thanksgiving in Connecticut in the late 1940s. The original photo has a thumbtack mark, indicating that Mead displayed it, possibly in the field. Richard and his family are not included, as they lived on the West Coast. Left to right, standing: Margaret Mead, Catherine Bateson, Edward Mead, Emily Fogg Mead, Philip Rosten, Jeremy Steig, Priscilla Mead Rosten, Elizabeth Mead Steig; sitting, left to right: Madeline Rosten, Peggy Rosten, and Lucinda Steig. PHOTO CREDIT: LEO CALVIN ROSTEN. COURTESY OF THE ROSTEN FAMILY LLC.

Rhoda Metraux, with her infant son, Daniel, and Catherine Bateson in June of 1949. PHOTO CREDIT: MARIE EICHELBERGER. COURTESY OF THE INSTITUTE FOR INTERCULTURAL STUDIES.

Margaret and her daughter, Catherine, who was by then a professional colleague, at a Wenner-Gren conference on ritual in 1973. PHOTO CREDIT: LITA OSMUNDSON. COURTESY OF THE INSTITUTE FOR INTERCULTURAL STUDIES.

Margaret with her daughter, Catherine, and Catherine's daughter, Sevanne Margaret Kassarjian (Vanni), born in 1969. PHOTO CREDIT: KEN HEYMAN. COURTESY OF THE INSTITUTE FOR INTERCULTURAL STUDIES.

began contemplating going back into the field. And, as always, she made new friends.

To Clemens and Mathilde Heller, September 10, 1952. Margaret first met Clemens as one of the organizers of the first Harvard Salzburg Seminar on American Culture in 1947, and the Hellers became firm friends. They lived in Paris and when Mead went there, she tried to make time to see them. This is a bread and butter letter, after a visit.

Dear Clemens and Mathilde,

I really meant to acknowledge your hospitality with something more customary than a telegram about earrings and a mechanical lion but you seemed to be so slightly oriented in space that I had no faith that letters would arrive at once.

NOTE on the lion, there is only one because I thought the twins could jointly *jetter a la lion* [sic].[36] I hope I was right.

It was a lovely visit, I feel caught up with all of you and with the different facets of your lives, with enough visual and auditory images of the children, up the pear tree, hid under the curtain in my room, wearing their sun helmets, to keep my reveries happily occupied.

As life stands at present I will probably spend a weekend in Paris towards the end of January, but I shall hope to see Clemens long before that. There is always room at Perry Street you know, for any number of people, any sized family.

Lovingly,

To Jim Mysbergh, October 3, 1952, from New York. Mysbergh had worked with Bateson in the OSS during World War II and had been a friend of Mead's since 1947. He worked in the Far East but when he was in the United States, he often stayed with Mead and later, Mead and Metraux, in New York City. He was one of the few people whom Mead wrote concerning politics.

Dear Jim,

It has been very good to hear from you and I have enjoyed all of it, the long letter on Burmese and Thai contrasts, the clippings and most of all a sense of you from that most delightful Burmese friend of yours whom you sent me. We became real friends in two minutes flat, an experience I haven't often had with Asiatics.[37] I especially liked the clipping on contrasts between American [and] Burmese religious views. . . .

At present our lives are all being pleasantly complicated by the presence of a troop of Balinese dancers; Colin lives in a flutter of articles; Jane, who has just come out of the hospital, takes them to the Metropolitan Museum, and I listen to the general racket. Cathy is a tall plump adolescent, very pretty and gay. Margo and Burt got married at Cloverly this summer, with Cathy as maid of honor and me as matron of honor, which is probably incorrect by somebody's book.[38]

As far as the political campaign goes, it looks extraordinarily close, with Ike continuing to believe that a political campaign is like a war, the main job being to keep your own team together and throw anything handy at the enemy, and the President losing his temper.

Philomena is back with me, working for her Ph.D.[39] I am just starting a new book—a sort of Latter Day *And Keep Your Powder Dry*, about America's place in the world, so any material on how we are regarded in your part of the world and any detail on mistakes that we make will be welcomed.

If you write any long letters, like the one before the last, do indicate whether it can be quoted from, anonymously, or by name, because you do put in such good observations.

Cathy and I send our love to you both,

Yours,

To Mary and Larry Frank, August 29 and September 2, 1953, while on her second field trip to the Manus in New Guinea. Mead and the Franks lived together in a cooperative household on Perry Street in New York for twelve years before Larry Frank retired and moved to Boston.

Dear Mary and Larry,

You should by this time have had two bulletins from me so be well filled in on the background of life here. And Marie passed along news of you all until she left for Europe, notably of Barbie's engagement, for which please give her my felicitations.

Sept. 2, 1953

I had hoped to get this note off to greet you in New Hampshire as you finished the summer, but when I had got that far, there was a spate of events, three births and two deaths, a handful of rows, and I had four hours sleep in two nights, and now I think this will have to go to Perry Street.

I have always had the trick of trying to find the people I care about in individuals in other cultures—who would Mary be, who would Larry be, born here. Mary would be Rosa Pwailep, the mother of the loveliest children in Pere, with room in her house for other people's children and generosity enough to give her youngest child to her husband's brother. Here the people who are good at bearing children, bear enough for two households, the people who are good at bringing them up, adopt all the motherless and fatherless, the old women who have rebuilt their houses out over the sea provide a refuge for the sick, the women in labor, small children at odds with the world.

And Larry would be Tomas, the young teacher who is clinging to the light of progress against all odds, who doesn't chase people away when he wants to sleep, so that his house is always crowded, the ladder is broken from hurrying feet, and fish sometimes disappear from the smoke rack into heaven knows whose mouth. Tomas sometimes gets fed up and says that he is sick of being the man whom all his sisters prefer to run away to when they quarrel with their husbands. True he is fond of these women, he understands them, but he does get involved in so much work on their behalf, can't some one of them pick someone else to listen to their troubles. The trouble with Tomas is he just doesn't have a bad enough temper at his command. And he has imagination which is a very rare thing in this literal-minded society.

And so it goes, continuous driving work, but I seem to thrive on it. My back sometimes gets tired from typing, I live a pretty extrovert life, getting gradually a very little thinner, and still preferring field work to anything else in the way of research—it is so much fuller. Ted[40] is developing an interest in introducing the procedures of structural linguistics into the study of culture.

I have had no news of you since Marie went away so I can only hope that all goes well. And my best love to all the Franks.

Lovingly,

Mead first met Australian educator Charles Bull when he traveled to the United States in 1950 on a grant from the Carnegie Corporation. He was a prominent expert on youth education who worked for the Australian Broadcasting Commission. Then in 1951, Mead and Cathy traveled to Australia, where Mead participated in a two-month lecture tour for the New Education Fellowship Jubilee Celebration, on the theme of "Education in a Changing World." Bull was president of the Australian Federal Council of that organization and one of the speakers on the tour. They remained close for the rest of his life.

To Charles and Marjorie Bull, May 9, 1954. Here, Mead recounted the delivery of a May basket with Cathy's poems to Léonie Adams. May Day was, throughout Mead's life, a holiday of special significance.

My dears,

This isn't going to be a proper letter but just something to say I love you, I've loved your letters and the pictures and the glow from last Christmas still spreads softly over everything which it touches. You both sound very well set in your new quarters, and I have pictures of Charles writing and Marjorie cooking and studying alternately, with very little doubt that any task in the world goes more easily if it can be interrupted for some sensory delight—a souffle or something solider.

Although my existence is terribly packed with events it has also never been as free from real worries or distresses of any sort. This May day which has been a day that my college group have always celebrated; one year we gave a maybasket to Edna St. Vincent Millay—and got caught in a narrow garden underneath a midnight moon and Léonie took her glove off to shake hands and flung it over the wall into the next garden. This year Cathy made a maybasket for Léonie, and put three of her own poems in it, and got herself caught and the poems read and came back with Léonie's new book—the first in 25 years—for me.

We expect to have to give up living in this house next year when the Franks retire to New England, but as Cathy says, "I don't mind leaving here so terribly as if we hadn't been so happy here." She has gone off to her first big dance-house party weekend; the boy is graduating and says Cathy: "I think Mummy it's hard enough on Brent for me to be so much younger than most of the other girls, so I am going to be very careful not to be original in any way. Otherwise it would have been fun to wear red stockings."

My young people are still in Manus and won't be coming down till early July. You remember their name is Schwartz, Ted and Lenora. I'll have them telephone Charles at the office.

The world is still a sort of welter of books, past, present, and future, reviews of this one and blurbs for the other, and when was the Italian translation published, while I try to think through the book on Manus which I really want to write. I have taken a cottage in New Hampshire near the Franks and we will have a composite household, Rhoda Metraux back also from a field trip, her six-year-old son and a Haitian nurse, Cathy, her best friend Martha, their assorted boy friends, and in between we will write books. Too bad about UNESCO but maybe the angels knew what would be better.

Very lovingly,
Margaret

To Englishwoman Pat Llewellyn-Davies, June 26, 1954, from Holderness, New Hampshire. Davies was involved in politics and her husband was an architect. The couple had become friends, and Mead was writing to let her know that her husband, Richard Llewellyn-Davies, was welcome to stay at Mead's home on a planned trip to the United States from England. Her attention to detail in the arrangements is typical and reflects Mead's own experience at being a guest in other people's homes.

Dear Pat, I hope getting a schedule with apparently no covering letter of any kind didn't seem too mad to you. Your letter came just as I was leaving for somewhere and I thought the most important thing was to get off word that I would be mostly in New York in September, so said "send my schedule" which is literally all you go[t]. It sounds rather like a railway timetable sent without request by a too eager railway.

As you can see from it, I'll be in New York from September 5 on, except for four days out at the end of the month. We will have plenty of room to put Richard up, I think reasonably comfortably. He can have a room and bath to himself, a key, and no obligations to be at home except when it is planned for something he'd like to do. If, of course, someone is paying for his hotel and there is no other way to collect that money, he might be more comfortable in a hotel. Ours is a real large-scale household of cats, children, adolescents who come in at terrible hours, alarms and excursions. . . .

Love to you all,

To Geoffrey Gorer, December 18, 1961, from New York, reflecting on how Marie Eichelberger had changed over time.

Dear Geoffrey,

. . . I've been thinking a good deal lately about the extent to which those who take as the principal charges in life their age mates or parents, never have any experience of weaning their affections except death itself. Marie, ten years ago, would have continued to relate to me till I died, but turning some of her attention to Cathy has meant she has been able to let Cathy grow up, then has weaned herself, and found herself new occupations and a new identity. . . .

To Charles Bull, August 15, 1962, on the S.S. Delos *off Ephesus, Greece. Here, Mead tells him the story of her walking stick.* [h]

Dear Charles,

Again—a moment to write because I'm traveling—and today I've chosen not to go ashore at Ephesus—and hear St. Paul's famous interrupted gospel reenacted. I'm in the middle of a Hellenic cruise with Geoffrey Gorer, and my ankle is playing up rather badly. So it's very fortunate it is a cruise and he can find other fellow Classicists to go ashore with. I basically detest sightseeing—so I don't really mind. The ankle is a result of ten days of very ascetic and tiring living at the great Cité Universitaire in Paris—where the distances were enormous and taxis never came. And as an addition to three sessions of meetings a day—Race and Ethnic Tensions, and Church and Society—for the World Council of Churches. I tried getting away three times a day to meet the melange of people that always seem to converge on Paris in the summer—I got thoroughly exhausted—or rather my ankle did.

I don't know whether I've told you about my stick, it's very tall and has a V at the top like this [drawing]—the kind Baden Powell always carries in a picture. It's really a lovely stick—keeps me upright rather than hobbling, and everyone free associates to it—it's a crook, a crozier, an alpine staff—a prophet's staff—a diviner's rod, a sling shot—something to kill snakes—and when they do tell you something about themselves—or what they think of you. Either way, it's fun and takes away any sense of being crippled. I'm doing a project these days on allopsychic orientation—and so have to think a lot about all the senses and the way each person is better at some than others.

But enough of all this. When are you and Marjorie coming to the states? I've lost track. I can't remember when I was disappointed that it wasn't "this year," this year or last year. And I do terribly want to know in time to plan for you to stay with me. But we do have to know ahead because so many people come—intermittently from different parts of the world and our guest floor—it is a whole floor, complete with garden, telephone, kitchen and bath—gets booked ahead. Also I'm away sometimes for two weeks at a time—even during term time—so please let me know soon—which year, which months to begin thinking about and keeping free.

I suppose Marjorie will want to see all sorts of social work things—not yet being retired and you—will you want to see Educational TV or something quite new and different?

. . . Life is in one of its smoother phases—for me—except for this blasted ankle which has slowed me down for over two years—perhaps to my ultimate benefit. I lost about the same—back to my N.E.F.[41] weight. Cathy has passed all of her exams for her Ph.D.—and next year will write her thesis on pre-Islamic Arabic poetry, no less—and the next year they will probably be out here in the Middle East while Barkev does his.[42] Gregory married again last year—a slight graceful girl with lovely eyes—who has a son the same age as his—who he has adopted. He's happy and cooperative—and I'm not worried about him. All my nephews and nieces are temporarily in safe harbor—altho Jemmy just recently cracked his skull riding a motor bike. I've no responsibilities in the world except friends and students and cherishing the life of the world—and the belief that there is enough love to go round.

This cruise will last 2 weeks—it takes in Istanbul and Athens, then I join Rhoda for a week in Sardinia—then to London for a "Pugwash" conference[43] with Soviet Scientists—then home on September the 7th and I'll hope to find a letter waiting for me,

>Love to you both,
>Margaret

To Charles Bull, May 21, 1964, from Emory University in Atlanta, Geor-
gia, where Mead was a visiting professor, on her upcoming plans and reflecting
the busy tenor of her life.

Dear Charles,

I always take such pleasure in writing you by hand, on far trips in far places, but as this letter must contain plans, I think I had better type it so as to be certain what I have said. I have your long letter of March 4, and now a note from Marjorie saying that you have been in hospital, and have come home to an even more restricted regime than before. I must say your previous regime didn't sound very restricted, with two long articles done, and I do look forward to reading them both. But it's sad that you have these interruptions in all that you want to do. I suppose the more you can accept them, the more you will be able to do in the end . . . [orig]—still I am sad for you both.

For my plans. I leave July 9 for Athens, to attend the second Delos Symposium, this one to be held, not on a ship, but at Delphi. My sister Elizabeth will be the member of my family I take this year; last year it was Barkev. There will be about two weeks between Delphi and the meetings of the international anthropological association in Moscow, then a week in Moscow, a trip to Tashkent—which they are opening to members of the congress, and then the long flight across India, Java to Australia. I want to have as much time as possible in Sydney—and I do have to get to Manus, do a couple of months work—and then get back to the United States. . . .

The plan that Rhoda and I would come down here for ten weeks and try to get our project written up was made over a year ago, but Georgia has proved to be an even more important place to be now that we have a Southern President,[44] and one who can stir the consciences of the people of this part of the world. I have been appointed to a small group of specialists who are directly advisory to the President—I find that trying to prepare materials for his eye is congenial in a way that it would not have been for any president in my life time. I have also been put on a special panel for social development advisory to the section of the state department which deals with international relations—our policy in the UN and in bilateral aid. These are actually the exact spots where I would like to be—not too much in the limelight and well within my own special area of competence. The war on poverty is providing a real sense of movement in the country.[45]

Rhoda and my life become more and more a kind of manufacturing outfit, as articles, reports, books, keep burgeoning in various forms and shapes. I am sending you a copy of my Terry Lectures—*Continuities in Cultural Evolution*—and a little later there will be a paper volume of collected papers. We now have plans for Rhoda eventually to take over the Iatmul on the Sepik River—Gregory's notes and mine so that there will be a junior successorship for all that material.

If that can be done—which will mean a trip to New Guinea for her—then all my big bodies of partly worked up materials will be taken care of: Ted, Lenora and Lola, all will know Manus, and there are several young people who know Bali. The anxiety which began so long ago, when I looked at the piles of untranscribed notes in my professors' cabinets, will be laid to rest, if I have successors to deal with each body of material.

I do hope you are going to continue to write articles which could be the basis of chapters in a book but which people will be able to have earlier, to read and think and work with.

Cathy will be teaching at Harvard again next year, Arabic and a course on linguistics. Barkev expects to have his thesis done this fall and will also teach next year, and then the next year they will probably take wing to some other university. They have their woodland place in New Hampshire for anchorage. Daniel goes to England this summer to visit the Gluckmans. Tulia has been having a wonderful time, running everybody like a queen while we have been away. There is a trace of your indulgence still apparent in the presuppositions on which the cats plan our lives—but only a trace— the kind of trace you must have left in your students' lives . . . [orig].

I will not be bringing any tasks this time, no film to hunt for at the last minute but only myself for a quiet visit, to appreciate the place that you

and Marjorie have been so happy in—a fuller ech[c]
into the country at Christmas, 1954. The years d[c]
good years! Be very good so that you will be feelin[g]
look forward to it so much.

With affectionate wishes to you both, in whi[c]
Margaret

To Mary Frank, March 17, 1966, following an [.]
psychological theorists.

Mary dear,

If you were here I would buy you some little li[v]
the day.[46]

Thank you for your long letter. I somehow feel [y]
point was when you said on the telephone that you so often disagreed with
the people you worked it [sic: with] and that I was safe to disagree with. This
happens so often in different ways; people decide I'm safe to hate, safe to disagree with, safe to scrap with during their analyses or their divorces or their
unfinished Ph.D's, perhaps because I am a combination of an authority figure and someone they basically know will not turn against them. You know
our friendship isn't endangered. Nothing could ever undo the long years of
affection and companionship and trust. But I did feel that something was
going on that neither of us understood and that had better be surfaced.

I hope you aren't working too hard. Liza[47] says Larry's speech to the Eastern Arts was very successful.

Lovingly,

To Sara Ullman, June 7, 1966. Sara and Allen Ullman, Mead's longtime
friends since the death of Eleanor Steele in 1933, were living in Spain.

Sara dear,

I was just sitting down to send you an invitation that read

Rhoda Metraux and Margaret Mead
invite you to a

HOUSE COOLING[48]
at 193 Waverly Place, June 11, at 8

when your letter came. I'm so glad that you feel they can get hold of your ills, and that Allen is painting again. This won't be a long letter because we are as usual up to our ears only more so. But if I don't answer it now, I never will, perhaps for months.

The house[49] is sold. We think we have found an apartment in the Beresford at 81st and CPW above the Museum but we won't be able to get in until November. Plans: Daniel [Metraux] goes to Europe and then straight to college. I go to a conference in Switzerland, and then a few days with Mathilde[50] in Scotland and straight to Montserrat. Rhoda goes directly to Montserrat for a month's field work; then we will stay at the new tourist spot and both write. I have to come back September 15 to begin teaching (I am givin[g] a course at Yale one day a week), Rhoda may stay in Montserrat longer. I wish there was some place in all this to fit in a visit with you and Allen, but there isn't. We are still terribly behind on our publication schedule.

Rhoda is better but tires easily. She has just had 17 days in Montserrat which she enjoyed very much but she came back to all the complexities of getting Daniel off and moving. Cathy and Barkev are going to the Philippines—for two years—on a Ford program from Harvard—with a year guaranteed after the[y] come back.[51] They have rented their apartment and their house and are very pleased with the world, although also terribly busy. Martha[52] writes us affectionate letters and seems to be very pleased with life. I try to keep up. Just the mail is staggering and it gets further complicated by things like being chairman of the NY Academy of Sciences membership drive . . . [orig] with all the not known mail coming back to me! Jane [Belo] is in the hospital at the moment but only briefly while Frank[53] is in Mexico. It was sweet of you to invite Marie [Eichelberger] and Philomena [Guillebaud], but I think they have probably planned every afternoon tea. Marie's brother has been very ill; she looks very tired.

My love, my dears. I wish you could be here for the House Cooling, but maybe you'll be here soon after we warm another.

Lovingly,

To Sara Ullman, August 30, 1966, from Montserrat, where she was visiting Metraux, who was on a field trip there, catching her up on the news.

Sara dear,

I hope by now you have been able to instruct the office what to do about the box of clothes. If not Rhoda will figure it out when she gets back. The Post offices of the world are impossible. I had left careful instructions about forwarding; Jimmy [Mysbergh] letters were NOT to be forwarded to the Museum. But they were, and ofcourse it was no use trying to send them back, so Marie had to get them from the museum and then Jim had to go to her house to get them. Up the Down Staircase!

This has proved to be a very felicitous choice of a way to spend a month. I've learned a lot about Rhoda's island, got a lot of work done, and had time to sleep and swim. I seem to be getting sleepier and sleepier, but maybe that is something I developed when I was at Geoffrey's living on his convalescent schedule. It looks as if he were going to be quite alright, if he just rests enough for another month or so.

The worst news however is about Jane [Belo], who has cancer of the throat. I've only heard from Dick Eells. She is in St. Luke's now and they are said to be going to operate. We don't know when. If you write her, write her at home, 315 W. 106, that ought to be the safest. It seems so unlikely she will have the strength to learn to talk again, which is, I suppose, what an operation for cancer of the throat means.

I also hear, via I don't know which grape vine, that Martha has bought a house in Portland, which sounds comfortably settled. Cathy and Barkev haven't got off yet. For the present, I think you'd better write them care of me, and when they have an address in Manila, I can send things on. Cathy is going to teach at the Jesuit University and Barkev is engaged in some big project about reorganizing business education in P.I. [Philippine Islands]. Two years there, with all their household goods shipped, or stored for them, and then a year back at Harvard. They have rented their house in the country to a charming sculptor who owns a truck in which he makes trips to New York.

Our house is not sold. But a great deal of the preliminary sorting and packing is done, so when it is we can move out gracefully. Meanwhile, it will be a very full year getting ready for Rhoda and my trip to New Guinea next June.

Daniel has had a wonderful summer, begun to speak German, taken in a bit of Scotland, and is all set for college.

I can't remember—because it seems only yesterday and also forever since you went away—whether you knew that the present Duchess of Argyll is a friend of mine. She was previously the wife of Clemens Heller, who was the boy who started Salzburg where Cathy and I went in the summer of 1947. So this summer I spent four days at Inverary, listening to the living ghostly voices of 9 earls and 11 dukes. It's a beautiful place, and although Mathilde is finding it almost impossible to run, what do you do with 80 rooms and with a garden where an order of 100 rose bushes just doesn't make a dent, she still seems also enjoying it. And so did I. And so did I. (refrain from old song. Trad.)

Here we have a modern hotel with beautiful little hexagonal cottages, with windows all round, glass jalousies, a view of the mountains and the sea and three minutes walk down to the beach to bathe. Even nicer than Crete was last summer.

Did you know I was teaching at Yale next year, I agreed to teach one course, with an expectation of about 50 students. And 600 have signed up. I now regret it but it's too late to back out. Instead I'll have to make some new inventions. Dear Sara, we miss you both so much. It would be lovely to be on the same piece of this planet again. Love to Allen, Lovingly,

To Mary Frank, September 7, 1966. Larry Frank's health was in decline and he died two years later.

Mary dear,

Marie has kept me informed all summer and I haven't wanted to add to your burdens by letters that had to be answered. So I have shared your ups and downs, renewed hopes and subsequent discouragements. This is just a note to send my love and say that I have been thinking about Larry, and all that you have been going through. In about a week there will be a bulletin letter for Larry and you. This is just for you. Let me know if there is anything at all that I can do. I get back to NY on the 15th, for my first lectures at Yale. Rhoda doesn't get back till after the 20th. I could always run up to Boston if it were useful. I suppose that Larry frets at being housebound. Would sending him marked copies of journals, or xeroxes of interesting articles, or newspaper clippings fairly often help break the monotony for him? I can easily manage anything that you think would be good, including long telephone conversations if they are something he tolerates easily. I am wondering how your days are going to be when you have to [be] away all day. Marie says that Colin and Robin[54] got off in a brand new car, safe for their long journey.

I've had a good summer, two weeks of terribly hard work in Geneva at the conference, ten very placid days at Geoffrey's house where he was convalescing from a mild heart attack, a week at the Duke of Argyll's castle in Scotland, (Mathilde is now a Dutchess), then a month here, in very pleasant surroundings, by the sea, learning to understand the field that Rhoda has worked in. That will all be in the Bulletin. Saturday I go to Western Canada to see the anthropologists at the University of Alberta and make a speech at Banff to the IBM executives, then back to NY. Don't bother to write a real letter—you always do you know—just a card to tell me if there is anything concrete I can do.

Affectionately,

To Sara Ullman in Spain, December 31, 1966, offering her condolences on the death of Sara's husband, Allen. The Ullmans' daughter, Martha, carried the handwritten letter across the ocean to her mother.[55] [h]

Sara darling,

We came home from Washington to Martha's telephone call and the terrible news—just a few minutes after Rhoda had read me your last letter. My dear. My dear—We are all so terribly, terribly sad, altho he's been such a precious part of life for so long—and to think of you all alone in Spain, and Martha all alone in Portland—and Allen gone from this world whose form and colors he loved so well and recreated so beautifully.

You both have meant so much in my life and so much in Cathy's life—it's as if the very fabric of existence were torn. Martha will bring this note to you.—and soon, I think, I'll be able to see you face to face.

So lovingly,
Margaret

Religion was a very important force in Mead's life, from the time she elected to be baptized into the Episcopal Church at the age of eleven. Her first husband was an Episcopal priest, and she struck up friendships with other clergy over the years. Richards Beekman was an Episcopal seminarian when Mead was put in contact with him early in 1967. She helped him find a job and a bishop to ordain him but was unable to attend his ordination. In gratitude for her help, he sent her the stole he had worn for the ordination. At the same time, he returned to her the flashlight she usually carried with her everywhere, which she had left behind when the two of them attended a conference together. This exchange cemented their friendship.

To Reverend Richards Beekman, November 14, 1967.

Dear Richards,

The flashlight and the beautiful stole arrived safely. It was extraordinarily dear of you to give it to me, and I shall treasure it, and then, do you not think that I should plan someday to pass it on to another ordination, in case I am fortunate enough to again be involved? I would like to think that this seems appropriate to you. One of the things that has always moved me most is the chain effect as one life touches another, through time.

I am sending you three copies of *The Catechist*, which you may not have seen, and which may interest you. Will you return them, please, after you have absorbed them?

I am still enjoying our ride together to and from the meeting.

You mentioned a book on friendship that I should read?

Sincerely yours,

Margaret Mead [t]

In a later letter, December 29, 1967, after he had sent her The Four Loves, *the C. S. Lewis book they had been discussing that dealt with friendship, she told him, in words that could apply to herself, "You have a genius for remembering, which is, I suppose another way of saying a genius for friendship."*

To Jim Mysbergh, October 6, 1970, from New York.

Dear Jim:

My office is filled with folders of mail I have never properly looked at and out of one of them came an old card from you, written long before you came home last spring, with a sad little message saying you thought somehow you and we were alienated over the state of the war and the world. I know you have seen Rhoda since and know that isn't so, but still it has been bothering us because we never saw that card and so never answered it. It just isn't possible, you know, that because you don't understand some point one of us may make or feel, or we seem not to understand some point of yours, that that can make any rift in our relationships to each other. After all, you've been a member of the family for 23 years. Do you realise it was that long ago that you telephoned me from the Penn Station, and came down to Perry Street?

The world is a terrible tangle, and the solutions are far, far to seek. But I don't really believe that anything will ever destroy our relationship.

News: Cathy's baby Vanni[56] is now walking and talking on the telephone, and generally being a delightful holy terror. I spent two lovely weeks with them up in the country, each of us writing half a day and chasing Vanni the other half.

Rhoda's VT schoolhouse is in working order. I went up for a weekend before going to Austria for two weeks in September, and Daniel and Judy were up there almost a month.

Gregory is chairing a conference on urbanization for a large group of biologists who know nothing about it! To see what they can come up with. So he and Lois and the baby, now two,[57] will be here in October, and Cathy is bringing Vanni down to spend the week with Marie while she goes to the meetings.

I am doing a terrible lot of teaching this semester, the semester to end semesters for all time, I hope. Work on the hall is speeding up and we hope to be able to open it by mid-May. Then I'll be able to go out to New Guinea for the summer months.

I was so sad to miss you, Jim, when you were here, and I've been very glad indeed at the news of all your plans.

Love

To Sara Ullman, March 9, 1972, carrying an old friendship into the next generation.

Sara dear,

I had a delightful visit with Martha, and I found that she had grown into a multifaceted person, many facets which I didn't know. There was so much to talk about and to share, as we moved back and forth over all the memories we share, the new things she is doing, new poems she gave me to read and references to old poems that she had read from my shelves at Waverly Place. I was re-impressed also with how sensitively Frank responds to Martha.[58] As you said, things did go very well on your last visit. They feel that too.

It was fun to sleep in the room which Martha has helped make comfortable for you, and to see how much she has made of her little house, with Allen's pictures[59] and a few things she has selected herself. I was wholly pleased and look forward to their being in the East next year, not just as a beloved child, but as a person, very much a person in her own right. Actually, I have only seen her at brief moments, in emergencies, or almost like ships passing in the night for years.

Life goes on here very busily indeed. Rhoda is very anxious to get back to the field, but she has a large number of tasks to do before the end of May when she must leave. I have piled up an incredible number of lectures, papers, and what have you, that have to be done. New secretaries are just finding their way around—and there are always emergencies—like getting off TWA and on another airlines next week. Probably no airline is really safe but it seems courting danger to go TWA at present. This weekend I will be home the whole weekend, working steadily, trying to catch up.

I do hope that you are going to find Mexico sunny, warm and congenial. As my plans now stand I leave New York for the summer on May 22. I may, but probably will not be, back before September—mid September.

Be happier,
Love,

To Léonie Adams, August 15, 1976. Mead preserved her own and others' letters and papers carefully for posterity. In this letter, she attempted to nudge Adams in the direction of writing an autobiography so she could have control of how her life story was told. They had been friends for fifty-five years.

Dear Léonie,

This is my last week of a rainy month in Vermont. I am visiting Rhoda [Metraux] in a tiny little remodelled schoolhouse which means that one person can't type while another sleeps, so when I wake up at night I read, and I've been reading Louise Bogan's letter[s][60] and inevitably reliving those early y[ears] in the 1920's particularly. And wondering about you. I haven't been writing for two reasons: one I don't know what I can do to make life a little easier for you, and also whatever I write displeases you. But maybe you would rather be displeased and stimulated, even if it is to displeasure.

I very much appreciated the notice of the Mass for my brother. He and his wife belonged to a little Japanese church, where the whole congregation had given him blood when he was ill before and really felt as if they had been responsible for his extra year of useful life.

. . . When I have a typewriter handy, which I can use at any time, day or night, I almost always have so much work to do that there is little time to enjoy just reading, or remembering, but here, with this accident of thin walls, I have had a real chance to read a few letters, and then lie in bed and think over those years when ofcourse I only knew Louise through her writing.

I wonder if you wouldn't be a lot happier if you wrote your own autobiography, the way you want it all said, and don't leave it for some biographer, no matter how good. Ofcourse, letters become an autobiography, but the selection is still very drastic. I have had to bring to final editing so many posthumous works and the choices one has to make are never sure. I know you have been putting your papers all in order, and that is good. Louise's letters to Ruth were included because I sent them back when I went through Ruth's papers before they were put in Vassar. But I didn't send Louise's letters to me back then, and the editor never thought to ask me. Ruth's letters to me about Louise, and all the rest of us, are NOT in the Vassar Library.

I don't really know how to see you. My friends the Redefers only ask me up every two or three years, and he is getting very frail, or was, when I was there last. And I gather you don't like to be descended upon and have to move all your papers in preparation for a visitor. Leah[61] is ill, having tests, and they aren't certain what is wrong. She doesn't sound well, at all.

If there was anything I knew to do that would make your life easier, or pleasanter, I'd try to do it.

Lovingly,

5

Colleagues: What Is
Important Is the Work

---

*Margaret Mead spent some fifty-five years as an anthropologist, from her gradu-
ate student days at Columbia University under Franz Boas until the end of her
life. The letters included here reflect her relationships with other anthropologists
whom she considered her mentors and her peers; with anthropology students
whom she helped or tried to help; the wives of anthropologists whom she encouraged to do anthropological work with their husbands; and with people from
other disciplines or ways of life whom she persuaded to come into anthropology
and do anthropological studies, such as Jane Belo, Caroline Tennant Kelly, and
Geoffrey Gorer. Many anthropologists or people whom she mentored became
personal friends as well, and that is also reflected in the letters.*

*Increasingly Mead's closest anthropological colleagues were those who crossed
boundaries, who attempted to integrate anthropology with some other discipline. By the mid-thirties, and especially after World War II, Mead began meeting people outside anthropology with whom she developed ideas. She interacted
with sociologists, psychologists and psychoanalysts, child development specialists,
educators, then later photographers, literary critics, specialists in emerging fields
such as cybernetics and ekistics, and anyone interested in new ways of thinking
about human problems and the environment.*

*With World War II, Mead entered the field of applied anthropology, and in-
creasingly her work became cross-disciplinary and focused on understanding the
current state of the world, including Research in Contemporary Cultures from
1948–50, then its follow-up program, Studies in Contemporary Cultures, through*

1951–52. The two projects together, she once wrote, involved about 125 different people, from 14 different disciplines. While maintaining her identity as an anthropologist, she worked with groups and individuals crossing many disciplinary boundaries.

After World War II, Mead also began a program of returning to peoples among whom she had done field work, to chart the changes that had occurred. This began with her return to the Manus of the Admiralty Islands in 1953, a people whom she revisited in 1964, 1965, 1967, 1971, and 1975. In 1967, she also returned to the Iatmul of New Guinea; in 1971 she briefly revisited the Mundugumor and Iatmul of New Guinea, and the Samoans; in 1973, she made a return visit to the Arapesh of New Guinea; in 1957 and again in 1977, on her last such trip, she revisited the Balinese.

By the 1950s, Mead's involvement with the new United Nations, through UNESCO, and WHO, and her work with the World Federation for Mental Health (WFMH)—which she called in a draft of her autobiography, "an umbrella term for all kinds of research from technological change to the reorganization of hospitals and the training of personnel for international roles"[1]—saw her bringing anthropology to non-anthropologists. In the mid–1950s she became a visiting professor of anthropology in the Department of Psychiatry, University of Cincinnati College of Medicine, a connection that continued into the 1970s. While beginning her revisits of places where she had done field work affirmed her identity as an anthropologist, increasingly her work became that of an anthropologist sharing her expertise with non-anthropologists, and interacting with professionals in other fields.

By the 1960s Mead was involved with the World Society for Ekistics, or the science of human settlements, and attending the Delos symposia associated with Ekistics that were sponsored by Greek planner and builder Constantinos Doxiades from 1963–72. She served on committees of the World Council of Churches, became a founder of the Scientists' Institute for Public Information, and became very involved in the work of the American Association for the Advancement of Science (AAAS), of which she became president in 1975.

In the last half of her life, Mead developed a vision of anthropology itself as an interdisciplinary science, collaborating with other social sciences to solve human problems and create a better world.

In 1922–23, as a college senior studying psychology and anthropology, Mead took a two-semester sociology class with William Fielding Ogburn—"The Psychological Factor in Social Problems"—which was concerned with cultural and historical factors in social change. It reflected themes in his 1922 book Social Change with Respect to Culture and Original Nature *(B. W. Huebsch, Inc.). In that book, Ogburn offered the concept of cultural lag, in which technological changes outpace cultural responses to them. Mead worked the following year as Ogburn's secretary, assisting him with his duties as editor of the* Journal of the American Statistical Association. *In addition to being an academic mentor, Ogburn was also a confidante of Mead's in personal matters. They remained close until his death in 1959.*

To Professor William Fielding Ogburn, sociologist and mentor at Barnard College, April 27, 1927. Mead had earned her master's degree in psychology before she moved on to a Ph.D. in anthropology.

Dear Professor Ogburn,

. . . Golly, I wish I could talk to you. I've got so many half-baked ideas which need rigid scrutiny. Of course this winter has mainly gone into just work, reading up for the Barnard classes, writing the book, transcribing my field notes, revising the storerooms, writing a Maori Guide leaflet, cataloguing new collections, etc. . . .

Do you remember my asking you once what my drive was and you said, "I don't know, Margaret, You just got drive!" Very probably you said it to mask some invidious thought but anyway I reckon it's true. It's certainly true as far as my drive towards Anthropology is concerned. All the others have a bug, and get an emotional kick out of Anthropological material, Luther because he hates all patterns, Ruth to absolve all individuals by proclaiming how strong the pattern is, Sapir to prove that the important thing is the mystic breaking of the pattern, Radin to prove that patterns are always broken, etc. But anthropology doesn't do that for me. It's just something to think with. I'd be as happy in psychology or any other social sciences, or anything that was good to think with. I've got a nice firm conviction that my drive couldn't be psycho-analyzed away, that it has no relation to any experience or any person, but is just a need for something to think about. . . .

My regards to Mrs. Ogburn.

Oh, we *do* miss you!

Affectionately,

To Professor Franz Boas, October 18, 1928, from Sydney, Australia, on some of the politics of Pacific anthropology. Boas was her mentor, the head of the Department of Anthropology at Columbia, and the most influential anthropologist in the United States. Mead kept in touch with him, writing letters on her field work, until his death in 1942. None of the letters to Franz Boas from Samoa are included in this book because they have already been made available to the public.[2]

Dear Professor Boas:

Reo and I were married in Auckland on October 8. We are spending three weeks in Sydney, equipping, inoculating and such. On the 31st we take the worst steamer in the Pacific up to Manus. Mr. Briggs, a professor of zoology here—he was in Manus recently—says that the native life is in pretty full swing; they even carve the big bowls in the interior. It seems therefore a good place to work. There are villages of as many as 200 people. . . .

I am having an amusing time with Radcliffe-Brown.[3] He approves of my work and once and so often tries to make out that it is contrary to the tradition of the American School, whereat I always come back by telling him it was your problem and I simply did as I was told according to the training which I had received. He's being very charming but he wants disciples and he's roundly annoyed at Reo because he insists upon sticking to his own culture[4] and not making up cosmic theories. For instance Reo found that the chief provocation to being killed by sorcery in Dobu was to possess more property than another man; the man whose garden flourishes greatly is in imminent danger of being done to death. Only in the kula can a man have temporary possession of a valuable and not be in such danger. So Reo suggests that one of the uses of the Kula is to give a man a chance to possess wealth and consequent prestige in safety, because it is only loaned wealth which must soon be surrendered. Brown replies thumbs down. Gift exchange occurs all over the world and for the same reason, this is just a local reason, therefore of no importance! It's really maddening. I am learning all I can about Tonga. He was there for some time.

I had a very good time in Honolulu. Gregory[5] was feeling aimable and says he'll give me field money for a couple months work in Samoa if I wish. I haven't given him my monograph[6] yet and anyway he is to hold it until I see whether I am going back to Samoa. The peerless Miss Stevenson lost a piece of the manuscript; it has since been recovered I understand but has not yet reached me. Dr. Buck[7] is a most excellent field worker. He combines [a] well trained mind with great powers of getting the native point of view. I cleared up a lot of points about the Maori through talking with him. I didn't stop in New Zealand at all. Reo said Best[8] has never recovered from that automobile accident and so it would be no use to talk to him. . . .

In a few weeks I'll have something definite about my new field to tell you. . . .

Affectionately,

Mead maneuvered behind the scenes in anthropological politics, here trying to choreograph a visit by E.P.W. Chinnery to the United States. Alfred Kroeber was at the University of California, Berkeley. He had been Boas's first Ph.D. student and was a prominent figure in American anthropology. To Alfred Kroeber, February 25, 1930.

Dear Professor Kroeber,

Mr. Chinnery, Government Anthropologist in the Mandated Territory of New Guinea, will descend upon your office sometime in the early weeks of April. He is on his long leave from Rabaul and is taking advantage of it to travel through this country, to get new ideas on anthropological research and method. I encouraged him to come, but I think that the anthropological gains will be rather otherwise.

Mr. Chinnery was a government patrol officer in Papua before the war, and was sent as [a] postwar student to Cambridge, where Haddon took an interest in him and published some papers on his material. These publications secured him his present position, which he will hold for probably the next twenty years unless he commits a murder.

Upon his interest and cooperation depends the work of every field worker who goes into the territory. Boats, guides, interpreters, escorts, carriers, etc. must be mustered through him. His judgment is taken by administration on the desirability of any project; he has to O.K. an ethnological collection before it can be taken from the Territory, etc. In other words, he holds the keys. It is not likely that he will do a great deal himself.

The government uses him for all sorts of routine and emergency jobs, such as investigating uprisings or a falling birth rate or a religious row. He is not permitted to do much steady field work, and his extensive general knowledge of a large part of British and German New Guinea is not corrected by a detailed knowledge of any one people. Although his position is secure, his prestige is at present a fairly delicate matter. The present governor and probably future governors will know nothing of anthropology.

I arranged, before I left, to have this trip explained to the governor as a grand rapprochement between American anthropology and Chinnery. So represented, on his return, with a bundle of contacts, addresses to learned societies, etc., it will be a very useful weapon for convincing the Governor of the desirability of Anthropology and of government aid to anthropologists.

Chinnery is an agreeable lecturer, I believe, and has had many good adventures. He sings native songs charmingly. Will you do what you can to make this little myth flourish. With all good wishes to you and Theodora.[9]

Sincerely yours,

By 1930, Mead was writing to Professor Boas more informally and less carefully than she had in the past. She wrote him this letter of July 16, 1930, from Macy, Nebraska, where she was doing field work with Reo Fortune among the Omaha Indians. Her project was to study assimilation among Native American women, which resulted in her book The Changing Culture of an Indian Tribe *(Columbia University Press, 1932).*

Dear Papa Franz:

I am sorry I didn't have your letter sooner, for it looks as if I were a thankless wretch not to have answered sooner. But I was all mixed up. When Ruth wrote about your news from Westerman and that you would be writing me, I thought it was from Germany and that you were already in

Europe. That seemed so far away that a letter from here seemed an uncertain quantity. And now I find that you have not sailed yet, but I fear that this letter may miss you, none the less.

The African news is grand, especially when it reaches us in the midst of such a broken down culture as this. It is not so bad for me ofcourse, because the more broken down and mixed up and illogical every thing, the more unadjusted the Indian woman is. Here it is a case of the Indian, living on rents and payments, having developed a perfect leisure class psychology, paying his gambling debts, scrupulous in his play obligations, scornful of work and of the poor "renter" who rents his land and slaves away upon it, European peasant style. Very little of the old life is left, but solidarity of race and language are still absolutely untouched. Everybody speaks Omaha by preference. I am not learning to speak the language however as it seemed an over-investment in terms of the time I had and the kind of problem. Also it is very difficult to get enough Omaha spoken to one by people whose English is perfectly adequate.

Thanks very much for the extra money for Reo. Work is quite expensive here, especially the endless contributions to feasts and the gasoline for everything is five miles from everything else. I have to close this to get ready to go to a hand game of the "Honeybunches," a new young girls' society of which this is the second meeting. I hope you have a very good time in Europe and get well rested for the Kwakiutl ordeal.

Affectionately,

Mead was never afraid to stand up for her work. What follows is an angry letter to Alfred Kroeber, May 1, 1931. He had written a review critical of Growing Up in New Guinea *(William Morrow, 1930) and Mead felt truly betrayed by his lack of understanding, especially since she had previously written him about her work on the material and had asked him to do the review.*

Dear Professor Kroeber,

I am writing to thank you for all the kind words of commendation in the review, and to answer, if I may, some of the points of criticism which you raised. I am tremendously grateful to you for the review because it has taught me how incredibly naive I have been in my reactions to previous criticisms. I have heard from here and there for years the accusation that I didn't present my data, that I didn't document my conclusions. But I thought this referred to my special problems only. I thought when I said that children's fights in Manus passed through a group, each child hitting a smaller rather than taking themselves out in combat between individuals or sides, that my critics felt that I should have published a list with dates, names and places of every such fight which I saw. I thought when I didn't do this my critics thought that I had generalized from too few cases. To discover that they actually thought me so lacking in method, so deficient in ethnological training as to be making flimsy generalizations without having done the kinship system or understood the economic arrangements or the religious ideas, was a real revelation to me. . . .

You have treated the book as if it were pretending to be a book of the Manus culture, instead of a book upon the social aspects of the education of Manus children. I should be the last to claim it as any sort of treatment of the culture, but I had thought that education was as legitimate a subject of special study as say Art styles, or age societies or medicine bundles. When the ethnologist presents us with a detailed study of the Sun Dance, we do not demand that they discuss the marriage rites unless there are analogies between marriage customs and specific customs in the Sun Dance. I was attempting to discuss this one aspect of culture in one Melanesian tribe, and to do so I spent months carefully observing and recording the behaviour of children.

To present a respectable ethnological statement of war one can put one informant a series of questions and find out all of the salient war customs, but to describe children's antagonisms and their expressions, events which have never received any cultural definition, a mass of detail must be collected before a single general statement can be made. When I describe what children know about kinship, I couldn't do it, if I didn't understand the kinship system and if moreover, I didn't know the relationship of every person in the village, so that when a child sulks, or begs, or runs away, I could explain the reason in cultural terms, if they existed, or hunt for personal reasons, if cultural ones did not exist. Knowledge of the behaviour of engaged children meant knowing to whom they were betrothed, who were their financial backers, how the kinship system was skewed to bring special children within the range of powerful individuals. When I state that children are ignorant of the details of such and such an aspect of adult life, how could I know that, unless I had checked the child's knowledge of an event against the actual event? Isn't education as much a proper subject for ethnological inquiry as pottery making or the blood feud?

When I wrote you that I had not "done the religion," what I meant was that I hadn't taken texts of the seances, and that I probably would not have been able to remember, nor could I find in my notes, all of the counts which the Usiai seer had dreamt in a dream to explain the death of Popwitch. Such details as these Reo collected in such minutiae that he has a hundred thousand words on the religion, form and function. But I never thought you or anyone else would think that I didn't know the religion and the social organization, nor that I wasn't continually collecting material on the various manifestations in daily life.

When there was a ceremony, I went to one house and Reo went to another—if there were two parts, or I stayed with the women's group. When there was a quarrel I sympathized with one side, Reo with the other. But you cannot spend four or five hours a day taking texts, tracing down the relationships involved in a seance the night before, etc., and also spend it observing children's play groups, supervising drawings, recording relationships between parents and children. Reo did most of the detailed recording of adult detail, just as I did of child detail, and we would talk over our hints and clues as to the meaning of institutions, or tabus, when we met at meals.

I think really you estimate my imaginative ability too highly. I couldn't have made up the culture, it only emerged from a tremendous amount of hard work.

The one thing in the review which I do regret most terribly is your comment on Reo and his work. He won his second year's field work in Melanesia on the basis of his work in Dobu which was as difficult and exacting a piece of ethnology as has been done anywhere. He had to learn the language without an interpreter from a people who loathed and detested all white people, and the basis of whose social fabric was secrecy. He selected the Admiralty Islands because they were ethnologically unknown. It was definitely understood that he was to do the detailed recording of the adult culture, I my special study for which I had received my fellowship. He was responsible to

Brown for a piece of ethnology, just as I was responsible to the Research Council for a piece of child study. He had done one fine piece of ethnology already. Yet your reference to his work sounds as if—I know you didn't mean it to—but it does sound as if his work would be inferior to mine. The coupling of his name with Kraemer—whose work is unspeakably bad—and the use of the "a" Fortune, sounds like a disapproving prejudgement of his work. Actually, there hasn't been a finer piece of ethnological work on the day-by-day functioning of a primitive religion published. It is being interpreted as a definite comment on the calibre of his work, which ofcourse makes things very difficult. He has always shared honors so scrupulously and generously with me that it makes me very unhappy to have had such a comment appear in a review of my work.

I did feel, even throughout sections of the review which appeared to me to disregard the intention of my work, that you wish me very well, and gave me of the kind of understanding upon which any enjoyment of work done is based. And that is most important.

Ronald gave me your message, thanks awfully, but we are going to have to go by Vancouver, the other way is much more expensive and involved. I am so sorry to miss your half year in New York, but we, without the benefits of the California climate, live in hope!

Sincerely yours,

Mead did not hold a grudge, however, and her relationship to Kroeber continued on a friendly basis, as many letters in the Margaret Mead Papers attest.

Radcliffe-Brown had taken a position at the University of Chicago and was teaching a summer course at Columbia. Mead and Fortune stayed in New York and took Radcliffe-Brown's course, seeing him socially, and getting to know him better.

To A. R. Radcliffe-Brown, November 10, 1931, on the S.S. Macdhui for New Guinea, after having been in Sydney, Australia. Mead brings him up to date on doings there.

Dear Professor,

Sydney was a shell without a soul; we really found ourselves very lonely there, despite the picture which hangs in the Cafe Claremont. All Sydney wonders, will you ever come back again? Raymond [Firth] is struggling to wear the mantle which you dropped but it hardly fits his slighter shoulders. Our converse with him was mainly fruitless controversy over your ideas, with Reo and I in the role of defenders of the faith, but I think Raymond has slight facility in handling abstract ideas. With Malinowski he feels that the individual native, with such and such a nose and such and such an ancestry, is the only real ethnological datum. Ian [Hogbin] is very charming, a true Polynesian in spirit I think. Raymond doesn't seem to take Ian's teaching very seriously, although Ian is plugging away, very dutifully. But Raymond has told the cadets and two or so of his star students that they needn't go to Ian's lectures. Carrie [Tennant Kelly] is still working on the illumination she received from you, but is not a popular student with Raymond, who says she asks too many questions and reads too many books. Raymond says he hopes to make *Oceania* a theoretical journal, but that seems to me a rather slim hope. As soon as you know where you are going to settle, you must start a journal, for how *Oceania* can be theoretical when its editors understand not theory, perplexes me. . . .

I am just finishing off the edges of the kinship system, which will then go into proof before it meets your critical eye.[10]

With happiest memories of this summer and very kindest regards from both of us,

Faithfully,

***

To Franz Boas, from Kenakatem, Yuat River, Sepik District, New Guinea, early in the fall of 1932.

Dear Papa Franz,

It was grand to have a letter from you and know that you are all well again. From other people I have been receiving rumours as to the miles you are walking and the miles of manuscript you are sending out in every direction.

Reo has been wanting to write you about the Alitoa language, but it is so complicated that all his attempts to simplify it into a letter defeat him, so he will just have to wait until we come home and tell you about it. It is a pretty language, though.

We are now doing a people called Mondugemoh,[11] who are on the Yuat River, a half day further up the Sepik and parallel with the Banara on the Little Ramu, or Keram River. We have heard rumours that Thurnwald is coming back here so we avoided going up the Little Ramu to do one of the other peoples there as we had originally planned, as Thurnwald may want them. The Lower Sepik people extend beyond the mouth of our river, and into it a short way, being one boundary with the Mondugemoh. Bateson (of Cambridge) has now put in about twelve months in the Middle Sepik culture, which extends—linguistically at least, over to the east and south borders of our Alitoa people.[12] We have temporarily given up the idea of doing one of the related Middle Sepik tribes, until we can talk the matter over with Bateson and explore the means of approach better. Travelling expenses up here are very high.

. . . The Mondugemoh are going to prove an interesting study on the effect of close inter-marriage and the breakdown of exogamy tabus upon the system of affinal exchanges which thread so many of these Oceanic societies together. We have only been here three weeks, so have only had time to do the grammar, the kinship system and a general census and survey of the culture, and it's not easy to tell what it is going to be like, but I don't think it will be a very long job. The linguistic processes are so simple that it will not take a large number of texts to illuminate them.

As we stay longer in New Guinea, I learn to manage things which I used to consider too time consuming to be combined with field work, so now we have real bread and an embryo vegetable garden. We really are very happy and comfortable, and we send our dear love, affectionately,

To Gregory Bateson, March 10, 1934. In the spring of 1934, Mead was planning a trip to Europe to meet up once again with Gregory Bateson. She wrote him, however, about a conflicting professional obligation which would mean delaying her trip. Her involvement with this program—the Hanover Seminar in Human Relations—and, particularly, her deepening personal and professional relationship with its host, Lawrence K. Frank, would prove critical in her interdisciplinary career. While Frank had, from his position administering Rockefeller funds for social science work, been heavily involved in funding culture and personality research since the previous decade, it is only in March of 1934 that Mead realized his significance in the field.

. . . I have been asked to take part in a specially fathered Rockefeller conference for which one representative has been chosen from each social science to do a piece of coordination work which will take about a month. . . . There is no one who can take my place. . . . As a duty to anthropology, I ought to do it, as it will mean contacts which may be used to get subsidies for a lot of people's work. I just discovered when I was asked to do this that Larry Frank, a man I've had some slight correspondence with for a long time and always treated most cavalierly, is the director of the Rockefeller Education Board—a fact to which everybody else has been keenly alive for years. He's a Turk,[13] but a good administrator and with good critical ideas. He's the man who got Sapir one hundred thousand dollars to run a seminar on personality and culture with representative[s] from every culture in the world. Well, if I were going to stay in anthropology in this country and had to plan for future work, plus the chance of having Reo at Sarah Lawrence and hostile and so losing my Museum job, why I couldn't afford to refuse this offer. It would be stark staring madness to do it, for it means all the difference between in [the] future getting any kind of project financed, and being out in the cold world. BUT THE DAMN CONFERENCE IS FOR THE MONTH OF JULY. It would mean that I would not reach England—or France—until the middle of August. . . .

It was important to Mead to make and keep connections with other Pacific anthropologists. To Australian anthropologist Ian Hogbin at the University of Sydney, June 25, 1934. Mead was in New York, and Hogbin was planning his field work among the Wogeo in New Guinea.

Dear Ian,

It was good to get your letter and to know that you are going to do Wogeo. It needs doing badly. I am sending you under separate cover all relevant publications . . . I'll send the whole lot to Carrie[14] to forward. . . .

I hope when you go out, through Rabaul, you will meet Judge Phillips who must have been away when you entered the Territory. He's a charming person, one of the most charming and most lovable people in New Guinea. Also I wonder if you have yet met Robbie, who was the joy of our hearts on the Sepik. If you see the Cobbs give them my best, also to all at Boram and the Masters of Muschu and Wallis, and anyone else who knew us and made our paths smoother.

I have spent the year on material culture, in a sense wasteful, but I learned as a child to eat the unpalatable vegetables first.

Thanks very much for having your works sent me. I'll look forward to them.

I saw ARB—RB[15] he is known as here—in the winter. He's situated in a specially aggressive graduate school there in Chicago and ofcourse he contributes his share. I felt as if I were living, all unarmoured, in a lion's den, although many of the lions were arbitrarily kind. Professor Boas continues to teach full time, and even takes his grandsons to Washington on a sightseeing tour and complains bitterly if fifteen hours a day makes him "tired."

I shortly go to New England to eat the bread of Mr. Rockefeller for a month or [so] amid pleasant hills, and prepare a new course of education for the bewildered young.[16]

Goodbye, Ian. I do hope you are in the field, and being happy there.

Yours,

To Caroline Tennant Kelly, March 30, 1935. Mead had met Kelly in 1928 in Australia. She was a friend of Fortune's who had quickly become Mead's friend. When they met, Kelly was writing and producing plays, and organizing Little Theater companies. By 1929, she had decided to become an anthropologist and Mead helped plan her training. Kelly became one of the first people Mead led into anthropology from another field. She struggled for several years, as she reentered school and was not taken seriously by members of the Anthropology Department at the University of Sydney. She did field work among the Australian aborigines and became involved in applied anthropology projects, such as reports on current conditions of aborigines rather than ethnological studies per se. When Fortune, Bateson, and Mead returned from the Sepik to Sydney, Kelly became Mead's confidante and Mead stayed with Kelly and her husband, at their home outside Sydney and in an apartment they rented in the city. Their letters are a mix of personal information and anthropological gossip.

Carrie dear,

Gregory [Bateson] is here—though at this moment he is in Chicago—and Malinowski is here, and Léonie [Adams] has chosen this weekend to temporarily take a vacation from her marriage, and altogether life is as complicated as usual, and thoroughly unreal, but very nice. Long tirades from Reo to Ruth because I don't write to him, and uncertainty about what Malinowski knows or will say are my principal worries, that and utter weariness because I have taken on too many tasks. However, I am making enough extra money so that a field trip next year is assured, I think, whether I get any outside grants or not.

Gregory has finished his long paper—as long as a book[17]—and when that is placed for publication he can start into the field again. My main problem at present is getting the Arapesh monograph written. As soon as I get that done, I can go, possibly by November or December, and he will probably go out—if we go to Assam—two or three months ahead of me. He is looking better than I have ever seen him and is very peaceful and gay and doing beautiful work. He's now visiting ARB and wires that ARB is very pleased with his new points.

Malinowski is just a little mosquito. Reo's exaggerated view of his importance is on a par with your exaggerated view of Raymond's[18] importance, it's overestimating the unimportant and disagreeable people who are at the other end of the square. I met him at dinner one night and gave a dinner for him the next night, and now thank God, he has gone to Chicago. His main interest in Gregory seemed to be whether G. was getting paid for his Chicago lectures, and if so whether he was getting paid more. Such questions interspersed with remarks that ARB was a bounder!

When are you going to send me some of your field work. I do want to see it, and I probably would have some hunches about it. Also you promised me a letter about yourself and there has been none.

I had a letter from Monte[19] describing Reo's two days in Rabaul. He told Ruth he had seen Diane, but evidently not Mira. I get so tired by his repetitious tirades.

Arrangements for legalities are proceeding quietly. I am going to get an identical divorce then either both are good or neither, which simplifies the matter. It would be so good to see you. With Léonie here with me I have such vivid memories of the refuge you offered me two years ago, and wonder, wonder, what I would have done without it. Love to Timothy and the Duchess,[20] and oh, do write, my dear.

Lovingly,

Mead had been introduced to sociologist/psychologist John Dollard by William Fielding Ogburn around Christmas, 1933. The next summer they attended the Hanover Seminar together and that solidified their friendship. He became famous for his book Caste and Class in a Southern Town *(Yale University Press, 1937). To John Dollard, April 1, 1935.*

Dear John,

Can you come down next Friday evening for dinner, to meet Howard Scott,[21] and Gregory Bateson, (an English Anthropologist who is working along the same lines that we are), and stay over night. I have already asked them because I wanted to be sure about having them for you.

The Institute to combine Freud and our other forms of enlightenment sounds very exciting, although at the moment I am very off all forms of cooperative work. The atmosphere of greed and envy and jealousy which hangs over this C and P[22] research simply sickens me. You said I could attend committees with low psychic cost, but I certainly can't handle having any power or prestige which other people want and resent my having. I am in a mood to resign from everything at the moment. I got carried away by the chance to get some good work done, but that is probably just a mirage anyway, and it isn't worth being spattered with this dank mud of envy.

I am writing to ask Gardiner[23] if I may have a look at your report. It may help to clarify my thinking. At present the whole phrasing of "Personality *and* Culture" seems ridiculous to me.[24] I can't see that there is anything except biological basis, the genetic process, and the culture, and each individual personality is a temporary phenomenon of these three contributory forces (?). [sic]

Let me know as soon as you can whether you can come or not. If Friday dinner isn't possible, how about lunch with Bateson and me on Saturday. I am afraid I won't be able to get hold of Howard for lunch.

Love to Victorine and Julie.[25]

Yours,

To English anthropologist A. R. Radcliffe-Brown, May 20, 1935, from New York, seeking his advice on how to define society and reflecting on why anthropologists and sociologists have trouble communicating.

Dear RB:

Thanks so much for your letter. I'll write you further about the point which you raised, later, but now I am in need of immediate help in another problem. In working over this Competition and Cooperation study, I find it necessary to define a society as over against a group of people sharing a common language and culture. I want a definition which will take into account the following points:

The group should have definite limits as over against all other groups, defined in such a way that every member of the society meets the requirement of membership and no outsider does.

There should be an administrative mechanism either formal or informal which acts upon every member of the society to regulate some aspects of his or her conduct towards that group.

The society should be permanent, not a merely temporary aggregation of individuals, acting for the time being, in response to a group of rules.

With such a definition one can distinguish between: the Ammassalik Eskimo in which there is no society, the individual is the largest permanent administrative unit; there is no sanction to prevent one man killing another except the fear of being killed himself by some other individual, that individual may act as kin of the victim but has no obligation to do so; there is no sanction to prevent parents killing or abandoning their children, or to make a man or woman stay with and cooperate with the spouse. Then next one hits groups of people like the Ojibwa in which there is a fixed connection between the individual and land, and between groups of related individuals and land, but there are no sanctions to effectively enforce ideas of kinship responsibility which the Ojibwa share, in ideology, with more integrated groups further South.

Then one gets the Arapesh, who can not be said to have a society, because there is no permanent group which is not so interlocked with other groups through individual ties that it is virtually impossible to make any clear distinction. It might be argued that at some previous period in their history the Arapesh localized patrilineal clan was a society, self administrative, and clearly differentiated from all other clans, absorbing into it all the wives of clan members. But this condition does not now exist, although the ideology possible for such a system does exist, nor has a wider integration on the basis of inter-clan ties, or the concept of locality—both of which exist—reached a stage where there is any closed society. Discussion of Arapesh has to take off from a point, as each clan is so tied in with each other clan, and each individual's actual social position has to be defined in terms of net-work of ties, each one growing colder as it grows more remote.

Then comes Manus, with each village a political unit as long as an individual lives within it, because of the institution of the hereditary war leader, and the strong religious sanctions for group cooperation; but the condition that each individual is free to leave that village and become a part of another village prevents one from discussion [of] Manus society, or Peri society, although a Manus village is a more permanent and better defined unit than an Arapesh village. All of these stand in strong contrast, say to the Omaha, with a definite tribal concept which distinguished sharply between the insider and the outsider and a definite administrative sense in which every Omaha was subject to some of the same sanctions, to which no outsider was subject.

You undoubtedly have thought a lot about this point and will be able to help me out. I am just writing now the summary of the C and C study[26] and I need advice. I hate to think of your writing the answer out in longhand. Couldn't you dictate it.

Last night I was asked to speak to the graduate seminar in Sociology at Columbia on modern anthropological theory. I had great fun doing it. I've been getting a lot of enlightenment lately about why sociologists and anthropologists in America have never been able to talk to each other, because the sociologists never faced the fact that the organization of society was a function of culture, and the anthropologists never faced the fact that the organization of societies was a problem. I have carried the two things along in my head, doing the traditional study of culture and doing sociology in primitive groups, without ever quite realizing the distinction, which you understand so well that you do not take the trouble to make it explicit. Gregory had come out at the same point in trying to work out the theoretical part of his paper. I have been trying to get it over to my students that in studying culture we use items of behaviour as units, and in studying a society we must [use] individuals, while in studying several societies, we use forms of social structure, and finally societies, as units. Do you think this is the way to say it? Only please answer the point about how to define this difference in the organization of society from Eskimo to Omaha first, if you haven't time to discuss both points.

Yes, please send more reprints. Ruth's and mine are getting worn out from too much borrowing.

The judgment on the Mundugumor is mine, not Reo's. I think they were as bad as that, just as he thinks the Arapesh were as good as that. It cuts both ways; and we both agree on the difference between the Arapesh and the Mundugumor and the depth of that difference.

Always affectionately,

To William Fielding Ogburn, July 1, 1935, on the work she was doing and ideas she was thinking about.

Dear Professor Ogburn,

Thanks so much for your letter. Your approval is still as very important as when I wrote that little first article on the Methodology of Race Testing, specially for your approval. And I am specially pleased that you liked the chapter on the Arapesh child because that is the section of the book that I am proudest of myself. But other people so seldom pick the part which one feels oneself is good. However, as you are partially responsible for all of my judgments, I don't suppose that can be regarded as an independent variable, can it?

As for selection, I agree with you that in any book for the general reader, it's a risky business. I am now writing my Museum monograph on the Arapesh and it's a great pleasure to be able to delve so much deeper than in the book. I don't feel there is anything in the fuller material which contradicts the book, however. But it is a great deal richer and more convincing. When the monograph is finished, there is one chapter which I shall be particularly anxious for you to read. It's going to be annoying to a lot of people, that monograph, for I have done a chapter for each school, that is each approach, a diffusion study, a structure study, a functional study, and Character formation, or psycho-analytic study—just to demonstrate that these are all possible ways of approaching cultural phenomena, each useful, although very differentially useful. . . .

We are really getting quite a lot of thinking done, Ruth, and Karen Horney, and Erich Fromm and John and I. Which is all due to you because you started John and me off working together, and that has been a tremendously productive partnership. John has a book on the criteria of the life history coming out in the fall which should be very very stimulating and helpful. Did you see Horney's article on Masochism in the July number of the *Psychoanalytic Quarterly*—done with anthropological problems specially in mind.

Theoretically I am also interested right now in trying to demarcate clearly the lines between what we may call temperament and the way in which basic temperamental trends may be institutionalized in a culture, and what we may call character. Character structures which are fundamental should be possible of attainment by any temperament, with, however, very different results in personality. And character structures should be characteristic of cultures. So far I can distinguish the character formation which works on shame—American Indian; guilt—our own and also Manus and Samoa; fear—Ifugao and Dobu; and "incongruity" (this is not a good word, but it's a rather new concept and I'm no good at word building), which is what motivates the Samoans, the Javanese, etc., in which the security of the individual is fastened to the perfection of the pattern of ritual, and the individual is threatened unless the pattern is maintained in all perfection.

I taught a course on the "individual in culture" at Columbia this semester and John taught a course on "the individual in society" at Yale, and the contrasts were useful. I think that I at last understand what the difference between a sociologist and a student of culture should be, something which I probably understood less because you always understood culture so well.

But reading Marene[27] has given me a lot of light, because if one takes his type of sociological problem and restates it as a function of varying cultures, it makes good sense.

I have also got a new hunch about the way in which homosexuality should be regarded. I think that it can be pigeon-holed with incest, as a form of expression which is socially dangerous, not because it does not lead to biological reproduction, (those arguments I think are very dangerous) but because it permits competition and aggression in the achievement field and in the sexual field to occur too close together, and so prevents effective cooperation. If incest is looked at as a way of maintaining the family structure without putting too great a strain upon the relationship between the constituent males, so also can a taboo on homosexuality. This would explain why homosexuality among males almost always involves a great difference in age, and an element of complete dominance and submission, often an exploitive element, while among females, there is a tendency towards age equality and a mutually exchangeable role, because we have already socially suppressed a great part of female aggression, and therefore a homosexual relationship with an equal is more tolerable. What do you think?

Horney and I have done a good deal of talking about women. It seems to me that the test of whether a woman has an equal chance with a man in any occupation rests upon whether she remains a love object after she enters it, equally with the man. Consider that angle to the statement "Even the best cooks are men."

Now see what an avalanche you brought down on your head by writing me that long letter. Aren't you sorry? Anyway, you don't have to answer this you know. And thanks also for your note welcoming Bateson. He got mixed up in looking for you, which was most unfortunate. I understand you're to be in Washington this summer. Have a good time.

Affectionately,

To her mother, December 20, 1935, from New York, on Lawrence Frank.

Dear Mother,

I am going to a meeting on the Sunday after Christmas at Bryn Mawr to which Larry Frank is also going and so I have invited him to dinner Saturday night. He is the associate director of the GEB[28] and the brightest foundation secretary in America, in fact, I think, one of the brightest men in America. It is through him that I have made most of my contacts in the last year and a half. It is not necessary to have any other people, just you and Dadda; Larry is rather fussed and worried by crowds or clothes, and I have told him that we had quite enough of the shabby intellectual about for him to be quite comfortable. I have said dinner at seven fifteen.

. . . It's so nice to be here for Christmas.

Lovingly,

Margaret

By 1936, Franz Boas was getting ready to retire as head of the Department of Anthropology at Columbia University and a search was on for his replacement. Mead had gotten to know W. Lloyd Warner well at the Hanover Seminar in 1934 and she encouraged him to apply. Later he would become famous for the Yankee City series.

To W. Lloyd Warner, January 14, 1936, from New York. In this letter, she coached him in how to conduct the various interviews. Mead had the habit of trying to manage things for colleagues and students toward what seemed to her their best interests, which sometimes they appreciated and sometimes not.

Dear Lloyd:

This is the way things stand. You should be asked to come East and talk matters over sometime in the near future. In conversations with the Powers that be your interest in contemporary studies in our own societies have been emphasized and that has somewhat alarmed our Papa whose heart is so wholly in primitive society. So when you talk with him be sure to keep that slant out and emphasize the importance of training students for primitive field work and getting funds for primitive field work. It will be better if he is happier, about it all.

Then if I were you I would go ahead and make arrangements to take your Ph.D. at once, just as quickly as possible. It hasn't come up yet but it will and it could be used against you. And the Australian material should be published even if you have to subsidize it, and you should be able to say something about the fact that arrangements are actually under way. This the powers that be care about—being the Dean et al.

There is also the matter of timing. If possible, it might be better if you planned to come the year after next. Ruth would be quite happy if she knew that the matter was definitely sewed up in a bag, and it would make the whole transition easier. The Papa would move out slowly, it would be infinitely easier for him, it would therefore be infinitely easier for you. And there is the point that the Foundation powers are pretty urgent about getting Newburyport done and done quickly. If you came next year that would worry them and you couldn't start Columbia off with a flourish of new support such as might be desirable.

Everything has worked out—I think—a million times better than I had hoped. I had thought the best we could hope for was five to ten years from now when some old stop-gap retired. We will all be frightfully happy if it goes through. With you here and John[29] going strong at New Haven we really ought to get somewhere. Ofcourse Ruth is definitely for you up to the ears. The Papa feels that he knows nothing about you; he is not at all antagonistic, just very anxious to know that it will be alright, and everyone will be safe in your hands. You can manage to convey that. He has been told that with the Harvard dep. what it was, ofcourse you had to go into the kind of studies where you could place your students but that your heart had never left primitive society.

I am sorry I had to write. Talking would have been better but I sail on Thursday. Don't come East or talk or show the slightest interest or knowledge of this, let them come after you. And always love to you both,

Despite her efforts, Ralph Linton was chosen to chair the Department of Anthropology at Columbia after Boas retired.

To psychologist Gardner Murphy, who had become a friend, June 2, 1936, from Bali. Murphy was interested in trance and psychic phenomena.

Dear Gardiner,[30]

In 1928 you wrote me a letter telling me how to tell whether a medium was really in a trance or not. I kept the letter carefully and I am sure I started out here with it, but now—I cannot find it. I haven't needed it all this time, and it arrived, as no doubt you have forgotten, after we had left Manus, so I never used the methods then. Now again, I am going to be working with trance phenomena and need help. I'll try to give you as full a statement of the conditions as possible, so that you won't waste your time suggesting methods which couldn't be used here.

All trance is sacred, and regarded as the entrance of a God. It would not be possible to set tasks to a person in trance, or to direct a strong light on their eyes, or anything of that sort. It seems as if it were a pretty risky business, the trances are sometimes very long, and there are forms in which the individual pushes a very sharp kris against the skin. Any failure in the state—we don't know yet whether the failure is being too much or too little in trance—seems to lead to wounds, etc. . . .

There are only certain individuals who go into trance, and there seems to be a wide range from the priest who simulates very mild possession out of which he gives orders and keeps a sharp eye on things, through the individuals who go into trance accurately, under prescribed conditions of: (1) breathing smoke, listening to a definite set of songs, or (2) participating in a dramatic performance in which a masked witch battles with a sacred beast; in the full play they are called in to kill the witch, and when they fail, they start stabbing themselves, with contorted faces and dionysian shouts, or fall down and roll in a terrific, apparently delightfully painful agony, (they learn to go into this state even when the mask of the witch is brought on the scene done up in a basket), or (3) by concentrating on some moving object as well as breathing smoke, either puppets which are vibrated on a wire or a hobby horse which the entranced man later rides about during his fire walking stunts. There may be more but these are what we know of to date.

In all the performances we have seen, the active states, which are either stabbing—for the kris dancers, or rushing about, eating hot coals and stamping on flames, for the hobby horse men, or dancing in time to music—for the little girls who have breathed smoke only, or breathed smoke and watched the puppets—these active states are interspersed with periods of limpness, slackness, semi-rigidity, etc. from which the entranced person rouses to follow the patterns of behaviour prescribed. . . .

In reading this over, I don't know whether I have made it clear that these entranced persons do very elaborate and highly stylized things, dance to complicated music, take parts in plays, utter speeches from the Gods, etc. They also occasionally do capricious and idiosyncratic things, insist on changes in the music, demand some object, etc. While the eyes may be closed, or open and staring, there is nevertheless a great amount of some sort of *attention* present.

I suppose you will have heard by now of my marriage to Gregory Bateson and that we are working in Bali together. I hope you are doing what you want to do, Gardiner, and working on mystical phenomena. Love to Lois,

Sincerely,

Margaret Mead

To Geoffrey Gorer, December 4, 1936, written from Bangli in Bali. Gorer was another of those, like Caroline Tennant Kelly, whom Mead encouraged to enter anthropology. Gorer, a successful English writer of popular nonfiction books, met Mead through a reporter for the New York Times *on a visit to New York in 1935, due to his book* Africa Dances *(Alfred A. Knopf, 1935). He had worked under Rivers as an undergraduate at Cambridge. She persuaded him to take up anthropology again and Mead and Benedict trained him in anthropological techniques in the winter of 1935–36. Then he set out to do field work among the Lepchas of Sikkim, which resulted in his book* Himalayan Village *(M. Joseph Ltd., 1938). He was a person, as she wrote from Bali, May 31, 1936, with whom she could have "those six hour conversations that were so effortless and so rewarding." Over the years he became a close friend. Here she scolded him particularly about his latest nonfiction book, written prior to his field trip.*

Dear Geoffrey,

. . . I am frightfully worried about the title of your new book. If as you say, it is a serious attempt to analyze that burlesque act and suggest what it may show about American urban working-class culture, then why give it a name which ear marks it as pornography? Why not give it a dignified title; heaven knows the material is explosive enough without underlining the explosiveness. I can't see how a serious student of "Our culture" can think that the publication of a book with a title like that is congruent with future work in cooperation with institutions like the Institute of Human Relations at Yale.

You already have to live down—from the narrow academic point of view—everything you have written, for assorted and dissimilar reasons of subject matter, treatment, political views and an acceptance of extra-sensory phenomena. Isn't that enough, plus the fact that your training has been unorthodox and that you have been—from the standpoint of the underpaid academic man—disgustingly successful, and have had much too interesting a life, without adding to your handicaps a book called HOT STRIP TEASE.

If you are going to do scientific work, you must have the confidence and acceptance of the scientific people. You know well enough how much my prestige has suffered through writing readable books that sold, and that with an orthodox training and post and a bulk of strictly dull and conventional publications to right the balance. I am saying all this in the hope that you will have time to perhaps change the title or at least add a saving subtitle of the order of "An analysis of a piece of cultural symbolism," or "seen from the cultural point of view"—anything to show that it is a serious piece of work. . . .

About personal relations and creative work, I wonder if we do differ so much. Only I have come to prefer the creativeness of a new idea to the creativeness of an indifferent poem, but I don't think that decreases my joy in a successful act of creation. But I have never felt creative work and personal relations as in the slightest degree exclusive, but rather as both reinforcing a total integration. I have never found that falling in love kept me from working; instead my mind has been stimulating and re-invigorated. Do you have a picture of just so much energy which if you used one way, then is not available for some other goal? Or do you think that love does not tap the same deep sources in the personality as creativeness, but instead somehow cuts one off from those sources. Bermondsey sounds very romantic. Have a good trip.

Love,
Margaret

To Lawrence Frank, March 28, 1937, in Bali.

Dear Larry, This, in answer to your Christmas letter, although it won't be as good a letter as I had planned. . . . I was delighted to get your letter and know that all your brood are handsome and well, and you much happier at the Macy Foundation than at the GEB. I still have in mind the conversation when you described to me—in veiled terms—the more satisfactory job which you might take, and I do hope it is continuing to fulfill all those hopes. It was also good to have news of the people of the various GEB projects with which I'd been associated. I'd had no news from anyone until I began to think it a dream that I had ever been working on any of it, and it occupying half my thoughts. Those who wrote either wrote personal letters neglecting the Commission's business entirely, or only giving a personal angle. And I *was* interested in the whole thing for its own sake and would have read even the 8th carbon copies of reports. However, I'm so busy it's probably just as well I don't have a chance. . . .

I get fits of baffled rage when I try to fit the material to people-from-other-sciences' questions and realise how completely their queries ignore the material itself, and what it can contribute; the problem is not motor development, but the specific effects of culture on motor development, for instance. I think that only by being very loyal to the material itself, and following in each culture, its peculiar twists and turns, can the social anthropologists really make any contribution to the whole character and personality problem. Anyway, Erik won't answer any of my questions. Meanwhile all the problems of should one take a life history, should one try to make a developmental study of a series of children, regardless of the cultural aridity of some of it, etc. keep cropping up.

I am getting interesting material on sex differentiation; men and women are differentiated here on a strictly physiological basis, with virtually no ethnological contrast, and a total lack of hostility, sexual inferiority, etc. The women are the most contented I have ever studied I think, the men the most lost in highly symbolic activities, cock fighting, art, gambling, playing with babies, etc. I think I shall have some material on why women do not excel in any but the oral arts, too. I wrote Erik about that, but no answer yet.

Meanwhile we live a curious life, ourselves very placid, while life swirls around us, amid texts, and bad cuts, and births of babies, and funerals, and just discovered hundred year old stone gods, and people taking worm medicine who must be medicated four times or they die, and so mustn't be allowed to escape, and Cinés of babies learning to walk, and little boys playing with butterflies and beetles on strings, and little girls of three walking with a pretense that they are pregnant, and a tame mad man who sleeps on the front veranda and makes himself headdresses out of the paper in the dust bin, and studies of the gestures of the deaf and dumb in relation to the standard gestures, and draftings of the local priestess pretending to be possessed by a deaf and dumb God who can only pronounce the ends of his sentences, and requests for marbles and coloured rubber bands, and collecting children's drawings, and getting dressed for feasts and praying in my home temple, and trying to decide whether—in an undefined new field, a mass of closely related material is likely to be more valuable than a series on more subjects. Love to you, Larry, write when you feel like it, and refrain from writing without guilt, and may life be good to you.

Lovingly,

Margaret

Mead met Helen Lynd through her work with classes at the American Museum of Natural History, and Robert Lynd at Columbia. The Lynds were the most famous sociologists of the 1920s for their study of a typical American small town (in fact, Muncie, Indiana), which they named "Middletown." They became good friends of Mead's. To Robert and Helen Lynd, October 12, 1937, from Bajoeng Gedé, in Bali.

Dear Bob and Helen,

It was a year ago today that I had Helen's long letter, so I have concluded that I am not to have one out of this vacation's free time, and anyway when I have *Middletown in Transition* and Bob's review to acknowledge, I had better mind my own P's and Q's.

I read it all, remembering *Middletown* well enough to have all the time the sense of the sequel with the same characters in it, and was properly refreshed and interested. I like the "trace" idea.[31] It's a long time since there has been a "social change" idea that was interesting. In a much milder way, it can be applied to cultures which, like Bali, suffer from periodic earthquakes and eruptions. The wholesale destruction of houses and temples serves to put a check on building, to accustom people to being in arrears, to make them feel that the Gods can wait another forty years, so that people who were born long after the earthquake will carry a trace of it in their social attitudes. It's a far step from Middletown to Bali, but even so Bali and the problem of studying it is nearer to your problems than any I have tackled yet, and so I stopped and meditated on method as I ranged through *Middletown in Transition*.

We have the curious phenomena here of the simpler, more primitive groups, having more emphasis on the village—or City-state—and less on the family, while the more advanced and urban Balinese emphasize the family at the expense of the community. We have footholds in three places, in this mountain peasant village, in an old Palace in the city where the Regent of this district lives, and in a priestly Brahmanic household in Batoean, a city which has long specialized in art and ritual. From these vantage points we try to work out samples and incidentally, when we move from one to the other, not forget the filter, or the long lens, or the medicine scissors, or the dysentery medicine, or the extra lamp mantles. It's an extraordinarily mad life; if you can imagine studying Muncie one week, Provincetown the next, and say Los Angeles the next for instance, with a touch of Kentucky hill towns, and each transition accomplished bag and baggage in a couple of hours. It means ofcourse a different vocabulary in each place, and even different systems of medicine depending on whether or not there is an iodine deficiency.

I was very touched by Bob's review in the *Nation*. So many people have written me about it. I was warmed by the friendly backing, and delighted as only an author can be when the reviewer emphasizes the points in the argument that the author thought important.

. . . I feel quite selfishly glad that I am missing the present period of war madness which must make everyone feel so hopeless and baffled. . . .

By the time [we] see America again Andrea will be in hair ribbons, and Staughton I suppose, in long pants. It seems a very long time.

My love to all of you,

To sociologist John Dollard, December 3, 1937, from Bali.

Dear John,

You *do* write such a good letter; you seem to be one of the people for whom dictation was invented. It gives me the sense of having really heard you talk.

About aggression; I wonder if one could use as figures of speech the difference between a frame which is full of explicit barriers, doors, locks, closed cupboards, things that can't be reached, things that can't be lifted, etc. and a frame in which there were no such things, where all the doors swung open to a light touch, and beyond the swin[g]ing door there was just another swinging door, until you got tired of wandering through the same rooms and came back—were not forced back—to the original room. Some cultures seem to arrange life in the first fashion, others in the second, and it must have very different results. . . .

About anthropology and possible cooperative training at Yale. With the present set up I suppose it is just impossible. But I hear that Edward [Sapir] is really very ill, and I imagine there is a real possibility of his not going back after his Sabbatical. In that case, I should think a complete reorganization would be in order and it might be possible to think about plans. I agree that it is stupidly wasteful to have to keep reduplicating departments. I haven't seen it yet, but I understand that Spier has raced to publish, way out of its turn, a violently hostile review to Comp and Coop,[32] written by one of Sapir's oldest friends, Wilson Wallis of Minnesota. This I take it is a two barbed arrow, one barb for me and one for Mark [May], with the possibility of Institute politics playing a hand in Spier['s] unusual zeal to get it published. If I hadn't learned so much from doing that job, I'd regret I'd ever touched it. . . .

About Ogburn. I had a long letter from him recently, strangely enough only talking about psychological things, the implications of Horney's book, etc., with never a statistic in the letter, and betraying as much serious interest in the psycho-analytic approach as I have ever seen him display. I think you really are very close to his heart and that actually he doesn't feel what you are doing as anything like as remote as you yourself feel it to be.

. . . My one student here—Jane Belo—has been being educated with *Caste and Class*.[33] If there were more they would be also. My father was tremendously impressed with [the book], Mother writes.

I wish, I wish you didn't have that hay fever. And it will be good to have a real talk again, about a year from now. Air mail will reach here mailed up to the end of February if marked "trans-Pacific." After that, address care GPO Rabual, M.T. of New Guinea. Love to Victorine and Julie and John Day and you,

***

Margaret Mead to former mentor sociologist William Fielding Ogburn, December 6, 1937, from Bali.

Dear Professor Ogburn,

For the last couple of months, I'd been wondering whether I might allow myself the luxury of another letter to you, but with vivid memories of my secretarial days, I hesitated to add to the pile of the things which you haven't yet done anything about but wanted to. Now with your letter, I can write. And such a good letter as it was—quite like a talk with you. . . .

About Horney's book; I was interested to see how seriously you took it. . . . Her lack of emphasis on the developmental picture is actually, I think an experiential lack; she knows very little about children, and she hesitates to deal with material she hasn't touched herself. And she shares with me a genuine distrust of reconstructed causes, and takes the Adlerian position—although she mightn't call it that—that a "first memory" is merely data on the present personality of the patient. One also has to take her feminism into account; I think that on the face of it the present Freudian position about female psychology is absurd because it first postulates the dominating importance of the body in forming the psychology and then only regards the female body as a distortion or echo of the male and allows it very little positive effect in character formation. . . .

The other current approach which interests me is Erik Homberger's.[34] I have never heard you mention him and I don't know what you think of it. There is a paper called—I think—Experimental Play Analysis, in a recent number of the *Psycho-Analytic Quarterly*—which gives a pretty good idea of his methods. He works with children, not with adults' past, and has I think great insight. He has not, however, grasped two important points: (a) that if bodies are the point, then he must elaborate the role of the female body, and (b) that cultural variation is wider than some of his premises.

I have tried the experiment of putting my developmental pictures from other cultures on the chart and I can put Arapesh, Manus and Mundugumor on, but Samoa, Tchambuli and Bali not. In other words, his premises are not inclusive enough to embrace the known range of cultural elaborations of the basic bodily picture—mainly because these three latter cultures don't work on zones at all, but work instead with the whole body, and an interchangeabi[lity] of body parts which is mechanical rather than libidinal. (e.g., where in Erik's scheme the mouth and the anus will be treated identically for libidinal equating reasons, in Bali an arm equals a penis because both can be waved, for instance.)

But despite the limitations of his whole scheme, the way of thinking in it, the translation of such notions as orality into *modes* of behaviour, is I think a great advance. It is perhaps not without significance that Erik is primarily a sensuous person, as John,[35] who is studying aggression-frustration, is mainly concerned with problems of the will. (I flatter myself that that sentence is practically Ogburnian!) And speaking of personality, as I figure Lloyd Warner out, he has, most unfortunately for everybody, identified intellectuality as feminine, while a kind of hale fellow well met organizationalism is masculine, and he has to spend a lot of time documenting the latter point. Drinking is also masculine! . . .

Do you remember little Jane Belo—fair and intense and very pretty who once, as an undergraduate, took your Psych. Factors Course. She was analyzed by Kemp, and had some work at the Sorbonne, and is married to a composer—Colin McPhee—who is making a study of Balinese music. They have plenty of money, have lived in Bali for years and know it well, and I started her in working seriously before I came out here. Now she is working hard; she has a secretary trained by our secretary; she sends me carbons of all his and her notes, and it gives me another eye and ear. The same is also true of another very intelligent woman who has lived here for five years and knows one village intimately.[36] I have trained her to record and trained her a secretary. This gives us access to more communities and to special things in each; for important things they are studying we go to at least one example and record it ourselves too as a sample. It really has all the advantages of an expedition and none of the jealousy and jockeying for position and quarreling over too-close quarters which goes with a real expedition.

. . . There is a great deal more that I would like to write but this is all the luxury that I dare take; there is an enormous amount of writing up waiting to be done before the notes cool . . .[orig]. I am so glad you are going to do something more on inventions. Please write when you can,

Affectionately,

Mead increasingly felt a responsibility toward anthropology students, but sometimes her help could be overwhelming, as she overplanned their training. Cora DuBois, who would later become the first woman anthropologist hired by Harvard, came to Indonesia just out of graduate school to do field work while Mead and Bateson were in Bali. Mead wanted to get her off to a good start, and she was frustrated by what she perceived as DuBois's resistance to her help. She expressed this in a letter from Bali to Ruth Benedict, Christmas Day, 1937.

Ruth darling,

. . . Now about Cora. She spent 33 days in Batavia and arranged to spend one and a half here, 33 days living in a hotel and calling on the wives of Dutch officials, etc. When we heard from the KPM, five days before we heard from her although I had written her how important it was for her to let me know so I could plan, that she planned to arrive on the 22nd and leave on the 24th, we were ofcourse pretty cross.

I had had a little house specially finished so she could have it to work in Bangli while Madé taught her Malay. (Instead she had to work through a Dutch speaking Javanese and is very dissatisfied with her Malay, though I think she knows a good deal but lacks confidence.)

I had bought her Balinese clothes so she would slip more unobtrusively into Balinese life here, etc. etc. We have planned and considered what she would get the most out of, literally for four months. I almost wrote her and said that if she was stopping at Bali for policy's sake, she'd better use that two days also in Batavia. But then I was much too busy to get around to it and anyway I thought perhaps she was shy and it would be a shame to discourage her if that were so. So when I knew she had only this tiny bit of time I planned it as well as I could.

We went up to Bajoeng specially to meet her, as all the arrangements there were the ones that would be relevant to her, photography, house structure, etc. To do that we missed a very important feast down below. I sent Bagoes and Madé to Beoleleng to meet her. Here in Bajoeng she saw a good big typical temple feast, and dressed in Balinese clothes, she was received very well by the people. Her manner is very good with these people, shy, but unfrightened of them and they accepted her remarkably well. She saw the whole show, Djero Balian in trance, children playing in the temple and how we record it, all the local dances, etc. Then later we had a sangijang dance for her, all complete.

Gregory showed her all the developing technique and how to cut her own film, etc. so that if she should decide to do any photography, she now knows the ropes. We showed her all our cooperative techniques because she has hopes that a Dutch girl will come out and join her in the spring. Finally late that evening, anent the Wallis review, (which [she] didn't think very unfavorable!) she said, well, she was convinced about my field work, that obviously I did speak the language, had the people in the palm of my hand, and an enormous amount of documentation. Frankly she'd been on the fence, it didn't seem likely that I could have found the things I'd said I'd found and probably she'd thought, I'd been influenced by preconceptions. But she hoped I'd publish a lot of this documentation.

So then I told her the story of Reo's offering to let Lowie text out our knowledge of Manus, and Lowie's immediately waving it aside, and then going right on saying we couldn't talk the language. And, I said, if you thought I didn't have the documentation and thought that it mattered, why didn't you ask to see it? Well, ofcourse she hadn't really been that interested, she said, and then said she supposed that she didn't want to admit that I could do what I could do because it was so much better than she could ever do. So, and I don't know where you go from there.

The next morning we took her over the village, then to Bangli, stopping to see Barongs by the roadside, and saw the Bangli set-up and then to Batoean, to see that set-up. In Batoean we had a grand Barong Tjalonarang show, with Rangda and trance and fine comedy, etc. Jane and Colin and Jane's secretary were there working on it too so she could see more cooperative work, Colin taking down the music. Then because she was enjoying the music so much, I asked Colin to get out his special classical orchestra in the evening and we went over there for dinner. The next day was picture buying day, and she saw the pictures as they came in and my group of Batoean children drawing. She also saw a lot of our Sebatoe togog series. She was a [sic: I] think impressed. . . .

Mead got angry, then got over it. She did not hold life-long grudges, nor turn against people to whom she had given her loyalty. A few months later, March 31, 1938, Mead wrote a glowing letter of recommendation for DuBois, and later supported her career.

To W. Lloyd Warner, January 11, 1938, from Bali, talking shop.

Dear Lloyd,

Some months ago I received a postal card from the *American Journal of Sociology* asking me to review your book . . . I enclose a copy.[37]

I have been down for three weeks with malaria tropica Dutch variety, and during my convalescence I eschewed my privilege of reading detective stories and read your book from cover to cover. It is literally studied with pencil notes of every sort, comparative data, queries, bright research ideas derivative from your treatment and so forth. I think it is a swell book, beautiful material and beautifully arranged. When we meet I want to talk [it] over with the possibility that your linear arrangement of materials may be suitable only for certain types of culture, while an interlocking gestalt arrangement may be necessary in analyzing cultures organized along other lines, with less of a cognitive and more of an affective basis, for instance. . . .

So much for shop, I know you hate it but if you will write significant books what can your poor friends do but enjoy them and be stimulated by them. By the way did you omit any statement of field methods, such as the degree to which you used the native language, the degree to which you moved about, etc., for the reason I suggested? Anyway I think it was a good suggestion. My life is not made any pleasanter by the muddle-headed and ill-natured reviews which I have been receiving.

We have had a grand two years; we have two other people with long Balinese experience working with us, and three highly trained native secretaries, which is the Balinese version of a literate informant. We have been developing a lot of cooperative techniques which may later be utilized for expedition purposes. For a big trance performance the other day we had four European observers and two trained native ones, two leicas and two Ciné cameras going alternatingly, for instance. I know you said I oughtn't to do any more field work, but I really think I can make as much contribution through developing field methods as sitting at home lecturing to students who don't like being lectured by women anyway! And it has been such great fun, the first time I've ever done any field work without storm and stress.

. . . As to the Columbia appointment you know what I think about it and I needn't elaborate. Perhaps it's worthwhile to have two good schools with different emphases and I don't feel so badly about it, from the standpoint of anthropology since ARB is no longer at Chicago and you have to hold the fort there. Still it would have been grand to have you in New York, and I am not pleased, aesthetically with such a successorship to Boas as the present one.

With best wishes to Mildred [his wife] and all the love in the world to you both, until we meet and drink again,

Love,

To Geoffrey Gorer, January 25, 1938, after reading the manuscript chronicling his fieldwork with the Lepchas.

Dear Geoffrey,

. . . Now about the preface by an Eminent Person. I think Haddon is the best bet. Seligman is stupid, muddle-headed, stuff[y], pompous, and fundamentally untrustworthy. (All very much between ourselves.) I wouldn't trust him not to accept it and then add some sly depreciating digs. He would be better than no one, even so. He is by the way, or will be by the time you get this letter, I think, at Yale—address Hall of Graduate Studies, Yale, New Haven. I don't think either of the Seligmans are fundamentally friendly to me—I've never met them—and the atmosphere of the anthropology department at Yale, is not going to improve their attitude any. I should suggest Haddon. After all you are Cambridge, and if you are to work in association with any University, it would naturally be Cambridge.

Haddon stands for dignity, sound work, and on the whole friendliness to people who enter anthropology from many different angles. He also is in sympathy with the psychological approach, although he doesn't make much of it theoretically. . . .

Now about the dedication. I was tremendously pleased that you wanted to do it and I understand fully why. But there is one aspect of your doing scientific work—whether in primitive society or our own—which is tremendously dear to me too, and which will be endangered if you dedicate the book to me. The number of people who are working from our point of view are very small, and perhaps because of the accident that I tend to mix personal and professional relationships up until they are indistinguishable, we form a group whose close personal ties are as conspicuous as our ties along theoretical lines. And that is a definite handicap to the advancement of our work, and to its availability to other people. I can't review Ruth, she can't review me, except for the public prints. If I raise critical points about her work in public, it would be regarded as a sign of personal desertion—not by Ruth, but by other people. The same thing applies for various reasons, to Gregory, to Reo, and to a less extent to ARB, to John and to Lloyd.[38] And I don't think it is good for the progress of science. It means that the people who really have some hope of understanding, evaluating, and either appreciating or constructively criticising each others' work can't do so. That leaves only the enemy, and leaves me with the kind of systematically stupid and hostile reviewing which I always receive, from [which] I learn nothing, and which helps nobody.

Now you are one of the people on whom I know I can always count for sympathetic understanding and constructive criticism; I don't really care much which it is, as long as it's one or the other. You will always pay attention and try to find out what I meant to do, and the chances are that you will always sympathize with my ends although you may think I have gone the wrong way about it to reach them. All this is very precious, and frankly I don't want to sacrifice it to the immediate honour of a dedication. Once you dedicate the book to me, you are less free either to appreciate or criticize my work in the places where they will do the most good. Such acknowledgements as you make in the introduction to *Hot Strip Tease* do not tie your hands nearly as much. It is taken for granted in this unpleasant individualistic competitive scientific world in which we live that people will ultimately hate anyone from whom they have learned anything, so the fact that you say you received inspiration or methods from Ruth and me is not a binding personal statement. But a dedication is.

Do you see Geoffrey? We have so many years work to do and so little to do it in. We all need both hands free. It seems a cruel jest that I [sic] the fact that I learn to love the people whose minds I respect most should actually hamper the group thinking done by that very group. But it does. And this is one time when I have a chance to choose that it should not be so. You will understand, I know, because you always do.

Didn't you tell me that you got your first interest in ethnology through reading Rivers and hearing about him when his name was a fresh legend in Cambridge? If that is so, and my memory hasn't played me false, I should think a dedication which made some acknowledgment of that fact would be very much in order, and would fit in beautifully with a Haddon introduction, the fact that you are Cambridge, etc. You don't, I know, have the time, the need, or the inclination, to waste time taking degrees and meeting academic requirements. But there are other ways of handling one's rapport with the vested interests. One of them is to start by standing, as it were, in the main stream of the ethnological tradition without over identification with any modern school or position. Then once established, you can go ahead. You have taken pains to make your Lepcha work conform to good sound standards, as well to do brilliant special work. It's worth reaping the benefits of that.

I quite see that you won't have time to send me your revised mss. if it is to come out in March. Anyway, it means the finished work will be more of a delight to me, and I think I saw it at the stage when I could be of the most use. . . .

Lovingly,

To her brother-in-law Leo Rosten, a social scientist and writer, May 16, 1938, from Iatmul, New Guinea. Rosten, known for his stories about fictional immigrant Hyman Kaplan, received his Ph.D. in political science from the University of Chicago in 1938, with a dissertation on the Washington, D.C. press corps. Published in 1937 as The Washington Correspondents *(Harcourt Brace), it*

was an influential early study of political bias in media coverage. Mead and Ros-
ten exchanged ideas back and forth for decades, remaining friendly after his split
from her sister Priscilla and Priscilla's eventual suicide in 1959. This letter echoes
a recurrent complaint of Mead's in her professional life: her anthropological col-
leagues did not recognize theoretical points when she made them.

Dear Lee, Disgraceful as it is to admit it, this is in answer to your letter
of Oct. 10, 1936, and incidentally to congratulate you on the intervening
high points, a son, the *Washington Correspondents,* Kaplan, et al. Your letter
was so full of points and also, I put off answering it until I'd written the article, and then in a way the article became a sort of answer. I am assuming
you take the journal. I had only a couple of reprints sent out here.[39] And
maybe you didn't approve. But your letter was planning a book at least. You
sketched idea after idea, approach after approach, any one of which would
have made a much longer article, and that with practically no material in it.
It brings home very vividly how sparse is the amount of material on which
theory in a realm like politics is based.

When I considered that it was my job—given the Princeton situation and
their plan of having a consulting anthropologist—to outline the type of detailed comment that an anthropologist could make, to give some sense of
the comparative method and of the use of material, I found that I couldn't
possibly make more than one point—the relationship between public opinion mechanisms and character formation. You probably thought it was
dreadfully slight, as against the canvas you sketched. I wish I could find out
why people never think I am making any theoretical points. Even Comp
and Coop has been reviewed as a contribution to fact without any theoretical implications. Is it because I do put in some facts or is it because I don't
capitalize the theoretical? But all this doesn't mean that I wasn't tremendously grateful for your letter, didn't file the whole fusillade away for future
stimulation, and then, with some sense of what I could do, derived from
your list of all that might be done, go ahead. I enjoyed the article on Roose-
velt very much. Weren't you to have done another? I have never seen it.

About the Balinese and matriliny as related to the female doll having gen-
itals and the male doll being sexless, I hope to heaven that secretary of yours
keeps carbons of your letters. It's a common amateur error—by amateur I

mean the expert in other fields here—to think that matriliny and attitudes towards the mother are linked. It is true that the stern, patriarchal father, probably can never bloom under a real matriarchate, but that is all that can be said. A society may be structurally patrilineal and yet the mother be the determining psychological influence in the life of the child, as in most of the United States, for instance. The point about the Balinese sexless male is that the penis is *detachable*, the vulva is not. . . .

Worth watching at present are two developments going on at Yale, the work of Erik Homberger who is attempting to generalize and put on a scientific basis the libido theory of Freud, in relation to zones and modes of behviour appropriate to each zone, and of John Dollard, who is trying to do the same sort of thing for aggression. Unfortunately they aren't working together, so one has to do the integration oneself.

Yes, I like the idea very much of our doing some collaborating later, and I would be interested in it in any way—either theoretical discussions, or common publication, or dual publication of two aspects, or whatever seemed to work out. If only we are ever together in one place for any length of time. This having as one's "life space" the whole world, can be very trying.

So now we will consider that old letter answered, although inadequately, and turn to newer things. I enjoyed the *Washington Correspondents* very much. I read it as soon as I got it; fortunately, I was just convalescent and so entitled to a rest with a book. I think you did a grand job of integration and I am enormously impressed with the extent to which you got their confidence and cooperation. It was surely a very difficult thing to do. Wouldn't [you] be interested in doing finer work within the larger setting, now that you have blocked out the main outlines, (by finer, I mean more detailed) or do you think more can be gained from doing other large problems, handled with about the same degree of detail. . . . I do think we will have something to thrash out when we meet about levels of material to be used, about the differences which probably makes you think my *Public Opinion* article too picayune, and makes me think your *Washington Correspondents* too large and open in outline.

I like your occasional pieces for the *New Yorker*, the Union Square one and the Gangster one, even better than Kaplan, which has also been enjoyed, in all its installments. . . .

Our best to you both. There is a bulletin which gives the current events on the Sepik River. And a mother's[40] sister's blessing for Philip (I had written father's sister—you can figure that out any way you like).

    Affectionately,

To Helen Lynd, June 18–25, 1938, from Tambunum Village, New Guinea.

Dear Helen, as usual I held the time when I would have time to write to you, like the carrot before the nose of the donkey, much too long. And whether this moment could be called "time" or not is a question. But Gregory is in bed with malaria with the fever just turning downward, and I find nursing, and keeping children out of the house so that he will be quiet, and dealing with small emergencies of purchasing food, following up squabbles over shot pigs, medicine, requests for ammunition from the shoot boy, etc. all a little too discursive activities to permit much systematic work.

We are now just nicely established to do this job of work; we have two comfortable adjoining houses with two mosquito rooms—just giant boxes set down on a little cement, made of mosquito netting—a certain amount of storage space, a bathroom out of which the water consents to drain and which has a plank shelf so that all the bottles don't upset; a veranda on which children can play with lighting suitable for photography, and a medicine place where I can get through the morning's complement of sores with some dispatch. The little outboard motor which is our taxi when we have to go the mile length of the river front in a hurry, is working; the houseful of boys have been reduced to some sort of a working unit, cooperation is not the Iatmul's long suite. The children are as jolly and friendly as the Manus, and my mosquito room is crowded with them all day, and my shelves contain lines of engaging clay figures of birds and fish, with real flowers in their hair.

We are a little tired of field work and a little anxious to get back to writing and talking with people again, but we sharpen our wits so much on each other's, that we don't feel as rusty as we might. And we find we love and hate New Guinea, whereas for Bali we have a mild, pleasant feeling, and sometimes a delight, but no higher emotion. Bali however is undoubtedly the place for a cooperative ethnological expedition—if such a thing is permitted to come off—and New Guinea as decidedly is not. . . .

Andrea and Staughton are already established in their newest photographs in my new children's gallery here, but I am sad to be missing all the reality of them. It's [good] to know that you are having the time to play with them. . . .

I hope we will have some of those theoretical discussions when we finally get to America. The thing that I have learned from Gregory is really the nature of scientific thought, the meaning of analysis. Most social "scientists" want to be scientists, for religious and social reasons, not from any genuine preference for the methods of science, as such. I think I myself, have a rather limited capacity for analysis, I am better synthesizing, but I do enjoy seeing a rule of procedure illustrated from chemistry, and then used to untie a knot in some cultural impasse.

In the case histories that come to me in letters from my relatives and friends, I have been impressed recently with[41] how important a conflict in ethos, between own family and in-laws, is in the attitude which an individual takes up about in-law relationships. It seems to me that much more than a matter of reconciling father images or compromising with incest, that if the child finds the home ethos incompatible, and marries into a congenial ethos, then, in order to successfully accomplish the identification with the new ethos, the husband has to be set up *against* the father, or the wife *against* the mother, as externalized symbols of the conflict within the personality. And the fundamental incompatibility within the home, may be due to temperament in the sense of "squares," or may be due to something quite other, as artistic interests versus academic, or scientific; desire to conform versus a premium on individuality, etc. The analyst would say that the individual must be analyzed, and the fundamental point in character structure must be attached, but I don't see how that will solve the problem of living between two incompatible ethoses, whereas a rational understanding on the part of both parties to the new marriage, may.

I am becoming increasingly interested in the whole problem of the relationship between male-elaborated symbolic culture, and women, who are biological[ly] fitted to handle only part of the symbolism, and who could, if given the opportunity, perhaps build up a new set of symbols which would be available to women of genius. It seems to me that there are only two promising roads: build a new culture which is bi-sexual in symbolism and which can draw equally well on the genius of both sexes, or go back to the position where women were virtually non-dependent upon using symbolic forms, and contented themselves with direct procreative and nutritive processes. The present solution, the induction of more and more women, into a culture whose symbols are at least fifty per cent alien, and 25 per cent antithetical to their own nature, and the development in the female character of alien drives, "masculine protests," etc. is culturally suicidal, perhaps, and very bad for women in any case. . . .

I am glad you are enjoying Erich[42] so much. He seems to have abandoned me entirely as too remote for correspondence, but then writing letters in long hand, in a foreign language for someone as busy as he is, must be a tiresome business. John and Geoffrey are my most faithful theoretical correspondents. Geoffrey is the only person I know who believes that I mean it when I say, write me about your field work as it goes along, and so all through his Lepcha work I had first hand reports to which I could react and make suggestions. But most field workers, including my own students, can't believe anyone off the scene can reconstruct it, and I find that I fail to give most people enough of a picture of the people with whom I am working, so that they can react on theoretical points. John just gives up, and says it's all too unreal. Perhaps the moving pictures will be sufficient to establish more of a contact with people like John and Erik Homberger, who do so much non-verbal thinking. We have 20,000 ft for Bali, and expect to have about half that for Iatmul.

When I get home I want to try to do some studies of my friends' children, especially of my scientific friends' children, by the same techniques of using leicas and ciné and a running record. Then you for instance, could read the account of a half hour in your children's lives, and see the pictures and the analysis, and then look at Balinese or Iatmul children handled the same way, and it ought to make a lot more sense. Just at present it seems to me to be more important to devise methods of maximal communication with half a dozen other people, than more minimal methods of communication with the bulk of workers in one's own field. Anyway, I seem singularly unable to communicate very much to other anthropologists as witness the recent *Anthropologist* reviews. . . .

I'll not attempt a new page. Love to Bob and the children.

So lovingly,

To psychoanalyst Erich Fromm, August 2, 1938, from New Guinea. Later he would be known as the best-selling author of Escape from Freedom *(Farrar and Rinehart, Inc., 1941) and* The Art of Loving *(Harper, 1956).*

Dear Erich,

I have not heard from you for such a very long time that I have begun to wonder whether those very pleasant contacts in New York were vivid enough to stand the strain of such a long absence and leave any sense of reality behind them. Helen [Lynd] writes of the great delight and increase of wisdom which she had derived from contact with you and that makes me happy for I envy you both. I realise ofcourse that my picture of my friends and their picture of me are necessarily very different; I move into a new world which involves human contacts only of a very different order and all my friends remain vivid to me, set against their natural setting, to be revived in my memory with a "I wish I could ask Erich about that," or "I wonder what Helen would say about this," or "How would John fit this angle of aggression in," etc. But to you, I have simply vanished into an unpicturable world, where I gradually fade away in a mass of blurred images of savage faces and native dances and palm trees. But I do hope that when I come home again, in less than a year, all the threads may be picked up again. I have been hoping and hoping for some articles or books of yours for I need all the stimulation and help I can get on this matter of character formation.

We are now working with the Iatmul, the people whom Gregory wrote about in *Naven*. It's proving exceedingly interesting for it is the best study of anger that I have ever been able to make. Instead of harnessing anger to a strong definite ego drive, as in Manus, which accomplishes things, they have made anger into an emotional experience which is in itself pleasant, reviving, invigorating and leaves them all in an active glow afterwards. It's one of those twists in character formation which are plausible enough once one says them, but which I wouldn't have spontaneously thought of unless I was formally working out every sort of possibility. And laughter and temper tantrums have virtually the same effect on them, both are cathartic, and they laugh when someone else gets something right, and when they themselves fail—a most unusual combination. . . .

I am enclosing a letter which there is a possibility that you may not care to read[43]. . . The letter is about the possibility of Bill Steig's being analyzed, and his being analyzed I presume by you. As to my right to write such a letter, I can only offer two arguments, that my sister Elizabeth's happiness is very very dear to me, and that it was I who sent Arthur Steig[44] to you so that if Bill's analysis follows as the result of Arthur's persistent urging of the subject, I shall feel that I was in some degree responsible. Now I know that you consider love as important, and that the fact that I love Elizabeth so much will therefore give me, in your eyes, some right to speak. She is not only my favorite sister, but as a child she came to stand to me as a sort of pledge of the beauty of life, because she was born three years after my older little sister died, and became in a sense that other baby reborn. From that experience I drew such a confidence that life holds a sequence of good things, that the most precious experience will not be lost, but will come again, that it has in a way patterned my whole life, and definitely surrounded Elizabeth with a special halo.

As her parents were both more occupied with other children, Elizabeth became to some extent my grandmother's child and to some extent mine, rather than a younger sister. She has been radiantly, beautifully happy with Bill, and anything which may effect that happiness profoundly interests me. You may think that I would feel that psychoanalysis would introduce a barrier between Elizabeth and me and that that is why I am worried, but that is not so. I should be perfectly willing to have Elizabeth analyzed by Karen, provided she and Bill both wanted it and did not feel it would upset the pattern of their marriage. Elizabeth is still very insecure, still carries the marks of a long traumatic illness as a baby, of left handedness which became symbolic of other difficulties, etc. If I had had the money I should have seriously considered having her analyzed several years ago, but I simply didn't have it, and I had to do the best I could with creating security for her by myself. So that it is not about her, nor about my relationship to her that I am worried, but only as to how anything that effects her husband may affect her happiness.

But although you believe in the importance of love, I know that you also believe [in] the importance of formalities, in a way of life where privileges are defined and limited in terms of definite situations and definite status relationships. . . . So if you, as an analyst, feel that I, as the sister of your prospective patient's husband [sic: wife], have no right to speak to you on the subject, don't read the letter. . . . I know that this is all a little melodramatic, but I am only trying to reconstruct the conditions which would obtain if I came to you in New York and reserve to you the same rights that you would have there. It is because I trust you that I am writing to you at all. . . .

With very best wishes, and the hope of seeing you next spring,

Affectionately,

William Steig was later analyzed, but by Wilhelm Reich, not by Erich Fromm. To John Dollard, September 19, 1938, while she and Bateson were working among the Iatmul in New Guinea.

Dear John,

I don't know why I always manage to find myself in the position of one for whom other people's duties are my forbidden indulgences. It used to be so as a child when I used to be punished by being kept home from Church. At any rate letters have definitely got into the class of indulgences because every waking minute ought to be utilized for typing up notes; I am always a couple of weeks behind in transcribing my notes. When the idea of studying what the natives do instead of what they say they do was invented, any sort of peaceful life for field workers was over.

I have heard from Geoffrey several times since he had his vacation for you and he seem to have had a beautiful time and gotten a lot of well-being out of it. He said it was the first time—his return from the trip—that he had felt thoroughly well and peaceful for a long time.[45] It always adds a lot to my sense of reality to think of people I know together, so it was good to have, first the news he was going and then the news he had been. But he is not an accomplished commentator on children so although he wrote originally that he was joining you and your "family," I don't know whether he saw the children or not and am no wiser about them.

I enclose a note for Victorine; we have run very short of stamps here and letters sent out with money have small chance of ever getting stamped.

I am trying very hard to get life history materials here, with an eye on the various problems which I raised in my April letter to you. . . .

I have just about decided that I shall have to abandon the methods which I myself have used for the last ten years, because other people seem unwilling or unable to use them and so are correspondingly unable to believe material based on them. Before abandoning them, in order if possible to keep my previous material from sinking even lower in the scale of anthropological evidence, I am trying to gather materials on the relationship between my old methods—special[ly] the use of the exceptional case method—and methods which possibly may gain more universal acceptance and use, such as our detailed photographic records, verbatim interview materials etc.

The chief difficulty with my present method seems to be that it depends upon making too large a number of continuous observations for any feasible method of complete recording; all day long, from dawn until midnight I am noting every passing person, in order to register the exceptional situation which needs following up, as well as writing down details from which to derive generalizations which will later then be used as a background against which to watch for exceptions. Forty people will walk by the house, say, while we are at breakfast, and I will only reach for my notebook once, to record that a father is walking on the woman's road, out of his own clan area, with a small male child, an exception to the general rule. But unless I have made critical but unwritten observations on all those 40 and every other 40, each day and every day, I can't seize my exceptions, or have a background for a statement that "Nemangke began to go about without her mother during the period when the household was split up owing to the birth of her co-mother's child." But there doesn't seem to be any way of making clear to other people how this is done, so I am enlisting Gregory's analytical mind to analyse what I really do do, and help me work out equivalent methods if possible.

Ofcourse, I think it's also significant, that most of the disbelief comes from anthropologists rather than from psycho-analysts or psychiatric social workers who are used to observing critically a great mass of material and only recording parts of it. In a sense your dictaphone records of analyses are attempts to deal with the same kind of disbelief and accusations of lack of control, etc. [I wonder] Whether psycho-analysis would ever have developed to the point where it was worth taking down on the dictaphone if there hadn't been people who were able to watch, criticize, form theories and discard them *as they listened to the patient* and not afterwards on the basis of verbatim reports. Even so your dictaphone is a more controllable technique than our photography because it is so difficult to get the event you want to record to happen in the right light, at the right angle, etc.

One of the most characteristic things about these Iatmul children is the advance-retreat vacillation in the face of anger. Where the Balinese child is either paralyzed or else runs, the Iatmul child screams, retreats, and almost immediately approaches the danger again. . . . But to get photographs which show this whole process, in each case, is an almost impossible problem. I can only hope for the moment to rely on the fact that if I report 3 unlikely things and have supporting pictures for 2 of them, perhaps the third may be taken on faith, that is on merely written records. . . .

Another interesting point in relation to anger and aggression here is that a state of sulks can be dissolved either by an outburst of anger or an outburst of laughter—either one seems to dissolve the tension within the personality. Laughter is, contrary to the usual practice over most of the world, when another person does something correct or when ego does something wrong. It is possible that theoretically these two should be separate. . . .

This debauch must stop. . . It probably is just a debauch at present anyway, a game that I play that I am communicating with you, out of this vast terra incognito.

. . . It will be over three years since I left. I realise very clearly what a contrast there is between the memory of people left in a known scene and the memory for people who have left one's own scene to go to an unknown one. While I steadily become misted over with a muddle of half-formed images of Bali and New Guinea, you all become steadily clearer and more etched in my cherishing memory of familiar ways of life. I hope I am at least bringing home some material which will make some sense to you. . . .

There is so much more I should like to ask you about but it's impossible to put it into a form that you could make sense of at this distance. It's only six months now and then WE CAN TALK. Think of it!

Lovingly,

To Eleanor Oliver, the wife of anthropologist Douglas Oliver, January 10, 1939,[46] *counseling Eleanor while the Olivers were in New Guinea doing field work. Mead believed in helping the wives of anthropologists to do anthropological work, and she did this several times during her career. At the time this letter*

*was written, Mead and Bateson were in Brisbane, Australia, heading back to
Bali for one final look at the culture, following a period of comparative fieldwork with the Iatmul in New Guinea.*

Dear Eleanor,

Your grand long letter just came a few days ago. I meant to have a proper
long time to answer it in, but meanwhile our plans have got into a frightful
whirl. . . .

About births: I'd let it go, in case you've not seen one yet. In nine cultures
out of ten they are not very significant, and the worry and fuss and loss of
rapport of trying to see them is very expensive. I've only seen two primitive
births outside of Bali, all the way through in my whole field career.

Your weaning material sounds very interesting. . . .

We are delighted that you are doing some work on gesture. Altogether it
sounds to me that you are not, as you say, an imitation anthropologist, but a
very real one. And I know how much the whole set up—the smells, the insistent natives, the screaming babies, the heat, the flies, the mosquitoes, the
other whites, the endless round with no escape, the feverish excitement of
mail days—all of it, gets on one's nerves. But, take it from one who longed
for civilization, that civilization at present with this war, spies, armament
mania, is no happy place to be. It almost makes one long for the bush.

Tell Doug, please, please to answer my letter and tell me whether he has
or intends to do any work on pidgin along the lines I suggested. Because
otherwise I'll publish mine.

Next address, care Kelly, Castle Hill, Sydney, Australia.

Love, sympathy, cheers, and all of the best. I wish I could write more but
the train leaves in a few minutes.

   Always,

To John Dollard, January 20, 1939,[47] *on the* S.S. Marella, *approaching
Java. Mead was heading back to Bali.*

Dear John,

Your long letter of November 30, written before you got my last letter,
reached me in Sydney, full of news of all of you which was very very welcome.

Especially since it seems to have been written without the stimulus of my intervening letter in which I most carefully raise no critical points whatsoever. Still the whole situation frightens me. I am probably one out of ten people in the world, who care most about your work, who have learned most from you, and am most willing to learn more from you. But every attempt that I make to learn more is met with your insistence that I am being critical, hostile, peevish, setting an intolerable pace, etc.

There have been different names for it through the years. When you say that we always get on better face to face, it boils down to the fact that even face to face, if I talk with energy and strong affective involvement in the theoretical position which I am discussing, you will also decide I am hostile—as at that conference the time Horney had us all together—but that when you accuse me of being hostile, I will burst into tears, and the tears will convince you that I'm not. And this isn't a unique response of yours; most people think I am hostile to them when I discuss anything important. And I don't know how to think without mustering every inch of feeling I possess to bear on the problem, whether it's my own idea or someone else's; the feeling acts like a sort of blow-lamp in the white heat of which, ideas come and combine quickly.

When I wrote you that April letter, I had been thinking, as hard and fast and furiously as I could, about the life history technique, trying to go all round it with the data at my disposal, your letter emphasizing it as the only type of evidence you would accept, Kurma's life history, what experience I had of partial life histories, what I could deduce from *Caste and Class* of the way you had used it. And as there should be, when one is really thinking about any technique in the world, even a very old and tried one, all sorts of unsolved problems, objections, provisos, etc. welled up in my mind. I sat down and wrote you a long letter, probably one of the longest letters anyone has taken the trouble to write you about your special technique, asking for help, asking how you would answer my questions, asking you to write certain things that people, who from this point of view were your students, and could make a student's claim, needed. Then I went on into the field and spent about a sixth of my time, trying out your technique, under of course, among other things, the additional theoretical stimulus which I had generated by writing to you.

Meanwhile, I waited, and almost prayed for an answer to my letter which would perhaps give me some additional leads which I could follow up on the spot. For in many ways the Iatmul are ideal subjects, verbal, expansive, willing to talk. (Their defects in low sense of own identity as opposed to the culturally usual, I described to you in a later letter.) But I didn't get any letter.

Because you didn't want to quarrel with me, and because my letter, highly charged with the emotion which I feel about *getting things right*, made you think I was quarrelsome, you very affectionately put off writing to me, until a general realization of our affection for each other could overlay the soreness which resulted from my letter. All of which I deeply appreciate. But, note this, John. If I hadn't written that letter, the work I did on Iatmul life histories would have been much much poorer than it is. And if you had answered that letter, I think, I believe the work on Iatmul life histories would have been better . . . [orig]. I haven't been able to solve the problem of how to use a deviant life history fully which I asked you to help me with, except in the way I solved it among the Iatmul. Therefore, I shall collect no more life history materials in Bali, because I don't know what should be done about it.

I can't think you've got such a large number of trained people who are not only willing but anxious to follow your lead, that it is worthwhile to neglect those you have. But it is all obviously my fault in someway; I can't think, either on paper or in conversation without arousing in my hearer a sense of aggression, which they feel as a response to my aggression, because I do not, by some set of postural, gestural, inflective symbols, discriminate between constructive aggression directed towards the solution of an intellectual problem, and destructive aggression directed towards the collaborator, the teacher, or the student, to whom I am talking. And yet, I do feel, perhaps without a right, that I do not feel aggression towards the people I am talking to, but only towards mastering the idea which I, or we, or they are talking about.

Geoffrey defines aggression as "the desire to master the environment"; if that definition were accepted any person who showed aggression in any respect might be correctly interpreted to be including in the environment his auditors, and they, objecting to being mastered, would respond with contra-aggression. That seems to me in a sense to describe some of your initial objections to working with me in Hanover;[48] you said I want to master *every-thing*. Yet there is a good deal of perfectly concrete evidence that I don't want to master everything.

I prefer my superiors to my inferiors; I try to learn other people's scientific idiom rather than to teach them mine (this is a good illustration; I do want to master their idiom so that I can understand what they say; but if I wanted to master them I would want to make them speak my language.) And I do like criticism, most most of all when it is directed towards the future, either towards revisions in a mss, or revisions in field work technique. That is, the statement, "I don't understand your work" is discouraging, but the statement "I don't understand your work because you do not state a and b and c," that is the kind of statement I value among all others, much more highly that [sic: than] praise.

Well, I don't suppose I should bother you with all this. Only if you were willing to think about it for a little, you might be able to solve the dilemma, or at least approach a solution for me. For it is a dilemma; if I don't muster a great deal of aggression towards a problem, I don't think as well as if I had; if I muster it, then every later expression of that thinking antagonizes you, and everybody else remotely like you. And Work loses, my work certainly, and I venture to think, in slighter degree, yours. . . .

Jane has been home collecting leads for the study of trance, and one of the things she has brought back is a mass of typed copies of (I don't know what form, published or not), Erickson's[49] hypnotic material. It has been like a strange dream coming across the accounts of demonstrations in which you participated. . . .

There was one remark in your last letter which I take strong exception to, the aspersion upon my powers of consuming alcohol! I call Lloyd to witness to them, he always claimed I was able to drink more than I [sic: he] did.

You bet we'll save an early weekend. I can hardly wait. It's particularly hard to be going in the opposite direction for the moment, even if it's only for a month. Love to all of you.

Lovingly,
Margaret

To her friend Jane Belo, February 2, 1939, in Bali. Jane had worked with Mead and Bateson for a year, in an agreed-upon apprentice role, then had gone back to the United States. She had divorced her husband, Colin McPhee, and come back to Bali to do some more work on trance. There was tension between them as Belo felt a need to be more independent in her research and she felt Mead was jealous of Bateson's attraction toward her. Belo put these points in a letter, and this was Mead's reply.

Jane Darling,

Yes, I think letter writing is a little better than talking because I seem to have so little control over the quavers in my voice. But I do cry easily, much more easily than you think, so you mustn't mind too much when tears do get into my voice.

1. About the pupil point. I don't think it's nearly as much of an issue as you think. You weren't my pupil originally. You were my friend who told me about Bali, and it was as my friend who knew about Bali and enjoyed working on Balinese stuff, that I wanted you to come out two years ago. Then you set up the teacher-pupil arrangement as a sort of pattern within which I would be free to criticise and direct and you would be free to expect criticism and direction—for a year. But in doing this it was only that you thought that would be a good way of learning, and I agreed. You called the tune, and I danced to it. But the year ended last February and I relapsed easily enough into the original position of friend and collaborator which I like best. If someone I care about asks me to play a role, either of teacher, or of pupil—I try to play it as well as I can, but I really like a face to-face equal relationship much the best.

Meanwhile you went home to America and found yourself surprisingly at home and certain of yourself in a world which had originally seemed alien but desirable. When I told everybody in Australia to read your articles in preparation for you[r] arrival in 1937, that only made you nervous. But this time you found you could criticise Bernard's thesis, and correct Abe if he made a naive point about primitives, and hold your own in psychiatric seminars, and so naturally going on playing the role of a pupil and an apprentice didn't make sense any more. But if you remember that it was always a somewhat artificial thing—something we made up to work inside for a year—and that the friend and collaborator and fellow stay-upper at Odalons are the real things, it needn't worry you any more, need it.

But just because I am willing to play any role you cast me in, I get cross if you think I am still being teacher when you have said that it was finished and I have felt that [it] was finished. As I see it, we are here now—Gregory and I—to do a few days special work of our own, and otherwise to act as your consultants and give you any help we can in planning the rest of your work. It's entirely up to you whether there is a lot of consultation or a little, and I feel perfectly easy about it either way. . . . And you know, Jane darling, really and truly, nothing that I could add to a plan of yours would give me one hundredth the pleasure I get out of looking at a good plan you've made yourself.

2. The point about my being jealous of your "femininity" isn't so easy to answer because you are less likely to believe what I have to say, as the point may be presumed to refer to feelings of a "lower level" than (1). All I can say is this. All my life I have chosen as friends whenever possible, beautiful women, and I have had some friends whose chief claim to distinction was simply beauty and feminine charm, simply because I enjoyed them. When I have loved people who were homely—as has happened once or twice only in my life—I have regretted their homeliness, because I didn't enjoy them so much, and I knew other people would be debarred from loving them often because the first impression would be unhappy. I have never chosen as a companion, a "foil," a woman less attractive than myself, and if possible, I have also chosen as close friends people whom I considered more brilliant, and more specially gifted. These historical facts you can document if you like by going through my photograph albums the next time we are together in some place where they are. It is a commonplace among my more critical friends that I am much too complaisant whenever someone is beautiful.

Second, I am much happier, and less likely to quarrel with you when you are looking lovely. I could name over with enjoyment the times when you have looked particularly well—the dinner you gave for the Sutherlands, the day you talked to DB; the night you came to the party at Jenny's; the night you saw us off in Sydney this time. And this is true for several reasons.

(a) I have an image of you at eighteen, which is perhaps my most vivid physical impression of you. When you look tired, or dress yourself carelessly, or don't feel up to being charming, that image is threatened, and I care about that image.

(b) Gregory will enjoy you more. You may think this sounds like nonsense but I know from actual experience that our contentment—his and mine—is threatened not by his admiration of a charming woman but by his exasperation with uncharming ones. Perhaps it is because I—to some slight extent—as a woman take on a little of the quality, good or bad, of any other woman who spends much time with us. However this may be, it is true that an unattractive woman makes him feel sulky about life, a charming woman makes him gayer and happier.

3. Your point about luring away men. It is a relief to know that you think this is a worry because it is so far from being one that I ought to be able to convince you easily. I think you are singularly lacking in any sort of "feminine" (in the uncomplimentary sense in which you say many men use it) attitude towards other women, in cattiness or competitiveness. I have always enjoyed seeing you with women—I remember what a pretty pair you made with Claire[50]—just for this reason. And I have always known that the danger was quite the other way round, that you might feel worried or responsible because Gregory gave you the kind of response that any normal man with eyes would. So don't you worry about that another bit.

Now I do hope this will make everything clearer. You see I *believe* everything people say to me—or am very likely to—and so I over-react to casual remarks which were only meant to express a passing mood, feel them as unfair, and have a fit about them. And that isn't good for either of us.

Lovingly,

Apparently this exchange of letters began to resolve the problem between them and brought their friendship back on an even keel.
To Ruth Benedict, January 7, 1939, mentioning this blow-up with Belo.

. . . I had a difficult time with Jane at first because along with her emancipation from Colin she had decided on wholesale emancipation from all restricting forces, and she projected it onto me, and was sure that I would resent her for having an idea of her own. That combined with the fact that she identifies intellect = maleness = father = me, drove me nearly crazy. Do you know that I have never had a single friendship with a woman in which I have been identified as a man? In a sense it's odd that I haven't when intellect and professional success, etc. are so male IOC.[51] But it is so, and I didn't like it at all. However, Jane's conscious character is very good, and if one pays attention to what she means with her conscious mind and ignores what one knows her unconscious is doing, it works alright. And she is convinced now I think, that I have no passion for satellites. . . .

Mead found psychoanalyst Erik Erikson exciting to think with, as part of this letter to him shows. Erikson later became famous for his books Childhood and Society *(Norton, 1950) and* Young Man Luther *(Norton, 1958). To Erik Erikson, June 23, 1939, from New York.*

Dear Erik,

. . . I wish you were in the East. Here nobody seems to be interested in the more complex manifestations of the libido; if they are interested in the libido at all it is all in terms of the death instinct, and otherwise one hears nothing but aggression and frustration all day long, with the kind of overemphasis on one set of factors that one finds in Horney's new book. I find I rely more and more on your kind of thinking and it's hard to rely at a distance and with only a few publications to fall back on.

By the way Erik, there is a usage growing up among analysts and perhaps other social scientists as well, of saying "anthropological" to mean "in other cultures" or "a product of culture." You do it once in your Sioux article, and I don't think it's helpful for clear thinking. I think the word, anthropological, should be reserved to refer to a discipline or a scientific approach, and not to the content of the approach. I know that other sciences do the same thing when they use the expression, "A Freudian symbol," which confuses content and approach, and I think that is equally misleading.

Gregory has a complicated problem about the abstraction which the children manifest in their statements of their difficulties, which he wants to ask you, but he wants to wait and talk it over when you come if possible. Remember we have a bed for you, and we do hope you will sleep in it because then we will be sure to see more of you that way.

. . . Please, please, please send me everything you write as quickly as possible because I learn so much from you.

Our love to all of you,

The Society for Applied Anthropology was founded in 1940, with Mead as a founding member and president. To Lloyd Warner, June 22, 1941, explaining the origins of the group.

Dear Lloyd,

. . . Have you joined the Society for Applied Anthropology? It seems to me important because you are the logical contact man in the middle west on the use of trained anthropologists on present day problems. For one thing, you've trained more of them than anyone else.

As you know it's in no sense a split-off or insurrectionist group. But the program committee of the Amer. Anthrop. refused my suggestion that we have a session devoted to the most immediate problem where anthropology could be applied—come next December. So we obviously have to create a separate place to discuss the things we are interested [in], at the same time ofcourse participating in the Anthrop. meetings.

I hope things are still going well.

Love,

To her friend, Caroline Tennant Kelly, July 12, 1942, about her war work.

Carrie darling—I know you have written five letters and had nothing but some parcel post packages of mixed magazines and it's very bad of me—like the years when I wrote five letters and sent cables and didn't even get the parcel post packages of old magazines. It's mainly that there is so much to write that I keep putting it off till I have time. And why haven't I time. Don't I just sit sometimes and play with Cathy? Yes, perhaps once a month. And don't I sometimes just fool away a whole hour when I might have been writing a letter? Yes, when I am so tired that even thinking doesn't make much sense anymore. The only empty spaces are between such packed hours that they hardly count at all. Today is Sunday. I woke up with cramps and with the determination to do no writing at all, so all the work I have done is to revise manuscripts, and look after Cathy, and even read ten pages of a detective story.

First, I will recite facts as briefly as possible in case this letter gets interrupted. I have a job in Washington, as Executive Secretary of a scientific committee that is advisory to the Office of Defense Health and Welfare Services.[52] It's a good job. I can put my finger into a great many pies, shape the program as I like, travel about the country, etc. The only rub is that it's in Washington and I only get home weekends. It began in February and for the rest of this year we kept the set-up we had; Gregory and the nurse and baby at home; the Museum office with Gregory's secretary for the Council on Intercultural Relations[53]—of which he is secretary—and two research assistants. He had to take over a lot of my jobs—such as being chairman of the section of the Academy, and a lot of things I'd been helping him with, ofcourse, now fell on him entirely.

I got back most weekends, when I wasn't taking a trip somewhere. Meanwhile, in between various jobs related to the war—on the propaganda side—he finally finished up the one Balinese book which is to come out for the duration. It's to have 100 plates, 759 pictures, sixty pages of text by me and 100 pages of captions by GB. It's the cream of the Balinese stuff and as the NY Academy of Sciences was ready to publish it, we thought we ought to try to get it out. Science is going to live on pretty thin fare for many years to come. That is just getting finished up now, final editing, glossaries and such. And Gregory now plans to make a study of the American army, going in as a draftee if he can get himself drafted.

The difficulty is nationality. . . . Alternatively, Gregory may try to go back to England for a few months and get material to use here in lectures to improve Anglo-American relations—which are very bad. His mother died last year after having had a very good year with a group of refugees—a musician's family whom she had adopted. They are living on in her house.

Meanwhile we are living this summer with the friends with whom we are joining forces for the duration. Larry Frank is one of our best friends, one of the finest minds in American science. He has six children, some of them half grown, and [a] lovely young Irish wife with one baby of her own—fifteen months younger than Catherine. They have a great five story house in Greenwich Village, and this lovely camp up in New Hampshire, and a farm. We are to have the parlor floor of the house as a separate apartment and Cathy will be absorbed into Mary Frank's nursery. . . . For the duration, the Franks and the Batesons will be one household, Mary will run all the domestic affairs and I will have no domestic responsibility at all. Cathy enters nursery school next September, where she expects to find lots of boys,—lots of girls, and lots of teachers. . . . It's hard lines to have been as good at absence as I have always been from grownups whom I loved, and then to fall in love with a baby, where time lost is lost forever. I have thought of trying to keep a diary for her but there is no time. The war presses too hard.

. . . This year has been a hectic one. I've only seen Phyllis[54] for a short time once. She turned up when we had an important and very confidential conference on and I couldn't keep her for dinner. Then Gregory gave her dinner once. So she hasn't had much to write you. You know ofcourse that Malinowski died. Radcliffe-Brown is in San Paolo, Brazil, got there after a terrific trip. Lloyd Warner, who is now a great power in the land, with the second volume of Yankee City out, is one of the members of my Committee. I have a Committee of eleven, and I am the Exec. secretary. Then I have 17 liason members from 17 government bureaus, and I have to orchestrate the combination. When I took this job—it was offered to me half an hour after Pearl Harbor—and the whole thing seemed like a repeat of the invasion of Poland, the knowledge that Gregory and I would be separated again—I had to give up my teaching at Vassar and Columbia. . . .

It's grand to know that you are doing the kind of jobs that really make sense to you. I'll send you out a batch of my Committee's stuff and that will give you a taste of what's new in our minds. We have developed a new method of opinion sampling, which is really anthropological. I have a crackerjack young assistant who worked part of it out whom I'm very proud of.[55] I'll also seriously try to gather together the reprints you've not had. Love to Timothy—our best to or [sic: you] all.

Lovingly,

To Caroline Kelly, August 12, 1945, as the war drew to a close.

Carrie darling,

A letter from Dorothy Colley brought me the news of your mother's death yesterday. I know how lost you must feel, especially as she seems to have preserved a good deal of her strength and determination until close to the end. It's the end of a terribly long period of devotion, of practically a whole lifetime in which you never made a move without thinking of what it meant to her, and in some ways it must be more like losing a child than a parent, but with the special poignancy of loss which I imagine only the loss of a mother can bring. Dear Duchess, it won't be the same without her.

In spite of my poor corresponding, I have taken *in* all that you have written. It must be marvelous that you now have your dip. anthrop. There is a sort of echo of your struggles going on here as Jane—who did not have a BA—is now trying to take a Ph.D. I think that one can give a discipline a very special quality by bringing people into it at different ages and types of sophistication, instead of just taking kids right out of the routine undergraduate preparation. Geoffrey Gorer is ofcourse an outstanding example, coming in at 32 from Classics and play writing, and Ruth at 33, from dancing and poetry! Your work will never be the same work it would have been if you had lacked those years before you went back to school.

I have a friend at present who is a grandmother—a very young one—and who, after years of being a lady bountiful member of boards of private charities, has gone back to school to take a degree in social work, and is, I think, going to actually invent a new profession, a sort of educational casework, which I have got very much interested in, in which you take children with restricted and contracted approaches to life, and help them expand, (dilate is the technical phrase used by the Rorschach people). It's not teaching and it's not therapy, it's really something new but of the order of case work.[56]

. . . I have finished my Washington job, very comfortably, and left it in a state of turning from a Committee on Food Habits, to a Committee on Living Habits, if the money is ever forthcoming. We rounded off with a Manual for the Study of Food Habits which should reach you in due course.

. . . I go back to the Museum and to all the punishment which every returnee who is not protected by putative exposure to death—will have to take from the people who stayed on in their jobs and did nothing during the war. In my case it will be especially heavy because Wissler has retired, there is a new administration, a new chairman of the department, Harry Shapiro, and everybody has a fixed belief that when I was at the Museum I ran all over the country as I have these last three years on this war job. So I am going to have to spend a year in conspicuous, officious dullness, always going out every evening with the crowd of rapid runners home, so that they will be convinced I am really working a dutiful seven hour day at nothing. I think I shall rather overdo it and perhaps at the end of the year they may be glad to let me alone again. Anyway, I need time to get back into the swing of peacetime types of activities, to get used to staying in one place and not carrying huge bags of papers every place I go. Is Timothy farming all the time, or only evenings and weekends?

. . . So life goes on, terribly full, of meetings, papers, manuscripts, committees, ideas. I had meant to write a book this summer, but now I want to wait and see what the atomic bomb is going to do to people's intellectual readinesses. We have made such incredibly unforeseen strides during the war, in theory and method, and we practically have to stop and catch up with ourselves. I am quite healthy, pretty plump, no more gray hairs than when you saw me and on the whole stronger. Reo when last heard of had gone overseas, reason unknown. He lost his job at Toronto. Sometimes I feel as if I had lived an awfully long time, to have a five-year-old child. Sometimes I regret the great skill I have developed for intimacy without complications, for affairs all conducted with mimeographed memoranda instead of flowers, for conversation instead of kisses, and sometimes I think it all makes one's step pretty light in spring.

Love to Timothy,

Lovingly,

To I Madé Kaler, November 9, 1947, from the American Museum of Natural History in New York City, responding to a letter from him asking for a copy of Balinese Character *(New York Academy of Sciences, 1942), the book they had produced from their Bali fieldwork, and enquiring about the other Westerners who had been there at the same time, including Bateson's mother, Beatrice, and her friend Lady Nora Barlow, who had visited the first Christmas Mead*

and Bateson were in Bali. He also asked for assistance in acquiring U. S. stamps
for a stamp collection.

Dear Madé,

It was such wonderful news to hear from you and that you are all well and flourishing. We have thought of you and talked of you and wondered about you so many many times during these long years, but first we could not write—because of the Japanese, and then we were afraid to write to you until we heard from you because of the disturbed political situation. We thought it might not be wise for anyone to be receiving letters from this side of the world. As you ask for the book, we are sending that too. Of-course we have always wanted you to have a copy but again we did not know whether this was the time to send it.

So you have four children. We have only the one little girl, Catherine, and she is *ampoen mekeotoes*,[57] and in the third grade of school now, very tall, with blue eyes but hair darker than the *Toean's*[58] or mine. She has seen so many of the Balinese pictures and knows many of the people of Bajoeng Gedé by heart. During the war we were both terribly busy. The *Toean* was overseas for almost two years, stationed in Ceylon, with short times in Burma, India and China. Since he has come back he has been hard at work lecturing and preparing a new book. The *Njonjoh Biang*[59] died in 1941, but the *Njonjah* Barlow[60] and all her children have come through the war safely. I saw some of them when I was sent to England in 1943. Mrs. McPhee is married again, and is now Mrs. Tannenbaum. We have heard that Mrs. Mershon has returned to Bali and had one letter from her. I suppose you will have seen her by this time. Mr. McPhee has written a book about Bali, especially about music, and is now working on a more technical book on the music. Miss Holt is now in the State Department in Washington, and her son is in the Navy. Neither the *Toean* nor I have changed very much; he got a little huskier-looking on the army food, and I got very plump during the war, but this summer I spent two months in Europe and got thin again.

I am back working at the Museum, and beginning to write another book on Bali, just about the babies in Bajoeng Gedé this time. It seems so strange to be working all day over the baby pictures of Karbo and Karsa and Kenjoen and realize what big boys and girls they must be now. I don't suppose you have turned into a photographer and could be persuaded to make a trip sometime and take pictures of them all for us, so we would know what they look like now. Or is there any photographer now in Bali who could be persuaded to go with you. If there ever is a chance for this being done, please tell me and I will send you the money.

Catherine collects stamps now—but of course without very much sense yet—and she is going to help me get you a set for your stamp collection. Is there anything else special you would like from this country? We would ofcourse love to see a picture of you. I am going to put a picture of Catherine in this letter, and perhaps one of all of us, if I can find one. Yes, here is one which we used for a Christmas card the Christmas that the *Toean* got back from overseas and it's quite good of all three of us.

Dear Madé, we are so glad, glad, glad that all is well with you and that you like your work and your children and are so interested in the future of your country. Please write us as often as you can.

Affectionately,

To Geoffrey Gorer, July 20, 1950, evoking a feeling of personal uncertainty in a period when they were very tentatively thinking about a possible marriage, and had been since the summer of 1949.

Geoffrey dear,

I've not written for a variety of reasons, mainly because it has been so difficult to clarify the plane trip and until it was clarified I somehow didn't quite believe in it. I still don't entirely. . . .

But meanwhile . . . [orig] I am coming over to see you, and not for any other reason at all. And so perhaps it would be better to get life very clear, both our immediate lives and how you feel about the prospect of War. I haven't anything more of the clarity you wanted than I had last year. Gregory thinks he wants a divorce,—"to improve our relationship." It's a pure piece of acting out, which has no reality for me, doesn't alienate me any more than any of the other bits of acting out. He isn't planning to marry anyone else, or change life in any way, and I feel still completely caught as Cathy's mother. To keep any sort of relationship going in which she can feel safe and happy still seems my major task in the world. It could alter, very materially, if I either got any sense of reality about the divorce, or if Gregory should marry again. Meanwhile—I am not being a martyr, but she seems a first responsibility. There isn't ofcourse anything new in this except a report on present status.

So if I come, it's only for a visit, only for now. There is still no future, except the sort of timeless haven which you have given me to anchor my spirit in . . . [orig]—where I feel eternally safe and cherished. That's one side, and the other is—the possibility of war, of war for years and years, of war which will take all our energies, and will, I should guess, be pretty much the end of our effective lives. If I don't come now, and if you aren't sent to this country— I suppose you might be?—we might not see each other for years—or ever. So you see, my dear, that we ought to be clear, that you think it a good idea for me to come, if I possibly can. I keep stopping this letter in the middle to cry—the idea of not seeing you just doesn't seem bearable, but it would be more unbearable to be out of step with your intent. So write me, and tell me what you think. I have one absolutely clear belief, that properly communicated what makes sense to one of us will make sense to the other.

[no ending]

The timing never became right and the moment passed. Gorer was at heart a homosexual, and as Mead wrote to Metraux in a letter July 17, 1952, from England, "It's trying ofcourse—intense feeling without any physical expression is so trying—and I'm more than ever convinced that is what it would always be." Mead and Gorer went through a rough spell of misunderstandings from about 1950 until 1952, as they tried to rebalance their relationship into friendship. After 1952, their friendship smoothed out again and from that point, they remained good friends for the rest of their lives, often taking vacations together and working together on articles and reports, or sharing ideas for a project that one or the other of them had taken on.

Mead decided that she needed to return to field work, after a decade of war work and studying cultures from a distance. She accepted an invitation to lecture in Australia and used the opportunity to research a good field site. To Professor A. P. Elkin at the University of Sydney, whom she had known from coming and going through Australia to field work in New Guinea, October 19, 1950, prior to her trip.

Dear Professor Elkin,

I was just on the point of writing you when your letter came, as I was waiting until it was quite settled that I could accept the invitation of the New Education Fellowship. I had ofcourse assumed that you would be in touch with what was going on.

My reasons for accepting the invitation were manifold, including a desire to see Australia and my friends there, after these long eleven years, but also a desire to think over new field work possibilities. I have been trying to decide what was the proper next thing for me to do—provided field work remains a possibility in a semi-warring world—and the suggestion you make was one of the possibilities I had thought of—to go back over the same ground, with new recording techniques, new insights, and the fifteen or twenty years of slow and violent change which have intervened. Alternatively I had thought that it might be a good thing if I tried to do one piece of field work in Australia, so as to make the Australian material more accessible to me, and round out my first-hand experience with the areas for which I am responsible.[61] At present I only know Australia through books.

I talked with Raymond [Firth] a little about this when he was here, and I accepted this invitation so that I could come out and consider possibilities with you, and other specialists there. Work in Australia would involve my going with other people—I should think some young couple among your experience[d] field workers. I imagine that at my age tackling Australian field work on my own wouldn't be too economical, and anyway I thoroughly believe in mixed sex teams to get the sort of material I am interested in. Now if I were to go back to New Guinea, I would also want to take with me probably a young couple, skilled in photographic and recording techniques. There, they wouldn't have to know the field conditions for which I think I probably still have adequate knowledge, inspite of all the changes in technology, but they would need the skills which I don't have—ability to walk distances, take movies, manage machines, etc. In either event, whether it was decided that I was to work in Australia or do a return trip to New Guinea, it would take a good deal of preliminary planning, and if I were to work with some of your younger people—which would be necessary for Australia—not as necessary for New Guinea—I'd have to meet them.

I had thought of this lecture tour as a reconnoitring trip only. I don't think I will be ready to go into the field again until September 1952 at the earliest. . . .

What I really need now will be some help from you in suggesting who I should plan to see in Australia and what would be the best way of getting a look at some aborigines of the sort Mountford photographed. I'd like to work with as simple a group as possible if I did work in Australia—though I don't even know whether the word simple is a congenial one there. The tour which the New Education Fellowship has planned takes me all round the periphery and there will be a spare couple of weeks in Sydney from August 12 on, in which I could make a flying trip somewhere if it isn't possible to dip in as I had originally hoped.

As far as revisiting former fields are concerned, I think the most profitable scientifically would be: Samoa, Manus, Arapesh and Iatmul. I had a chance to reinterview natives from Mundugumor and Tchambuli. Both were, even in 1932, small and very badly shot cultures. Although they might repay a quick visit—especially with a moving picture camera, and perhaps a census, work boy history, etc. I don't think they would repay prolonged work. In fact Mundugumor was a doubtful field site in 1931[62] and we only spent a short time there.

I find it's very exciting to be thinking about field work again after all these years, and I only hope that the world stays peaceful enough to make it possible. The heavy commitments in government research which I have been carrying since Ruth Benedict's death all wind up this summer, and the Australian trip, on which I am bringing my daughter who will then be eleven, will be a kind of celebration of the end of an epoch—I hope.

With personal good wishes,

Sincerely yours,

Margaret Mead [t]

Mead's book Male and Female *(William Morrow, 1949) had come out in 1949. Former student and anthropologist Jules Henry reviewed it for the* Journal of Orthopsychiatry *and panned it. To Jules Henry, December 14, 1950, disappointed in his review of her book, and trying to avoid a public confrontation with him.*

Dear Jules,

I came home the other night at midnight after an interminable day of work and meetings to find a marked copy of your review in Ortho on my table. My first reaction was pretty close to despair. I expect people like Nadel[63] to find some difficulty with the parts of my work which use psychoanalytic concepts but not you. I felt as if all my efforts, all the time I spend trying to do what seems the next step, all the work I put into careful appendices and descriptions of steps in thought and everything were all quite useless. I sat down and wrote a sad note to you and a refutation to the *Journal of Orthopsychiatry* because there is always the chance that quite a few people who would understand what I was talking about would be quite put off by your review, and the group of people who prefer making misstatements about other people's work to doing any of their own would quote it as proof of something, etc. But I haven't sent it.

I hate writing refutations, especially to a review by one of my friends and collaborators. In the case of Wallis' review of Comp and Coop his misstatements were administratively significant, involved the project rather than me myself, and I felt I had to. Thurnwald's review besides being personally malicious and instigated, gave such a false impression, bolstering a misinterpretation with misstatements of simple fact, that I felt that had to be answered. But you and I have worked together in Ortho for years; I have been so delighted with the job you were doing there in forging links from anthropology to the psychiatric field, and I can't bear to come out in that journal of all others with a statement that you have completely misrepresented me, absolutely misrepresented my position, bolstered your misstatement of my theoretical position with what looks like—but is not—a great scholarly oversweep of my work for the last twenty-five years, etc. It surely will help no one for me to do that. Yet it has to be done in some way.

I care about my work for what it may mean in the world, I have been more continuously conscious probably than almost any other living anthropologist of the responsibility involved in each decision to publish one thing ahead of another, to go to one field, or even one meeting rather than another. This afternoon for instance—although I am both sick and busy, as I haven't been able to eat anything for two weeks because of auromycin poisoning—I am going to [a] legislative hearing to talk about aging because I think aging is an important issue at present—not because I have ceased to be interested in infancy! Three years or so ago Reinhold Niebuhr wrote a review which totally misrepresented my position in the *N[ew] Y[ork] Times*. If NOT had been inserted in every sentence it would have represented my position quite accurately. This was pointed out to him by the editor of the series and he wrote a retraction. Now I have no editor to do that, for *Male and Female* ofcourse stands on its own feet.

But I am writing you first to ask whether, if I take the time and the trouble to demonstrate to you that you are wrong, you will—if convinced—write a retraction. Ofcourse I know it depends on whether you were genuinely mistaken, or whether this has just hit off some political or psychological blind spot of your own which makes it impossible for you to perceive what I have said accurately. Or you may be angry at me for some obscure reason which I know nothing about. I have no way of knowing. I know I failed to get to the last two Ortho meetings, but the failure was genuine enough. These have been three pretty terrible years, stretching from Gregory's leaving in the spring of 1947, through Ruth's death and my having to take on all her responsibilities in addition to my own, through my mother's year-long illness and death last February, and finally divorce and Gregory's remarriage. One of the reasons that I in the end cut the Ortho. meetings out when I was practically flat on my back was because that was a place I felt I was needed less because you were carrying the ball there, and didn't seem in need of help. However if there is some irrational element of a personal or political nature back of your misreading, I don't expect to be able to convince you by facts. But if you really did misunderstand, then I am sure I can.

You see it's not a question of what is the correct interpretation of Arapesh data. If you had written a review presenting a different interpretation or questioning my interpretation on theoretical grounds, although I might have felt that a theoretical discussion would be interesting and fruitful, a refutation would not have been called for. Your review consists of misstatements about what I have said and believe, supported by wholly false statements about changes in position. I imagine the general reader will accept my word rather than yours as to what I believe and what I have done, especially when I can if necessary document both up to the hilt from published material. Because believe me I can, with serial quotations through time, going back to the 20s. So little has my position changed that I have just lifted a large part of a paper I published in 1930, to use in a chapter in Alexander's new textbook.

WE KNOW more than we did in 1925, a great deal more. In 1925 psychoanalytic theory, topological psychology and learning theory were all in a very rudimentary state as far as applicability to our material was concerned. Cybernetics and psychosomatic medicine didn't exist. Work on imagery and the primary process was rudimentary. Through the years I have tried to follow these various developments and use them as they became more available. . . .

I do *not* have a deterministic scheme, I do *not think, believe* or *say* that child training methods cause culture. The first section of *Male and Female* is devoted to a discussion of HOW children in particular societies learn their culture, with special reference to the role played by differences in primary sex characters. The section is called "The Ways of the Body," not "The Origin of Culture," a subject incidentally on which I believe it is pretty unscientific to write, although one may ofcourse discuss the roots of particular cultures in other cultures, in internal changes which can be identified in some instances. I explicitly state in the book that I am using material from a distance on discussions of the body because that is easier for Americans to assimilate. So I drew my material on this particular point from 7 cultures. On all of these cultures except Bali, the general social structure had been published and considerable material on Bali had been published. Gregory's statement about the relationship between character and social structure in Bali, was finished in 1947—the ARB volume was delayed until 1949 in publication. I had published these social structure statements, I had edited Comp and Coop which stated the relationships clearly, I provided a careful bibliography of all of this material. . . .

Now how about it? You will realize that it will be much easier for me to write a flat refutation to Ortho. which they will ofcourse publish. It may take a long complicated discussion to convince you that you have misstated my position. It's a question of whether you think it worthwhile. This is in no sense any sort of blackmail. I do not intend to establish any sort of institutionalized quarrel, I will continue to back you up for things which I think you want to do, or which other people want you to do—in the Michigan case I didn't ask you whether you wanted it or not as it certainly [did] you no harm to be backed for the job in any case. I regard old friendships as quite irreversible. If it is too much trouble for you to try to understand what I said or if you feel that in any case you would not wish to make a retraction, I have no choice except to send in a refutation. In that case you may get angry at me, but I shall feel that I made every attempt to prevent it. This is not a case of the pot calling the kettle black, for I do not write irresponsible reviews.

It must have [been] almost twenty years ago—in the summer of 1930 in fact—Radcliffe Brown said, when I was fuming about some misinterpretation—I think that time it was the review Malinowski inspired which said I didn't understand Manus kinship!—"Nevermind, in 20 years your work will be recognized." My "work" then consisted of *Coming of Age in Samoa*, through the first publications on Manus. And I said: "But that is no comfort at all. The important thing in a developing science is for work to be used when it is written in order that we may reach for the next *steps*. I don't want to become a classic, later, I want to contribute something now to the work of other students."

These attacks on the work we are doing at present, these continuous misrepresentations and misstatements are going to do harm, and harm of a subtle and serious kind, not a gross kind, but harm that in the end may mean that whereas in World War II, many millions of lives were saved through a proper handling of the Japanese, the same thing may be made impossible. In the work we have done on Russia, the first attack was that swaddling doesn't exist in the USSR. When the USSR published pictures of swaddling babies, and even a manual on how to do it, illustrated with an English translation, the attack shifted to the statement that we had claimed that swaddling produced Russian culture, when all that was claimed was that swaddling provided a paradigm of how Russian babies learned their first and important learnings about Russian culture.

So, how about it? Please answer this as quickly as you can because I should like to get it settled one way or the other. Adrenalin which comes from anger against impersonal forces or even unknown persons can be stimulating and useful, but not when one's friends are involved.

Christmas love to Zunia and your young daughter.

Yours,

Mead later published a reply to his review in the American Journal of Orthopsychiatry *20, 4 (April 1951): 427–428.*

*In 1953, Mead returned for a restudy of the Manus people, with a young
American couple, Ted and Lenora Schwartz. Mead published a book from this
restudy,* New Lives for Old *(William Morrow, 1956), chronicling patterns of
change in Manus culture in the post–World War II world. This field trip began
the pattern of serious restudies and brief revisits to her former field sites by
Mead and various colleagues to document changes over time that characterized
her later life.*

*Mead's interests within anthropology continued to go in directions that were
seen as on the fringe in the late 1940s and 1950s. In that period, she became
interested in the work of a young student, Ray Birdwhistell, on nonverbal communication or kinesics, what would be called "body language" in the 1960s,
when Birdwhistell's work was popularized widely.*

*To Ray Birdwhistell, October 7, 1952, from New York, responding to his let-
ter with the news that he had suffered serious medical complications after hernia
surgery and had nearly died. Birdwhistell had suffered damage to his parasym-
pathetic nervous system during the ordeal, and Mead—who was fascinated
throughout her life by sensory differences between people—was concerned about
the possible effects on his special sensory gifts and also about the role that these
gifts may have played in his trauma.*

My dear,

My first impulse was to say Thank God for modern medicine and my second to wonder whether modern medicine had not been originally the demon
of the piece. It must have been an appallingly vivid experience for you who
are so keenly conscious of your body already, and I wonder what it is going to
mean—whether it will add to or detract from your kinaesthetic awareness. It
is such a precious gift and one you were using so fully—perhaps—one is
tempted to wonder, whether this was a violent protest from resources which
were being overdrawn. Have you thought of that?

Anyway, it would seem worthwhile to know, to watch, how your memories go, whether you have a strong impulse to forget these nightmare weeks,
and if you do, or if you don't, what happens to your memory for movement,
your own and other people's. I'm not suggesting violent explorations because

they would be hazardous with the possibility that this is some sort of desperate search for respite, but just a very gentle exploration around the edges of self-consciousness, to see how it feels.

. . . It's worth considering whether it might be worthwhile to give your body a vacation and just use your mind for a while; perhaps what you say about reading is part of it.[64] How active are your muscles when it comes to incorporating written material? Can you let your eyes do the work? It certainly seems clear enough that the very sensitive work in the psycho-cultural field is done with people who are *liable* to this sort of thing—which wouldn't ofcourse prove that doing that kind of work caused it. Ofcourse it may also just be that you wouldn't be alive if you couldn't call on the cooperation of your body in the way most people can't and that all the malign-seeming symptoms were protective in some way.

Now, please, please, be careful. Is there anything I can send you to read which you can't get there? Suggestions about categories of stuff to read? Problems to play with? Which parts of you have to convalesce? We missed you at the meetings in September, an intensely harrowing affair because of Bob's[65] death the week before, and I'd begun to wonder just what you were doing. The more I think over your thesis, the more I am convinced that it's a major step forward. . . . Lloyd [Warner] concurs. We must get it published soon.

I will try to get down within the next two months if it is at all possible. And be careful. *Interdependent* is a lovely word.[66]

Love,

To her friend, the sociologist David Riesman, April 8, 1953, taking issue with his characterization of her approach in the book Sex and Temperament (William Morrow, 1935) *in a recent paper.*

Dear Dave,

It's always such a pleasure to have a meeting with you and Evey[67] because—in addition to the moments themselves—they ramify so. The copy of your Franklin lecture came on time for me to read and get myself oriented. Everyone there was delighted with it, but I heard one point which I think throws light on your own expressed dissatisfaction. You see, you talked "about" your books and your theories—from the outside, practically, taking for granted that the audience knew them, or perhaps I should [say] you talked "around" them. Then in the question period they had a chance to come to grips with them. I think you find the paper unsatisfactory because not enough of it is given to stating your position—it's too elliptical and indirect for a new audience. You have to assume that they do not know your theory, even if they do—that is, give them a clear brief summary to hang the rest of their thoughts on. *Then,* you can talk about your general feeling about problems, etc.

Now, for details in case you use present paragraphs as they stand. I don't think it's fair to make me into a nameless representative of a mistaken emphasis, in your discussion of approaches to sex typing in the South Pacific.

. . . What I actually did was to pose a problem—how does sex membership provide a peg on which cultures can hang the process of socialization, and then I went and looked in cultures selected for other reasons, to see what happened. The book grew out of the material, and its main emphasis is on the ways in which temperamental cross-sex characteristics can be specialized to either or both sexes, including the members of either or both sexes which do not have the temperamental proclivities to start with. All of this was based on the methodological point that you must try to get the learned factors clear before you tackle the constitutional ones. I *never* argued that culture was all powerful against sex. . . .

I imagine that when you look at it again you will see that what I was talking about was not that "culture was all powerful as against sex," but that there were temperamental clues to personality, themselves cross-sex, which could be assigned to all members of either sex or to both. Once having got that point out of the way, I could go on to explore the extent to which difference in sex, biologically given, was intractable to forms of social patterning, etc. . . .

I am sending you . . . another copy of the *Moments of Personal Discovery* reprint, and . . . my obituary article about Ruth [Benedict]. I will be sending you in a few days a draft of the Bob Lamb obituary to see if I have the points about his relationship to you straight and if you have any other criticisms.

Edith Cobb was very touched by your sending her the batch of reprints. From me, also, many thanks. None are read yet so no detailed comment yet. Life is, as you might not guess from the length of this letter, very thick at the moment but free of viruses.

Love to you both,

To her former student Wilton S. Dillon, October 9, 1961, in Ghana, where he and his wife had gone to do field work on the life histories of African intellectuals. As an older student who had served with the U.S. Army in Japan, Dillon became fast friends with Mead and Metraux upon taking one of Mead's Columbia University anthropology classes in 1951. When he went to Paris the next semester, Mead introduced him to her friends and colleagues there and suggested projects utilizing his interests and talents. Soon Dillon—and subsequently his wife and then their son—became an integral part of Mead and Metraux's extended network of kith and kin.

Dear Wilton:

I've a note to myself for two months to try to write a catching up letter and each time the weight of unwritten matters seemed too heavy to lift with the time available. Now I have your letter about a variety of professional matters—and no current address—so we will call the Fund and find out. All your points duly noted.

I realize so acutely that I am so far behind—only a telegram for your Ph.D.—never a word to Virginia about the upcoming baby[68]—no acknowledgement of all your notes along the way. It has really never been as bad as this. Now it will all have to be wrapped up in one big packet of love, congratulations, and felicitations and how we miss you both.

This summer has been frantic, first with the trip to the USSR, on the basis of which I am now launched into the development of a theory of cross-ideological communication, which it seems to me is our next need, even though we have not, of course, learned to use what we know about cross-cultural communication. But the need for the next step is urgent. It opens up a whole new world of research. I suspect that in addition to cross-religious and western democratic-eastern democratic (sic. Free world and Communist world), religious-secular, etc., we may need to re-examine the whole question of cross-racial communication once race becomes embedded in national behaviors as it is pretty sure to continue to be. You might give this some thought while you are there. I've just scratched this so far, got into the WFMH statement (the second Roffey Park International Study Group on Mental Health, written a draft of a paper which I am circulating to a lot of people for comment on US-USSR communication, with the possibility of using a cybernetic vocabulary, which, when criticized by the appropriate cyberneticists, should appear in *Science* and the *USSR Science and Life* simultaneously.)

I haven't thought about race at all, but I think, for example, that consciousness of race difference reappears when every other difference—education, way of life, etc.—is ironed out, and this should be important. With an educated African, who has full mastery of English, charm and manners, I have a vivid—probably because quite guiltless—consciousness of race. Where in situations of minority status or wide educational and cultural differences, one attempts to ignore race, in order to deal with individuals, or to deal with culture—as, for example, in New Guinea—this compulsion towards ignoring race vanishes under conditions of similarity and equality. This will mean that habits learned in one context may be inappropriate in another, and it may well affect the same people—especially Americans. Alonzo,[69] for example, believes that race should be ignored in favor of all the other variables of individuality, culture, education, etc. What would happen to this belief in Ghana, and what would happen if he maintained it, if he altered it, if he never questioned it, etc.? This should be a familiar area for you to think about after your working on the size of the Japanese.

After the WFMH Roffey Park meeting, I came back to NY for six weeks, worked furiously on the ubiquitous Terry Lectures, then back to three delightful weeks in Paris, three days in Athens making a speech, and twelve days traveling about Sicily with Geoffrey [Gorer], thinking about a reorganization of European studies of cultural depth, including surviving peasant societies. Now we are up to our ears again. Rhoda sends her love.

And my dearest love to Virginia,

Love,

In 1957, New Zealand–born anthropologist Derek Freeman, of the Australian National University, wrote to Mead asking for an assessment of one of her students who had come to his department on a grant. This exchange was the first correspondence between Mead and Freeman, a correspondence that lasted off and on for over two decades. In a letter to anthropologist Morris Carstairs, July 4, 1962, who had asked her opinion of Freeman, Mead had characterized him as "a brilliant man with definite dangers toward a messianic and over-systematized point of view."

On November 10, 1964, Mead and Freeman met in Australia to discuss her 1920s research in Samoa. In Freeman's own researches in Samoa he had come to different conclusions about Samoan sexuality and culture than she had earlier. In a seminar the following morning, they had a heated public exchange over the differences in their views of Samoan culture, the repercussions of which were far-reaching.

Freeman wrote Mead on November 11, after the seminar, expressing his distress over their public altercation and hoping that there would be no continuing bad feelings between them as a result, even though he expected to publish conclusions at odds with hers. Their correspondence was generally cordial until her death and included his compliments on Mead's work as well as questions about it. Freeman eventually published volumes after her death claiming errors in her work and sparking a controversy in which other anthropologists, in turn, challenged his claims.

To Derek Freeman, December 2, 1964, from New York, in response to his letter of November 11, 1964.

Dear Derek,

As a matter of fact I rather enjoyed the seminar, although I think it is unfair to expose students who have not been analyzed to that kind of interchange. However, you asked for it, and you got it. It didn't offend my personal susceptibilities except on the grounds of taste and estimates of what other people can stand.

I have been thinking over our discussion and I think your identification of the defloration ceremony as a key plot element in Samoan culture is illuminating. This is quite aside from how seriously, or actively, the forcible seduction of virgins has been pursued in different parts of Samoa and at different periods of culture contact. This may be a problem which can never be resolved.

It is also a question of plot—as a basic cultural theme, in addition to plot in the Roheimian sense of the key infantile culturally patterned experience that partly determines the character structure of members of the culture. I do not have any material that would show (on Ta'u in the 1920's) the defense of virginity as the basic training given a girl or the destruction of virginity as the basic aspiration of a boy. But I do think that we must account for the Samoan's extreme touchiness about status and his violent response to insult, and about insult that seems extraordinarily slight or symbolic. If one said that the basic Samoan plot in pre-contact culture, with later, perhaps pathological, manifestations in various stages of post-contact culture (which would include both the West Samoan mothers' obsessive visits to the hospital and the use of chicken blood in some defloration ceremonies) was a desire to take something irrevocably away from one's rivals and a determination not to [allow] one's rivals to take the same thing away from oneself (that is the daughter's or *taupou's* physical virginity), this would make a highly charged obsessive center (a *cultural* as compared to an *individual* "complex"); see my paper, Mead, Margaret. "A New Framework for Studies of Folklore and Survivals." In *Men and Cultures: Selected Papers of the Fifth International Congress of Anthropological and Ethnological Sciences, Philadelphia, September 1–9, 1956*, ed. Anthony F.C. Wallace. Philadelphia: University of Pennsylvania Press, 1960, 168–174.

Such an intention would fit the records of pre-contact social behavior and the behavior that Samoans abroad, particularly in Hawaii, have been reported to show over the last twenty years. It is less congruent with actual observed behavior on Ta'u, in the 1920's, where the tendency to soften and rationalize conflict permeated the whole testimony on the past as well as contemporary behavior.

It will be up to you to find the analogues (in the child's relation to the breast) of this pervasive attitude toward one's own irreparable loss through the defloration of one's own daughter and one's insatiable desire to deflower the daughters of others. I agree analogues should exist. In cultures in which I have studied infancy and childhood in detail, I can show just such precise relationships, but in Samoa, unfortunately, I did not study infancy or early childhood.

I don't know what "bailed up"[70] means, but I mentioned by way of apology that I had not read your thesis, and I was unprepared for your response. Anyway, what is important is the work.

Yours sincerely,

Margaret Mead [t]

To Caroline Tennant Kelly, Mead's Australian friend, February 5, 1965, from New York, regarding all the things that Mead was doing outside of anthropology.

Caroline dear,

Thanks for the pictures. How they bring it all back. The New Guinea ones now go into a new file for Rhoda's new project. We got the grant, so that means a four-year project, this year on comparisons between Montserrat and Manus, next year she's [to] work on Iatmul notes, mine and GB's, the next year 1967, go up the Sepik for a spell.

Ted has been playing up and changing plans as usual, but I have finally I hope pinned things down so that it is certain that I will be out there next fall, although not certain whether I will meet TS in Australia or in Manus, depending on the way his plans work out. Apparently Lola won't come home until July.

I've been put on the steering committee for the new World Association for Ekistics and it looks as if I would probably be going to Delos Three next summer.

In two weeks I go to London to give two lectures at the Bartlett School of Architecture and to attend a Transcultural Psychiatry Conference. And so it goes.

I am trying to get two books finished at once, *Family*, the photographic one with Ken Heyman, and *The Wagon and the Star*, a book on American community initiative to be used overseas. Meanwhile yesterday a bathroom rug slipped out from under me, and I fell across the bath tub. It feels as if I had broken several ribs, but the X rays say I haven't.

. . . In writing the section of Grandparents in the *Family* book I realised that the great function of grandparents is to make children realise that something existed before their parents did, which in turn will make them able to take in the idea of the future, of all that will exist after their children have lived and died, how in fact the depth of the future is a function of the depth of the past. I said this in a way in *Continuities*, but now I feel it more sharply.

At present, you somehow haven't much sense of the future—only of finishing out your own life. You have never talked about grandparents—were they all left at home in England?

We watched the magnificent Churchill funeral all day on TV—all the forms of communication available in the world were used.

I am so anxious to know how things are going, what you have done about the flat, what Oscar's "plans" were, what you hear from Jimmy,[71] how the negotiations with the University of NSW have gone? So write me soon.

Lovingly,

To Ethel John Lindgren, a Cambridge-trained anthropologist who was originally a friend of Bateson's, May 14, 1965. This letter highlights Mead's concern throughout her career that no anthropological research go to waste. It was the work that mattered, above all else.

Dear Ethel John,

. . . The last time I saw Reo, we tried to find you at some point where he was convinced you would be—"right down the next street or the next"—and you weren't! I haven't been to Cambridge now for several years. Sometimes it seems a good idea to see Reo and sometimes I seem to reactivate old conflicts. On the points in Dobu, I think he was shocked to find another form of marriage in Dobu and suddenly took everything back, but it is a pity. On the whole I think Ann Chowning's review was important in keeping the status of his field work intact. And, of course, since then *Manus Religion* has come out in a handsome paperback, with, thank the Lord, no introduction or changes. Anyway, thanks for whatever you contributed to getting *Sorcerers of Dobu* reissued. I need it for teaching; it's really one of the best, most compact monographs we have.

I enclose a wedding picture of my daughter and her husband. She now has her Ph.D. from Harvard and is teaching there—as Mary Catherine Bateson, professionally—in the Department of Middle Eastern Studies—Arabic, Arabic poetry, Middle Eastern languages—and she is working on her third major manuscript. Others, not yet out.

Gregory is in Hawaii at the Marine Research Laboratory,[72] working on communication among dolphins, very glad to be studying animals instead of schizophrenic families. He made a major contribution to the theory of schizophrenia in his conception of the "double bind" which is already quoted without attribution.

My main problem is time—there just isn't enough of it. I have two field projects going, and a third just beginning: (1) Studies in Allopsychic Orientation—creating a cross-cultural model for orientation in time, space and relations to the strange and unknown; (2) A Study of Cultural Systematics in New Guinea—field-directed by Theodore Schwartz. Part I, an areal study of the Admiralty Islands. Field work will be completed this autumn when I make a second trip—first in this project was in 1964; (3) A Study in the Cultural Structure of Imagery, field-directed by Rhoda Metraux, who will pick up on G.B. and my work on the Iatmul, do field work there, and take over that great bulk of unpublished film stills, notes, etc. This will complete the successorship on unpublished field research. Ted Schwartz is the custodian of all the Admiralty research, and the Geertzes[73] will be of Bali. Then I will feel reasonably secure that no work will be wasted for lack of successors who can use it.

Meanwhile every paper I write draws on too long a life and takes too long to document. I am trying to find a way to stop this—my last book, *Continuities in Cultural Evolution*, is supposed to put a stop to it. At the moment I am doing the *International Encyclopedia of the Social Sciences* article on Incest, and finding it too interesting for the length of the piece! The summer will bring a trip to Greece, for the Delos Symposium III, and then in September to Manus again by way of Iran, I hope. Last summer I went to the Moscow conference and then across Soviet Asia. You don't say what you are doing—except I gather the reindeer are still going strong.

It is always good to have news of you.

Affectionately,

Margaret Mead [*t*]

dictated by Dr. Mead
signed in her absence

To Caroline Tennant Kelly, February 8, 1969, as Mead's life became more and more crowded with work.

Carrie dear,

The reason you don't get letters is because I always put them off "until I have time to write a real letter." Then it gets to be so long that only a catalogue will do justice to the state of the world and I am repelled by the idea of a catalogue—and there we are? Or here we are.

Anyway, this will have to do as a sort of letter, as I should be: writing an abstract of a paper for a Nobel Prize Foundation symposium to be held next September, typing out a list of Waddington's[74] bright remarks at a conference last month in Chichen Itza, [and] selecting a piece of the MAN AND NATURE lectures that I am giving at the Museum in March (this is an honour as they are usually given by eminent outsiders) for *Natural History*. And [I must also] decide whether or not I will do a brochure on sex for college students, and make final arrangements for a student who is having trouble getting permission to study a nomadic people in Kenya, and find a committee for Bill Mitchell's exams (he's the boy who was in Iatmul in 1967 with Rhoda and me), and finish plans for a conference on Floating Universities on Monday, and most of the weekend is gone because I have the afternoon gone (Floating universities), tomorrow morning, vetting the illustrations for a set of children's books, late afternoon a seminar which is going badly, and evening to look at the pictures that Ken Heyman brought back from Bali. (This for a new book he and I are planning on Four Families—Mexico, Bali, Sicily [which we did this summer]) and Vermont. Three are now done. And so it goes.

We have been having all sorts of disorganization at Fordham, where I am supposed to be becoming professor and chairman of the social sciences in a new liberal arts college, (which is currently more or less abolished), my Pacific hall is supposed to be getting done but we haven't got it back from the construction people and meanwhile the girl I have got to run the thing is getting restive and wants to go to South America, so I may be stuck for an assistant. Classes at Columbia begin on Tuesday. We are rushing all our film work through as Ted's project finally ends in May. Life is continuously interesting, too crowded and works if no one gets sick. This week I was in bed for two days with some sort of gastric virus—not the flu—which really emphasized the way my life is organized not to get sick. Two days out—and the roof starts to fall in. Living so close to the Museum, as we do now, means a continuous carrying things back and forth to the house. Fortunately this year we have a hippie, who only wants to save enough money to go round the world and is very amiable and he does lots of odd jobs. When he leaves I am going to find another; all the rest have classes and exams and careers to worry about.

Rhoda is working intensively on her Iatmul tapes, and do[es] the duplicating herself. We at last are going to have Iatmul in sound as well as in words and visually, and they have changed very little. My NET film was finished—yellow hat do you remember[75]—and I presume sooner or later will get to Australia and you will see it. It's been a strange experience—to have everyone one had ever known simultaneously watching a film that itself spans forty years—cousins, schoolmates, people's friends, relatives. But everyone has liked it, and thought it good for anthropology. Craig found some wonderful early films on the territory, and also very good war films of the landing in Manus. They did the cargo cult by reversing a positive to a negative of a real film of unloading a supply ship, which was very effective. . . .

Barkev has a three year apt at the Harvard School of Business; Cathy has a grant this year to work on linguistics and children's learning. Daniel is in Japan, and so is his fiancee, so this Christmas we were childless and planned to go to New Orleans on the way to Dallas (for the AAAS meetings) but Rhoda got the flu so we stayed very quietly at home. My quietest Christmas, for everyone thought we were out of town. So now, what is your news my dear. Your note to Rhoda just mentioned a new enterprise in housing research, and the farm, so that is all I know.

Lovingly,

To Geoffrey Gorer, May 13, 1973, while on her revisit of the Arapesh of Papua New Guinea.

Dear Geoffrey,

I am in one of my fits of wanting to keep life tidied up and not have me and all my ideas on the same plane. This trip has brought a lot of things into focus. I have been thinking . . . [orig] idly . . . [orig] about what you said about the impersonality of hate. And when I was looking at this settlement, where all my old Arapesh village descendants live, as a closely related group, and the rest of the people are people without ties to each other, loosely grouped as Chimbus and Tolais, so that payback deaths can be anyone to anyone, I realized how much more hateful it is to be killed by someone just because one is in a category, and how it makes one hate any member of the category of those who attack one as a member of a category. To be mistreated by one's close kin carries with it all the complexities of ambivalence, of bound aggression and bound libido.

Do you remember you once thought I was depressed because I had found I could hate, but then you were wrong because the only kind of relationship which can arouse hatred in me is when I am bound to another person's weakness. And I've been fortunate and not asked to hate because in most cases I have not been trapped in a kind of dependency I can't bear. It seems to me . . . [orig]—to go back to my previous letter, that what friendship is about is that it can be a complete and unambivalent commitment to a limited relationship, which just because it isn't like marriage or total companionship, doesn't imitate a parent-child relationship and doesn't contain, as I think probably all marriages that last do, an element of hate.

I've emphasized before that one of the good things about the family is that it is a weaning device, that the dependency of the child, no matter how good the relationship is, makes the child want to leave and be free. If marriages were made terminable, perhaps that dreadful love, hate dependency could be reduced, as it often is for grown up children after they get away from home. All this somehow fell together as I was read[ing] Forster's *The Longest Journey*. It's odd how a book, and a piece of field work, and a theoretical point all inform each other.

Love,

To Geoffrey Gorer, February 1, 1976, from New York, reflecting on the ending of her presidency of the American Association for the Advancement of Science and other work outside anthropology.

Dear Geoffrey,

The house is very still, Rhoda is still asleep, with the two cats competing for space on her bed—she is sleeping in the living room because her German friend, Hannah Palmerston—the friend who worked with her at OSS, do you remember—is visiting. I am acutely conscious of how long it has been since I have written you a proper letter. Although the pressure won't be quite off until the AAAS meeting is over the end of February—when at least my presidential duties are over, still life is a little more relaxed than it has been. . . .

The last two months have been difficult ones. I got back from Africa with the first bronchitis I have ever had, and quite deaf from an 18-hour plane trip with a terribly bad cold. The deafness has cleared up and I am not coughing as much, but my hair is still drab and dead, and I tire too easily. Possibly it was some kind of virus. . . I stayed in bed a lot of the time, and by New Year's was ready for a double psychiatric stretch—Menningers and Cincinnati. Now I am back from that . . . my presidential address is off to *Science*, and I have only the illustrations—slides, film tape—for it, to arrange for the Boston meeting. I will send you off a preprint as soon as we get it. I think I have said something quite new and good, but giant steps to oneself, may seem mousy to others.

There is to be this *fest* for me also at that meeting, to which I understand you were preliminarily invited in case they could raise money. I'd have loved to have you there, but I would have felt rather ashamed of money being raised for something like that when everyone is desperate for money. It was all done behind my back. There are also all the political possibilities of trouble at Boston, the National Labor Caucuses—who attack JR Reese, (for psychological warfare), Rockefeller—anyone, [Ronber][76] and advocate fusion. They also attack me. The Vice Pres. is to be the speaker and I will be on the platform. Boston is still in the throes of busing riots. Also we have our own annual bunch of leftists—always the same people, getting older each year, Science for the people, with a raised red fist.

The president also has a lot of ceremonial hospitality duties. I decided to set a good example for wifeless women, by having my young secretary—Barbara Sperling—who is warm and charming, play the role of wife, and set a good example to the groaning women around the world. But there is no doubt that people need wives—or a lot of servants. . . .

I am also involved in a big report on the dangers of proliferation of plutonium which means taking on a very formidable nuclear lobby. However it is really a lot easier than fighting the AEC, for Americans trusted the AEC,

Congress was intimidated by the physicists' jargon, but they know industry is self interested. We get bombarded with films by GE and Westinghouse—Progress for people—showing how better electric lights—specially made to conserve energy! stop mugging by lighting streets and stop accidents by better headlights on cars—and then at the end—jus one little plu[g] for a nuclear pellet—which will do it all, everything a split level house needs for 6 whole months—in a tiny little nuclear cylinder. Every effort is being made to destroy all the solar energy effort, and to associate nuclear power with good things, from saving the developing world, to saving individual lives. I am also being heavily importuned to take part in the initial presidential [campaign] but I am sticking to my position, no participation till the party choice is made, then limited support on issues and perhaps some kinesic coaching about TV.

This letter is getting too long. . . .

Love,

Less than two months before her death Mead was still making plans and thinking about travel to far away places. She wrote to Gorer, September 10, 1978, on the possibility of meeting him in Rome for Easter.

To Caroline Kelly, October 9, 1978. Mead wrote about the possibility of coming to Sydney, Australia, in late November or December to see her for a week en route to Manus for Christmas with her colleagues Barbara and Fred Roll.

Dear Caroline:

This is a very tentative inquiry of something that may or may not come off, depending a great deal on your circumstances at the moment but also on whether I am well enough to travel. . . . There is a theory of my doctors that if I stay in one place for a month with no travel or engrossing events, I should be well enough to travel by the end of November.

. . . I could spend a week or so with you sometime between the 27th of November and the 17th and 18th of December. Barbara and Fred would drop me off at some cross-point on their way to Japan and I would join you in Australia . . . It isn't necessary for me to go to Australia, but it would be a fine time for a good visit with you and I don't know when I would have another opportunity. I am trying to plan my life for five years ahead, but the future is necessarily uncertain. . . .

Love,

P.S. Write me a letter as well as cabling because all forms of communication seem very unreliable.

Although Kelly cabled that she welcomed the visit, Mead replied by return cable on November 3 that her health did not permit the trip. Mead died of pancreatic cancer on November 15, 1978.

6

Growing Family:
Kith and Kin

❖

This chapter brings Mead's relationships full circle to her extended family, both by blood and by choice. Mead kept in touch with her aunts and many cousins, even second and third cousins. She also had ongoing relationships with her two godmothers, Lucia Bell Cabot, who helped her to channel her own money discreetly to Columbia and the American Museum of Natural History for field trips in the 1930s, and Isabel Lord, as she proofread and indexed several of Mead's books through the 1940s.

Mead's own daughter, Mary Catherine Bateson, was born in 1939, and her delight in her daughter and her growing relationship with her is reflected here. Besides her daughter, Mead had ongoing relationships with many children for whom she became a godmother, and with some who came into her household. During World War II, Mead and Bateson arranged for two young girls, Philomena and Claudia Guillebaud, to leave England and come to the United States for the duration of the war. They were the children of their friends Claude and Pauline Guillebaud, and Philomena was Bateson's goddaughter. Later, after the war, she took in Margo Rintz, the daughter of another friend. When Mead and Metraux took a house together at Waverly Place, Metraux's son and Margaret's godson, Daniel, lived with them. Mead's granddaughter, Sevanne Margaret Kassarjian, born in 1969, was an added delight.

Mead had always looked forward to having children, but had been told by a doctor while in her mid-twenties that she probably could not, due to a tipped uterus and a tendency to miscarry. To Ruth Benedict, February 13, 1937, from

Bali, where she was working with her third husband, Gregory Bateson, discussing her attitude toward having a baby.

Ruth darling,

. . . We are now prepared to have a baby, if God wills it. The extraordinary thing is that I find that the mere complete willingness to have a child, the final removal of all practical objections, is apparently quite enough to make me quite peaceful. I do not feel as if my heart would break if I didn't have one, nor do I have any of the same sentimental tendencies to feel strongly when I hold a baby in my arms. If I don't have any children, there will be compensations in the form of greater freedom for work, etc. But I do not think I will ever again be unhappy about it. Evidently the thing that I couldn't stand was ever making the decision against children in contexts where they were possible. I wonder if part of the reason I am happy isn't that I set my heart, in the end, only on goals within my control. If my goal were to have a child, I would stand the risk of great frustration, with the strength of feeling that I have, but if my goal is really—as it seems to be—to do everything I can to try to have a child, then I cannot be frustrated. I know you usually say simply that I have an enormous power to rationalize and to say I didn't want what I didn't get, but I wonder if I ever really set my heart on external things. Even my desire to do good work is to do good work in intro-jected terms, it's not a desire to external recognition which could of-course be thwarted. I admit I was surprised by this feeling about the possibility of having children. I had really thought that once the decision was made, I would be in an agony of alternate hope and disappointment with the possibility of genuine frustration at the end. And it isn't so. I feel as peaceful as if every hair on my head were lying just in the place God meant it to lie. . . .

Believing in kinship as something that was not only determined by blood ties but also chosen, Mead nurtured and relied on her extended adopted family throughout her life. Mead had selected Lucia Bell, daughter of the church rector, to be one of her two godmothers when she chose to be baptized into the Episcopal Church at the age of eleven. She remained very important to Mead throughout her life.

To Lucia Bell Cabot, November 22, 1937, from Bali.

Dear Godmother,

I almost did say "dear Lucia," as the gap between our ages seems to be narrowing down so mightily. Your letter reporting everything running smooth, and enclosing your Christmas check has come safely. For the Christmas present, we thank you very much and will buy a lovely textile with it—textiles are the things which tempt us most because there is really no good scientific reason for buying them, and they are so very pretty.

And thanks also so very much for your continued help with engineering my chaotic career. . . . It was good to have your account of Mother and Mother's account of you, by the same mail, and mutually complimentary. . . . You have always been a symbol of "someone who never says anything disagreeable about other people" to Mother, you know, and she always added that she thought you had a very accurate idea of what people were really like!

Life goes on here in the same incredible style. The Balinese have [a] complicated calendar full of good periods and dead periods. On the good days, and at the good periods, everybody has feasts, ceremonies, elopes, cremates and what not, and one dashes wildly from spot to spot trying to keep up, with very little sleep, and—thanks to the motor roads and to the fact that we have three separate establishments and roots in about three others, with great changes in climate—on the scorching beaches one day, in the chilly mountains the next. We live in a motor car almost as if it were a houseboat, carrying stools, cameras, lamps, typewriters wherever we go. But for the first two seconds when I wake in the morning, I am never sure where I am, or what has wakened me, dogs, beetles, the Regent's fighting cocks, or the priest's bell. . . .

Our love to you both and best Christmas wishes,
Lovingly,

To May Dillingham Frear, January 6, 1939, from Sydney, Australia. Frear and her sister-in-law had been classmates of Margaret's mother at Wellesley, and "Mother May," who lived in Honolulu, smoothed Margaret's path on her repeated trips through Hawaii to do fieldwork in the South Pacific.

Dear Mother May,

My good angel has again been kind and arranged for me to come by your door. . . .

I hope very much that you will be there, for it is a long time since we walked together on the edge of the crater and I told you how far my life had got then. And I should so much like for you to meet Gregory. The last three years have been the happiest and most productive of my life; we work together so smoothly that we hardly have to take time to explain in words what we are going to do next. Now after three years of work, and before plunging into work again, this trip across my America which he has never seen is to be by way of a little holiday.

. . . My dear, I hope my angel will be kind and have you in Honolulu when we go through.

So lovingly,

To Beatrice Bateson, December 29, 1939, after the baby's birth. Beatrice had given birth to three sons; Gregory was the only son still living. Mary Catherine Bateson was her first grandchild and the only one born during her lifetime.

Dear Beatrice,

I am sorry not to have written you sooner about your granddaughter, but I have been a little slow getting to the typewriter, and my handwriting isn't up to conveying much except thank you and merry Christmas.

Catherine is continuing to thrive in a most exemplary fashion; she has gained 10 ounces since I brought her [home from the hospital] six days ago; she has gained steadily ever since she began to eat. Aside from an attack of malaria, I had an extremely easy time, and the baby's head is perfect; not a bit of moulding.

It looks to me as if she had her grandfather Bateson's head, Gregory's brow, eyes and mouth and chin, and his hands and feet. Her nose is the most uncertain part of her and at present looks a little like mine. She has Gregory's long fingers and holds her hands now just as he does, and just [as] he does in the baby picture of him with his stuffed cuckoo which you sent me, flexed down from the wrists. She has Gregory's quietly determined temperament also.

She is being entirely breast fed. I have a nurse here at home for these first two weeks at home so as to be sure I am thoroughly on my feet before I undertake her care, but she is so little trouble it is hardly necessary. I have made a rapid and easy recovery.

It's sad to think you won't be seeing her soon, though I don't know how much small babies interest you, and maybe by the time she can walk and talk, the war will be over. Ofcourse I am not hurt by your decision not to come. I have always wanted you to do just what would make you happiest, and the plan was made when it looked as if England would be a much more uncomfortable place for you to be. . . .

I do hope Gregory is going to be glad of his decision to come back. I am rejoicing in the thought of seeing him and having him see his baby without letting myself count upon his being here permanently, in case the war should take a turn that would call him back. . . .

I am enclosing a copy of the birth announcements as a curiosity; they are the conventional thing in this country, but Gregory said they would not be needed in England. We have no newspaper here which everyone reads, not even within one city.

A letter from Madé[1] says "Dear Mistress, will you please write and tell me where the *Tuan* is. It drives me nearly crazy to think he may be at the war. I have not written before because I was afraid you would not want to hear from me with your head full of the war. Because of this bloody war I am losing my job in 1940, for there are no tourists coming to Bali."

Elizabeth Raverat[2] came to see the baby. I wanted a representative of the paternal line to report upon her, someone who could be regarded as less prejudiced than her mother.

Affectionately,

To Margery Loeb, a friend from the Department of Anthropology, American Museum of Natural History, January 18, 1940, about her new daughter.

Dear Margery,

. . . I am just beginning to get life organized, with a table so placed that I can work and hold the baby at the same time, and such necessary points.

And I still get very little sleep at night. She shows no signs of skipping her two o'clock feeding, and she doesn't go to sleep easily after eating so it means being up around an hour and a half in the middle of the night and only very short spells of sleep. But she is gaining beautifully—in fact the pediatrician's only criticism is that my milk is too rich and she is gaining too fast. She weighs ten pounds 12 ounces now (birth weight 7–13) and she isn't quite six weeks old. I find her very charming and very moving, and enjoy her as a person. I don't feel I am learning much about babies I didn't know, but a lot about styles of bringing up American babies model 1940, and a lot about maternal psychology! . . .

To Nora Barlow, in England, April 18, 1940, from New York City, after her daughter's birth. Lady Barlow was, Mead noted, from the generation between Bateson and Mead and Bateson's parents. Mead saw her as an extended family member of Bateson's and as a part of her own extended family.

Dear Nora,

These last three months I've been composing letters to you in my head but they never got on paper. The baby wore her lovely soft shawl all through the bleak days of January and looked very snug in it. Now it is put away for her little brother. Not that there is any little brother definitely on the way, but I am trying to plan definitely to have another baby within two years, however grim the world looks.

It's been heavenly to have Gregory back. He enjoys Catherine so much and handles her as [if] he'd spent his life among young babies. He is, I think, very placid and content to be here, and feels that his whole position is thoroughly unambiguously clear, that England didn't want him and so he might as well get on with his own job. There is ofcourse the possibility that if this country is drawn into the war, social scientists would be used over here. He keeps that possibility in mind in making plans. But there is a pretty sharp contrast between the way in which his English colleagues deprecated his originality and the number of people here who are appreciative of it—and that makes a great difference. He did a great lot of new thinking during those hectic Autumn months; in fact, he is getting so used to having at least one good new idea a day that he is almost blasé about it and just takes them in his stride.

I've been incredibly lucky all through; the baby was born so easily and I've been able to nurse her (US for breast-feed) entirely, and she's been a healthy bouncing infant from the start, very tall and very jolly. Also I have a household out of all proportion to my finances or my deserts, a well trained English nurse who also does my housekeeping and alters my clothes and with it all adapts to my peculiar ideas about the care of the baby. All this because I take in also her fifteen-year-old daughter. It works extraordinarily well. I went back to work half time the first of March, and next September I'll go back full time. I have to do a good deal of lecturing out of town but I just take the baby along with me and make the lecture committees provide a nurse for her. She has no fixed habits and high adaptability already, and already drinks from a cup quite nicely. Taking her away with me gives me a chance to have her all to myself every week or so; an odd rhythm with a nurse but she seems to thrive on it.

We see your niece Elizabeth occasionally; I think she is getting a little bored with life and the latest war news has upset her rather, but Gregory made her mount pictures all last Saturday afternoon and that had a soothing effect. I don't think she has enough to do to really fill her time.

How is Jeremy? I know you're enjoying him probably as much as I enjoy Catherine. Don't tell Beatrice that the baby hasn't fixed habits or she'll be sure she is being brought up all wrong. We're getting a lot of light on the way Gregory was brought up. I'm sad not to have been able to come to England this year and not to have you see Catherine while she is small. But someday the war will be over. And meanwhile, all our love.

Affectionately,

As the bombing of Britain began during the war, families started to try to move children out of British cities and into the countryside. Those for whom it was possible tried to get their children out of Britain altogether, sending them to friends, relatives, or willing people contacted through organizations. Mead and Bateson wanted to help friends in England, but their daughter was just an infant, and both expected to be involved in war work. They decided they could do it with the cooperation of Mead's parents, since Mead's mother had wanted to take in children affected by the war. To her mother, June 22, 1940, setting out the details.

Dear Mother,

We talked over the question of English children with Dadda yesterday, and it seems pretty clear that girls will fit in better. So. We will try to bring over the two Guillebaud girls, Philomena, aged 13, and Claudia, aged about ten. Their father is a lecturer of Economics at Cambridge, a Fellow of Gregory's college (St. John's). Their mother is an Austrian, and was a very dear friend of Gregory's parents. Philomena is Gregory's Goddaughter. We will assume any legal responsibility that is necessary, and will pay for clothes, doctors' bills, etc., if you and Dadda can give them board, room and laundry. Philomena will be in High School. Dadda thought the public school might not be suitable for Claudia. However, we'll wait and see where she would be in school and then perhaps we can get a scholarship for her. If the children get here this summer they can come to us in New Hampshire until school opens. They seem to be charming children; they are old enough to look after themselves, and well worth saving. Dadda didn't want to obligate himself for more than one year, but seemed thoroughly pleased with the idea of having two little girls—and no boys—around the house. . . .

Lovingly,

To Claude and Pauline Guillebaud, August 1, 1940, from Holderness, New Hampshire. Mead wrote this letter shortly after Philomena and Claudia arrived.

Dear Claude and Pauline,

We are a very happy household here with your children fitted snugly into the big corner room overlooking the lake that we have been keeping for them. By the time that Claude's letter came we were getting so impatient and run-out of any ideas of more that we could do to facilitate their landing safely, that the letter was a real godsend. I don't know when I have ever seen Gregory as excited as he has been over their coming, and I've been waking up in the night going over lists of games and books and plans. I am going to be very envious of my mother's having them all winter; my father meanwhile objected to our plan of keeping them this summer, he is so impatient to have them under his roof. My one apprehension was that they might be very silent and then I wouldn't be quite sure what to do about them, but they both chatter to us happily and make it possible to establish a quick give and take at once. Claudia is darling with the baby, whom she handles as if she'd had a younger sister all her life. I've gone over their scholastic attainments and sent the record off to my mother so that final arrangements can be made for their schooling, and as soon as that is settled I'll write you the details. I am planning to follow your advice and give Philomena an easy year. She's enormously ahead of her age group, probably more here than in England, and she is a slip of a thing. They seem to me to have a kind of basic sureness which means that even such an adventure as this should bring only good to them.

For dinner last night we had a chicken. "Claudia what part do you like?" "I always have that at home." GB. "Both joints or just this one?" "Just that one." Drumstick deposited on Claudia's plate. She attacks it manfully with knife and fork. MMB. "Do you like it cut for you?" Claudia. "No thank you." Phil. "I think, Claudia, that perhaps you could forget your manners and . . . [orig]" with a glance at me. Then full permission to pick up the bone, while Phil. commented on the over-relaxation of manners on the ship where there were no mothers. But they are quite firm with each other and between them will keep their standards flying.

In fact, we love them.

Yours ever,

To Claude and Pauline Guillebaud, February 20, 1941, six months later.

Dear Claude and Pauline,

We were distressed by your last letter suggesting that you got so very little and such very late news from us and from the children. We have written in fair detail. They started their music lessons in November, for instance, and have had them steadily ever since, with an Austrian woman as their teacher so that they can keep up their spoken German. I hope you approve of that. There is every danger of German being cut out of the schools as it was in the last war and I thought this a good precaution to take.

I am simply amazed at the completeness with which they have settled down into my parents' home and into school. Both schools are delighted with them. Claudia is in the top form of her school and will go to Philomena's school next year. This means that they will go and come together but be in quite different divisions in the school. I spent a whole day at Philomena's school a couple of weeks [ago], lecturing to the school and to separate classes and interviewing her teachers. They are all thoroughly delighted with her. It is fortunate that we have been able to place her in a school with a good deal of flexibility and imagination for ofcourse it is difficult for American teachers to understand the difference in standards.

The English system is to put a child somewhere where the work is so hard that top grades are impossible, the American to place a child where according to its intellectual ability it will do almost perfect work. But the school has accepted Philomena's lower grades in her better subjects as a sign that she feels she knows the stuff and will not be bothered doing the examples. That accounts for her lower record in math. This term she is being moved into a much higher form with very much older pupils to see if that makes her work harder. But I found absolutely no criticism or desire to change anything in her. Her home room teacher is a charming sympathetic young woman and altogether it's a good set-up.

The school is a big roomy old country house; Philomena's home room was once the chapel and has lovely lines, an open fire, and green from every window. She clung to going to school by tram and bus for a long time, but now she is going by car with other pupils which means that she doesn't have to leave the house so early in the morning. Both children are growing like weeds, visibly if I only see them every three or four weeks. Philomena has not yet begun [to] menstruate, but is visibly developing, still very scornful of boys. They get on remarkably with my parents, who are after all of the grandparent generation, and they have oriented their whole lives and all their immediate plans around the children. Each time I go over to Philadelphia I try to find out from every source if there is anything wrong because it seems almost too good a record to be true—but there it is. It's a delight to us to get off a train and have two bright streaks of red tear down the long station platform with open arms. One of the things I admire most is their easy happy affectionateness.

I am on my way over today again to do some lecturing in Philadelphia which is the way I pay back somewhat for the children's scholarships . . . This letter has been delayed, first because we thought we would get more cabled news and second because Gregory is so very busy at present working on the morale set-up, that he barely takes time to eat. He has been fortunate enough to get into a key position to be maximumly useful over here.

One copy of this goes air mail, one boat. Some packages sent by Pauline to the children have come through I know, but I don't know details. The duty was slight, something like $2.50. In making things do allow for their increasing size. I have told them to send you measurements—almost hourly.

Yours ever,

To her mother, August 8, 1941, after Philomena and Claudia Guillebaud had been in the United States for a year.

Dear Mother,

Sorry not to be writing oftener but we really are impossibly rushed and sink into bed at night dead tired. We plan to go down the first of September and the children can stay at Marie's until their school opens. If that is before you get back I presume that it won't hurt for them to be there with Dadda and Nelly for a few days, and they can travel by Phil's[3] old route until you decide whether you want to make any new arrangements. . . .

I go back to your long letter to me. I quote: "I realize in the same way that I cannot say: "Why so insistent upon candy when so many children in the world are being starved, let alone getting any sweets . . . [orig]." However I remember a little girl who gave all her allowance to the Church and such things." What you fail to remember is that you did say things just like that to me, and in my presence often. I didn't develop a social conscience spontaneously but because I heard social attitudes expressed all around me. Now these children have been reared in a very different setting, with a much lower sense of articulate social consciousness. Pauline is working now from eight in the morning till nine at night, but she hardly mentions it except to apologize for getting so few clothes done. What contributions Claude and Pauline make, the children have been told nothing of. It won't hurt them at all to hear more about social responsibility.

The other day, both Gregory and I lost our tempers a little over table manners and said abruptly they had got to stop this and that or eat by themselves. I thought I had been much too tough but the astonishing thing was that both children were *more affectionate* the rest of the day. They can't bear to be urged, reproached, suggested to, in fact they aren't accustomed to American democratic upbringing. A clear sharp order makes them feel much more at home. They both say—and incidentally so do all the other English children who are written up in articles—that they prefer direct orders.

This summer they are working very well. The three children do all the dishes, and then Audrey[4] cleans the house while P and C practice. I have had no difficulty with the practicing since I got it slotted into the whole household in this way. They are fined without emotion for every delinquency—if the dishes are not perfectly done, all three are fined; if beds aren't made—but they have been!—clothes left in living room, don't go to bed on time—5 cts fine at once. But it must be done absolutely without any emotion or any sort of moralizing—the combination I think they would resent. I don't plan to bring up Catherine in this way, but then I will be able to start her off. These children's characters are already laid down and our job is to find ways of continuing the emphases which their [upbringing] has started. At the table, a quick incisive—"take your elbows off the table—" said the way one chases a Balinese dog, almost without looking at it and without interrupting one's own conversation, has had enormously good effects.

What they can't bear is preachment, exhortation, or really reasoning about points which they admit are valid. Their argument would be: "Why talk to us about table manners. WE KNOW we should have them, but we are lazy and careless and forget and it's grown-ups' business to keep us on the tracks." I suggested putting them alone behind a screen to eat by themselves if they didn't pull up sharp, and Claudia looked absolutely terrified. "We'd eat like pigs!" She especially feels that she needs the continuous moral support of adults. We have settled her eating so slowly too, I think. I'll tell you about it when I see you, it's too long to write.

About the extra expense for me, I can manage it, don't worry. I don't think the children have ever considered the problem of who is paying for what. They assume that they live in an ordered world where grown-up people know what they are doing and have made adequate arrangements among them. . . .

The other thing which is, I think, a mistake is to compare them with American children—especially with me. Both are very young for their age—I think even in English terms, and English children are younger than American. . . .

This letter was originally designed to tell you I could keep them in September until school started, but it has become an answer to your long letter. I must get to work.

Catherine flourishes—high social development. Spoken to by a visitor to the hotel. C. "Who's that?" Visitor. "Can you say Mrs. Wright?" C. UH? Visitor. Can you say Mrs. Wright? C. Howdy do Mrs. Wright!

I am writing Dadda weekly reports on her.

Lovingly,

***

Here follow two of the weekly reports mentioned above. The first is to her father, July 23, 1941, with anecdotes of her daughter Cathy, then nearly twenty months old.

Dear Dadda,

When her sand toys came, I told Catherine that "Grandma" had sent them but she insisted "Bompa" had. And so, it turns out, he had. They are most suitable and much safer for her than tin ones. She is now her blooming self again, and looking very well with her front lock of hair cut a little short as it was getting in her eyes. I brought her a smart slack suit from Hollywood and she looks darling in [it], strolling about the hotel grounds, looking for "pretty owers" and "pretty ladies" and "nice pretty birdies," and mooing like a cow—heard across the valley but as yet unseen except in books. You should see her shake hands from a high, slightly limp wrist with a sweet, slightly bored, "howdy-doo." I have taught her to feed herself this week; she gets a little weary in the middle of a meal and smiles sweetly, "bye by dinner" or, raising her voice, "Nanny! Take away" or leans forward very confidentially and whispers, her face just touching mine, "you eat it."

Are you still planning to go up to Wellesville? I hope Mother got off alright.

Love from all of us. Whenever your name is mentioned Catherine remarks "Want to see Bompa" very clearly. If you wanted to come up here, you could stay at the Hotel and be quite independent of our household, but see as much of the baby as you wanted.

Lovingly,

To her father, July 29, 1941.

Dear Dadda,

Yes the puzzle and plasticine arrived safely. The plasticine is still saved for the next early day and so far she only consents to take the puzzle apart and resists all who would put it together again.

Last night she woke up at ten, shouting for "my book" and stayed awake until one—much to Nanny's distress, very sociable, and greeting each showing of the flashlight with "Hello sun! Nice sun!", trying at midnight. But today she is as fresh as a daisy and no signs of her debauch, but airily puts on Phil's bathing cap and announces either "I'm going swimming" "I'm going to the country." One of her favorite haunts is a patch of tall hollyhocks, where she can watch the bumble bees going in and out of the flowers. "My bees! Want to see my bees. (hands raised to be lifted to a level with the flowers.) Hello (pronounced Hiy) bees. There nother one. Where's nother one?" "There's one." "Hello one!" "That's the bee's dinner, Catherine." "Hello bee's dinner." Yesterday she convulsed us all by going all about the room looking for "Who." "Where's Who." "Going to find Who." Her "p'ease" said in a soft sweet voice is irresistible.

Lovingly,

P.S. Asquam Hotel data enclosed.

To her distant cousin Elizabeth, January 28, 1943. In the 1940s, Margaret began to make a concerted effort to cultivate personal connections with extended family members related by blood. She became close to some of them and developed relationships that lasted until the end of her life.

Dear Cousin Lizzie,

I hope you don't find typewriting rude, but I have typed for so long that my writing is quite disgraceful and completely illegible. I was so pleased to hear that you liked my book. Grandma didn't live to read even the first one you know, and she never had a chance to judge whether after all I had mistaken my career. She wasn't quite sure that I hadn't. I wrote her long letters from Samoa but she never saw a line in print.

We were so delighted that Dadda finally made the trip out to see you last summer; he had been daydreaming about it for years, and he did enjoy it so much. I still hope that some time this winter I may make a trip in your direction and be able to stop off. I usually have to make such close connections to get back to Washington that [it] is hard to make a side trip, but I am hoping that perhaps in the spring I will be able to make it. You have always been so real to me, seen through Grandma's eyes, and I somehow feel as if John and I had grown up together, even though we haven't met.

I don't know how much you keep up with all the members of the family so perhaps you have never heard of the visit I had with Callie—whom I had never met before either—in Cincinnati last year. It was like a dream to talk of all the people who have always lived in my imagination. I remember how outraged I felt when Cousin Lida married Em Eiler (? I don't remember how to spell it) son [sic]. That old rail fence off which Em pushed Aunt Lou cropped right up in my mind.

Perhaps Dadda told you something about my very tall English husband, who likes America so much and sees things about us which only an anthropologist would notice, and I am sure that he told you about my daughter who is now three. Last week she insisted upon being taken to Church. I had taken her into Churches occasionally, at Christmas and Easter, to see the flowers, but I thought her too young to sit still. But she insisted and sat as quiet as a mouse for forty minutes before she asked, "Are you tired, Mummy?" She only spoke once before, to breathe, "Isn't Church pretty?" When we came out, she asked: "Was I quiet?" "Yes," I said, "You were very very quiet and good." "Now will you take me to hotel!"

If I am lucky, I'll get to see you before the year is out. And I'm so proud and pleased that you liked my book.

Sincerely and Lovingly, your first cousin once removed,

To Nora Barlow, April 20, 1943, from 72 Perry Street, New York City, on the complexities of the extended wartime household that she and Bateson and their daughter, Catherine, shared with the family of their friend and colleague Larry Frank and his young wife, Mary.

Dear Nora,

Your letter about my book[5] made me feel very happy. I have never quite learned to enjoy English under-statement, but when I get such a warm response as you gave, then I add to it the belief that this is probably understatement too, and feel positively exalted. We are more interested right now in trying to improve Anglo-American communication than in anything else—it may decide the next election here, and so indirectly the Peace. A great deal of planning goes into backgrounds for radio scripts and moving pictures and articles for Transatlantic, etc.

Our highly complex life goes on at such a pace that I sometimes wonder what it would be like to not be in a hurry again. We live in a great old brown stone house in Greenwich Village, with some friends who have six children, the wife being the step-mother of five of them and the mother of a child a year younger than Catherine. Larry, the husband and I, work in Washington. The eldest son works on a night shift. The Hungarian maid has a husband in Africa and does most of her work at midnight and then sleeps till ten A.M. Gregory and I have the parlor floor as [a] flat of our own—though we lived in one room of it all winter because heating is so poor now—and Catherine sleeps four stories above in the nursery next to the young Irish mother who is the psychological center for it all. That was the thing I found it hard to adjust to; I have been the psychological center of every house I've lived in since I was four. In fact I am proud that I managed it because a home, no matter how large, can only have one center.

Catherine comes down and wakens Gregory in the morning, and sometimes he sees her for a while in the evening. I get home for most weekends—it's four hours commuting on [a] very crowded late train each way. When I don't get home, it's because I am in Detroit or Topeka or Birmingham. Gregory's main job is analyzing German propaganda films and my main job is food and altering America's food habits to fit the emergency conditions, but in between we write reports on every conceivable subject and finish conferences late at night in the dimmed out Penn. station, just before I take a midnight train somewhere. We are both astonishingly well and still buoyed up by the feeling that the peace is not lost yet.

Catherine is still said to be more English than American, thanks to her nurse. What a combination of Irish foster mother and nursery school will do to that I don't yet know. She is a blue eyed golden curled child, tall and sturdy, very peaceable but subject to sudden attacks of intense, unreasonable and unquenchable desire for the motes in a sunbeam or a glimpse of some city with a strange name, like Buffalo. She has—on the whole—Gregory's mind and my emotional make-up, although she knows how to sulk like Gregory as well as how to lose her temper in a quick meaningless flash like me. Questioned as to whom she looked like, she said Mary Catherine Bateson. "And who does She look like?" "Mummy and Daddy and HER." She asks philosophical questions which delight Gregory's heart, such as "When you hurry is it too early or too late?" The principal sacrifice the war is asking from me so far is to see nothing of her at her most delightful period. She herself, I think, is none the worse for my absence.

There is just a chance that I may be sent to England for a couple of months to do some speaking. I am mainly anxious to do [it] because there are so many things I need to know in this job of trying to interpret England and America to each other. At present various interested people are collecting data on how many non-high brow groups I have addressed at your side of the water [for] fear I will be too high brow. I think I should like to come very much, but it will depend on what else develops. Gregory is now developing a plan for what he calls "culture cracking" units, to work out the main themes in modern European cultures. If that went through, I'd probably have to stay here and help organize it.

Affectionate regards from all of us,

Mead did go to England in the summer of 1943 and saw the Guillebauds. To Claudia Guillebaud, August 25, 1943.

Dear Claudia,

I have been slow to get a letter off to you about your parents because I have kept hoping for a long time to write in and the long time never comes. So here goes for a short letter, with lots of details to be told when I get back.

Your father is so like you—even to the tone of his voice—that it was startling, and I felt warm and comfortable with him at once. He is tired from all his heavy work with trade boards, but I don't think you need to worry about him. His rooms in St. John's are lovely big ones which your mother has furnished beautifully, and he has bought a new bright splashy modern painting which goes over his mantelpiece. He spread all your pictures out in a row on the window sill and looked at them and again and again, saying "So that's Claudia!" and then he would tell me another story about you.

Then I went up to see your mother, who looked just as surprisingly like Philomena, and was much slighter and gentler than I had expected from all this talk of her magnificent rages. She has a nice little house with a real garden, hedged with artichokes, and two real friends who work at the same big hostel, and love her and look after her—one of them in particular, so you need not think of her as too alone, although she is ofcourse, simply living for the time when you and Phil come home. Then they will give the tenants in your home a month's notice to quit and you will all move in there again.

One thing seems very certain, your father has been much easier in his mind since he could be sure that you both and your mother were a long way off from the bombs which don't fall but which might fall, and it makes him happy to know you are safe.

You asked me to find out about his next book—it will be the edition of your great uncle's writings.

As for you, when you get home you will go to the Pearse, and be happier there, I fancy, the harder you work this year. Isn't that a more [positive] note to end on! I think I convinced them both that you are still very much you—which is what they wanted to know most.

Lovingly,

To Bateson's aunts, Florence and Frances Hermia Durham, August 16, 1944.

Dear Aunts,

Your welcome long letters to Cathy and me have just arrived, and as I find the only time to answer a letter is immediately I get it, even if that means the answer must be short—here goes some sort of reply. Cathy is up in the country so will have to dictate her reply later. Her head is still full of your garden and the donkey so she will be specially pleased to have news of him.

I can't tell you very much about Gregory's job, as it isn't the kind that is talked about.[6] When last heard from he was at Kandy. Poor Cathy, when she explains that her Daddy is in a town called "Candy," people think she is joking. He is enjoying his work very much and as a side line making useful observations on British-Indian inter-relationships. I was afraid they would exasperate him, but instead he is enjoying analyzing them, but he is beginning to think that there has been some drastic change in the public schools in the last twenty years, if he can judge by the younger men who go out. That may merely mean he is getting middle-aged.

I'm glad you found the Penguin. I would have sent you one but I got the idea you had had the American edition. I am sure I sent it. Iso and Hedwig have some spare copies stored at the Mill House, if you want them for anyone.[7] My observations on the relations with the Americans have been embodied in a good many pamphlets, etc. official and unofficial, and I am on the whole quite well pleased with the results of the visit. . . .

This summer I have been directing a six weeks summer school at Wellesley College, and it was anything but a vacation, although it was cooler than Washington, and I had Cathy with me the whole time, and Philomena Guillebaud up to near the end. I sent Claudia back early in the summer and Philomena is now on the high seas—a lovely young creature whom I should have been glad to keep for the whole of her education. She enters Girton this autumn. So now our household has shrunk to only Cathy and me; my parents, who have had the Guillebaud children in winter are moving to a tiny house as my father retires this summer, and I have only perches here and there but no real home. There is no telling when Gregory will be back—he's been gone six months. But the war has been going on so much longer for us than for most Americans that sometimes I feel as if I were talking of the earlier war. It's been hard to pull my thoughts away from England this summer—with the doodlebugs[8] so badly reported in the press that we never knew just how terrible they were—and get on with the business in hand here. Cathy will answer her own letters when I see her next. Keep well and soon you will be living in a world in which free movement is possible again.

Affectionately,

To Philomena Guillebaud, in England, December 12, 1944, from New York.

Dear Phil,

As Christmas gets closer we all get more lonely. It's hard to realise it was only a year ago that you and Gregory built blocks on the floor on Pine Street. Mother says that she gets so lonely—even in their new little house—and longs for you back again. She has probably written you about the house, which is really just right for them. The stiff little colonial furniture looks right in the low ceilings and lit by the small paned windows of the tiny little house, and it's just of a size for two people not to rattle in. Father has taken an interest in furnishing it and Mother—after spending forty years trying to make him "take an interest" isn't sure that she likes his taste. I have managed to divide ourselves up for the holidays as follows: Christmas Eve with Marie, Christmas breakfast with the Franks so that Cathy will have other children to open her stocking with, and Christmas dinner at home. New Year's we go out to Jane's. But we are a very small family indeed. Mother sent you a batch of pictures of you and Cathy, which Jane took last summer. I have the rather too brilliant picture of you framed in my office, and I carry the set with Cathy around.

My life has suddenly gotten more comfortable—physically speaking. Marie vigorously attacked the Perry Street domain while I was away and in California, and got the brass shined, the bureaus redone, etc. and Mary had put curtains on the glass doors, and there is a gas grate in the front room— we used it for Cathy's party last week. Whether the fuel supply will be adequate to actually use the gas much remains to be seen, but it's pleasant to be able to use it once in awhile. And in Washington I have half a house, no less, in Georgetown, one of the tiny little remodeled houses, very neat and clean, with white curtains and morning glory wall paper, shared with the wife of a man who was Gregory's boss and is now overseas, who has a car and drives me to work every morning. All quite exceptional luxuries for me.

The news from Gregory continues to be good; the latest is a picture of him patting a mongoose. He wrote me in September the elephant story he wrote you in June, and began the story "the other day . . . [orig]." He just has no sense of time at all. Wellesley people all ask for news of you. Chick is to be director next year, I am bowing out and feel like the mother of six daughters who has successfully married off one at least. Audrey came down for Cathy's birthday looking conspicuously handsome with great savoir faire. She apparently has a "gang" where she works and is really happy with a group of girls as she hasn't been for years. Staughton[9] also came to the party at Cathy's request and was gaily friendly and unembarrassed by Cathy's outspoken devotion.

Christmas love to all of you, dear, and we are lonely for you,

Lovingly,

To Philomena Guillebaud, on the same day, December 12, 1944. Margaret wrote two letters to her, the one above and the one below, responding to Philomena's frustrations at readjusting to her home and family. Philomena was about eighteen years old at this time.

Philomena, my darling, I have written two letters in case you want one to share about in the family. I was so unhappy about the state in which you wrote your last (!) sic, only letter to me . . . [orig] or the only one which I have received. I know you wrote in a fit, but the thing that worried me most was the tic. Is it really a tic, and how bad a one? Perhaps you had better write more and more violent letters to me and get it out of your system.

Part of it you know, is that you and your mother are so much alike. And part is the inevitable conflict between radical youth and conservative age, such as I had with my father over everything, labor, international affairs, race, etc. Funny, he has now come completely around on the Negro question, whereas even ten years ago he was impossible. But I used to feel as if I were battering my head against a stone.

As neither your father nor mother have written me a word since you got home, I can imagine that they are in some [degree] feeling the repercussions of adjustment also. Then you lack your mother's almost inexhaustible driving energy. I imagine that is one thing which perplexes her to see someone so like her and yet so unlike. It's rather like the difference in energy between Marie and Priscilla. Marie, a near invalid but indefatigable, Priscilla not ill but always nearly tired. I don't think I ever pointed that out to your mother but it might be a point worth making.

Probably it's all worth thinking out as well as blowing up about. Anyway, write me any time you want to blow up. After all I have seen a good deal of both your mother and father and will probably know what you are talking about. And always I love you and miss you. You were a great delight to us and we would like to have kept you always.

Affectionately,

To Nora Barlow, July 8, 1945, from the Franks' summer home in New Hampshire.

Dear Nora,

. . . No word from Gregory for almost three months, and his last letter sent by hand was on the eve of some rather doubtful sort of expedition into what my insurance agent used to call "them interiors" with a very doubtful shake of his head. His boss—here at this end—has a young and pretty wife who has been working for me, and she has set her husband finally to cabling in search of his whereabouts but all to no avail. I am not very badly worried, yet, because really bad news travels fast, but I do feel cut off.

As for myself I am beginning to feel the whole impact of these three years of utterly disorganized living in suitcases and in between farewells, in other peoples' houses trying to adapt myself to other peoples' ways. There is still another month of it, in addition to a month when I teach at Vassar, and then we cut the cooperative household more or less in two, and I turn my first floor into a "self contained flat" and take on the care of Catherine, not very exacting, but continuous. It's a funny civilization we've built up in which people feel they oughtn't to live together unless they are kin or in love, that a multiple family is somehow an insult against nature. Cathy is such a definite child that I don't think it has hurt her never to be the center of the household, but sometimes I wonder. She's a good and philosophical child, with quite heavenly hair. Not that curly hair hasn't become a liability these days, never looking twice the same, when the fashion is all for setting up a hairdo and buying one['s] shoes to match. . . .

Affectionately,

To Philomena Guillebaud, May 10, 1946, from New York.

Dear Phil,

I feel very guilty over this dreadful long lapse in writing especially after you had such a burst of being a good correspondent. You can add up the reasons yourself, that Gregory wanted to write, that maybe we were coming to England this summer, etc. However, we have now definitely decided that we won't come until next year. We have taken a cottage near the Franks— on the road to Meredith, and will have three months in New Hampshire. There seemed no use in adding ourselves to the complications of life in England, especially as we have so very much work to do.

This winter has been very full and rather hectic. Gregory took several months to shake himself back into civilian life and get into a scientific groove again. Cathy had to get adjusted to having two parents—she alternately killed each of us off in imagination, but then came to the decision that the one she kept would marry again and provide her with a step-parent so she might as well keep the pair she had! She has been kept on a sort of assembly line, sometimes at a friend who has another child the same age, sometimes at Marie's, etc. when I had to leave town. Gregory can feed her but finds two braids a little trying.

There is still far, far too much to do. Although I travel very little, still engagements pile up. Our concern over the state of the world exceeds our concern in 1940–41, as the situation seems far more dangerous to the whole of mankind. That means that certain sorts of meetings and plans have to take priority over every other sort of engagement. Meanwhile I made the dreadful mistake of deciding to try to be active on what Washington calls "the local level" just to explore what it was like to move down from the "national level" to the "local level." I have found out, a[t] tremendous cost of time and energy, that the lower and more local the group in which one works, the greater the likelihood of one's being more skilled and experienced than the others, and therefore the greater demands upon one's leadership and time. All this was tried out on Cathy's new school, an experimental cross racial, cross class, mixed religion and nationality school. Now I have to stick by it at least through this year—then never again. I return to the national level and the company of my peers where I work with others equally or more competent to do a job than myself. And maybe get some sleep.

. . . We still live in happy symbiosis with the Franks, borrowing from each other's ice box, and swapping children. There was a horrid few weeks when the city threatened to pull the whole block down to make a play ground, but that is now past. Housing is terribly tight in New York, and we feel very lucky and practically placed forever. But we have hopes—after this year—of spending our summers in the British Isles—perhaps in Ireland, and perhaps running a very small intensive kind of social sciences summer school, to which you would ofcourse be a great addition.

The threat of Father visiting England is now definitely past. I know you will be glad to know. He is engaged in some new teaching, short courses for bankers, and has decided that the book he planned to write would not contribute much at this time. I imagine you pretty thoroughly dreaded his coming, but it was a day dream that I thought would do him more good than it would you harm. I hope I was right.

We have had Margo for vacations this year. Frances is in Greece for UNRRA. Possibly Margo may join us for the summer. She is a more contented visitor than you ever were, just curling up silently with a book, plus cigarettes, gum and other forbidden pleasures, but not as much fun otherwise to have around. Her being here makes us miss you more. She may live with us next year, if Barnard is the only college she can get into. It is almost impossible to get into a resident college. She had intended to go to Wellesley.

Wellesley goes on this summer with Chick as director, and I go down for a piece of each session. Then Gregory, Cathy and I are to spend a week at Vassar together.

Now a detail or so. I sent you a food box for Christmas. Did you ever get it? Then the end of March I sent you both some nylons, which were taken over by a friend and mailed in England. I hope they got there. If I was perfectly certain exactly what you wanted, e.g. material for some garment, or suit, blouses, underwear, etc. with size, color, etc. I could get them for you, but everytime I think of shopping I am pursued by memories of how difficult you were to please, plus your mother's annoyance with the things Claudia brought back. I imagine it is no longer against the law to send specifications. I think Mary would enjoy shopping for you, or Claudia, and if you are worried about asking for things, you can pay us back in England. Either way is alright with me, only I don't like to think you need anything which we could give you, easily, if we knew just what to hunt for. Shopping conditions here, except for stockings, are about as they were when you left.

We weren't very excited by your "good second," which means ofcourse that you are still vacillating as to whether to be an intellectual or not. The middle course is not very satisfactory—which is what you seemed to be saying, and why you felt in such a muddle. If you don't want to be an intellectual, have the courage of your convictions and *don't* be one. We love you very much and we wish you were here to make life seem more complete.

Lovingly,

To Gregory Bateson, May 28, 1947. When her marriage to Bateson began to come apart, Mead was concerned that their daughter not become a casualty of the breakup.

Gregory, this is about Cathy. She woke up this morning saying "When am I going to see my Daddy?" I said I didn't know but sometime soon. Then she said, "I want to send him a post card." (enclosed. The choice was simply from a set I took home yesterday so that she could send a post card to Steve, and not her primary choice for you.) When she came upstairs, she stood in front of the double picture, and said, "That's Daddy when he was a child and that's Daddy grown up." I asked her if they seemed like, and she [said] yes, but different too. Then I showed her the earlier picture, the imperious one with the lily, and told her that was the one I'd always had to think about what my children would be like. She said that the lily one looked like the stuffed bird one, and the stuffed bird one looked like the grown up one but the lily one *did not* look like the grown up one!

She wrote the postal card herself, and you will see it follows the pattern that other people have written to her on post cards. (I was standing at the window spelling it out and watching for the taxi, and I let my voice break in spelling n-i-c-e, as I had my back to her. She was at my side in a second.

"Why was your voice different then? It sounded so sad?" So you see what thin ice I have to skate on every minute.)

We had a conversation this morning about the air raid demonstration and she said it didn't make her think of bombing, only flying. Then I asked her if she'd heard any talk about war with Russia. (This was a lead question because I know she has.) "Yes." "What do you think of it?" "It's all nonsense." "What do you think the people who don't think it's nonsense, think about it? "They think the Russians think that we are not going to help them get things for their children that they need—like tooth brushes!" How's that for good schismogenetic analysis?

So, I think it would be a good thing if the next time you come in town you could see her. Friday is a holiday and she will be at Marie's then and also Saturday and Sunday. You could take her from Marie's without Marie wondering particularly, more than she is wondering now. I am trying to keep Marie's knowledge down to just wonder as long as she is in contact with Cathy this spring. If you could spend an hour or so with her, perhaps help her load her camera (films and camera are on her shelf in my bedroom) and take her first pictures, and think over the fish, etc., it would be good. I think she is simply lonely, and simple loneliness should be easy to arrange to still.

mm

To her daughter, Mary Catherine Bateson, eight years old, August 10, 1948, from England. When writing to children, Mead was always sensitive to their age and tried to write in language they could understand, filling the letters with details tailored to the interests of the child involved.

Cathy darling,

Aunt Marie sent me a whole bunch of snap shots of you swimming and diving, and the one watching the chipmunk which I liked best of all. You look very tall in all the pictures.

Aunt Mary has come back from Ireland and she and Uncle Larry have gone away for a week in the country, and I have come down to Uncle Geoffrey's for four days in his cottage. On the train coming down, we saw something very interesting. Far up on the sides of the hills, you can see the pattern of a great white horse—several of them. People who lived in England long ago must have made these patterns by cutting the top sod off the white limestone underneath. Grass and bushes and trees have grown up all around them, but there they remain. Perhaps someone—country people who have remembered, do a little weeding. Ask you[r] Daddy to tell you about the temple in the woods which we saw in Java, temple for gods who had been forgotten long ago, so people said. But in front of the stone statues we found fresh little offerings of flowers and leaves. The statues were like the three little stone ladies on Aunt Jane's[10] mantel piece.

There was an enormous storm in England two nights ago. The wind howled mightily, the oats were beaten down in the field, many of the flowers were ruined before they bloomed. But it is all so wet and green that they have fresh blooms on them already. Have I ever told you about English roses, which have a kind of glowing light in their petals so that you can still see them when it is almost dark.

Yesterday I drove past the place in the park where you went with Philomena and the little Spanish girl last summer, and played in the short fat yellow children's boats. Do you remember?

The conference that Uncle Larry and I have been to is all finished, and all the people who didn't know each other two weeks ago but who have become friends by living together like a family for two weeks, said goodbye sadly. The young priest went back to Ireland, and the Dutch to Holland and the tall young Swede who had sharp blue eyes that looked out very coldly and clearly and a face that never changed no matter what he felt, went back to Sweden. And I went back to London and then to Somerset and found it very restful not to be among so many nationalities at once.

Outside the window now there is a great bush of bright pink flowers which we would call hibiscus in America, which would look lovely in your hair. Your letters sound as if you were having a very happy time, swimming and picnicking and reading poetry. And there is a whole half of the summer left, or almost half. So be happy, my darling.

Lovingly,

While Margaret's godmother, Lucia Bell Cabot, gave her advice and helped her to funnel field trip money, her other godmother, Isabel Ely Lord, became a trusted editor of her books. To Isabel Ely Lord, November 15, 1948, discussing changes needed in the manuscript that eventually became Male and Female.

Dear Aunt Isabel,

. . . I have done a good deal of the reorganization since I talked with you and I think now what the manuscript needs most is cutting up some of the too long sentences, and elimination of some illustrations, perhaps. You will remember it is to be done in English spelling and pronunciation. I suppose that glossary as well as index may be left till the proof stage. Will you mark the words which you think need glossary statement, including obscure geographical words, unusual technical terms, as well as native words used more than once.

I have to lecture every night until Saturday, and I will be in Chicago all day Wednesday, but I will plan to be able to come over pretty quickly when you need me.

It gives me such pleasure to put this manuscript in your wise hands in which I have such faith.

Lovingly,

To Philomena Guillebaud, July 24, 1949.

Phil darling,

You don't know what a great pleasure it gives me to picture you at Salzburg, walking along the paths on which I used to pursue Cathy, or pace furiously with Clemens,[11] or just wander by myself, or listening to lectures under the cupids. Have you discovered the little secret stairway in the library lecture hall, opposite to the cupids, which turns on itself? I'd love a long letter about it all, but if you are feeling lazy or busy, don't bother, for I shall see you in October and that is not too far away. I am going over to give 4 weeks of lectures at Birmingham, two a week on successive days and the rest of the time I can go a visiting. I'll undoubtedly spend some time with Geoffrey in London, I mean to go to Paris over one weekend, and in general get a little caught up with the world.

I spent the day in Philadelphia with the family. Mother had another slight stroke which rather undid the damage done by the last one, so her aphasia is almost gone, she is more herself, but also more restless and impatient of convalescence. She says she had a very sweet letter from you but that you didn't seem to know she was ill. She may just not have read your letter straight—but did you get my letter telling you she'd had a stroke in January, a pretty bad one involving her right hand and speech? Sometimes she is extraordinarily gay and sweet, at others rather bitter. Father, in taking over more responsibility, has started to grow up a little.

I hope you have told the Kluckhohns[12] who you are in Gregory and my life, for they are great friends of both of ours. I wonder if there is anyone there whom I know. If there is, give my remembrances of 1947 to all of them. My love, my dear, and I'll see you in October, if no letters get written before then.

Gregory and Cathy have just been climbing California mountains, now Cathy is back at Cloverly, Gregory and I each back at our respective and continent apart jobs.

Lovingly,

To Mary Catherine Bateson, ten years old, August 15, 1950, from London, England.

Cathy darling,

Your picture is smiling at me from my bureau, and it makes me happy to know that in about a month you will be coming home. But you and I have a lot of vacation things to do between then and now. You have another ten days of camp, and then the trips and arithmetic with Daddy, and the flight across America. I have five days in Paris, and a flight to Italy, a speech to make in Italy, four days with Clemens and Mathilde [Heller] going to a beautiful French fiesta in Brittany, and then the flight back to the U.S., and then ten days in New York, and a meeting in Michigan before you get back. It doesn't really seem as if all of that, including for you a year's arithmetic, could get tucked into such a little time, does it? Oh, yes the Russian report has to get revised in that time too.

Uncle Geoffrey's new house[13] is now very much set up with a great number of animals. He has a charming dog named Meg who looks like a Thurber dog and gets very easily attached to people. There are two cats and two other dogs, besides the cows, pigs, ducks, geese—everything in fact except horses. . . .

Philomena is coming over in September, will get there about a week after you get home. She is going to go to Columbia and learn Russian so we will have a grown student around the house again. She is going to bring *two* guitars so that you can learn to play duettes with her.

Mrs. Gorer has given me a lovely real jade bracelet which is too small to go over my knuckles. I am just hoping that you'll be able to wear it for a year or so. Then it will have to become a buckle.

Have fun my darling,

Lovingly,

To Claude and Pauline Guillebaud, October 12, 1950, from New York, after the arrival of Philomena.

Dear Claude and Pauline,

This is a bread and butter letter from a hostess, or something equally unallowed for in our society. But it is to tell you what a great delight it is to have Philomena here. It's really my first glimpse you know of any permanent rewards of child rearing, and I feel I do a sort of 1/6 of the clucking over this particular chick, and at the same time send my vivid thanks for the other 5/6s. I feel so sharply the contrast between Philomena as she was when she left America, and this delightful, gay, poised, responsible girl who is helping me turn this house from a rather forlorn nesting place for a mother and child, into a full going household again. She takes hold of everything with such a will and such energy; Cathy and she appear to be entranced with each other. Cathy's one desire is to have something happen so that Philomena has to sleep in her room, and Philomena has dreams of my going away and leaving her in complete charge. My secretary is enjoying working with her—she works in my office a couple of days a week—and altogether life is very rosey, and I thought you deserved a proper thanks for lending her to me this year. I am especially pleased with her gaiety in a group playing the guitar . . . [orig] which is *echt* Pauline. Her economics professor pauses in the middle of his remarks if she frowns to ask her father's daughter anxiously, "Isn't that correct?"

And Cathy's thanks and mine for the two charming spoons. Whenever we have a big ham now, Cathy dutifully scoops my marrow out for me.

Affectionately, and thank you,

To Claude and Pauline Guillebaud, January 31, 1951, from New York.

Dear Claude and Pauline,

I know Philomena has written you about the divorce and Gregory's remarriage, but I think perhaps you should have a letter from me also, if only to live [sic: leave] the way clear for easier reference in the future. I had so many, many letters to write that when Philomena said she wanted to tell you, I said to go ahead.

Having her here has made it a great deal easier for me in many ways; her unshakable affection for Gregory, her warm ties to Cathy, and her tie back into Cathy's English inheritance, combine with her sweet responsibleness to make 72 Perry Street seem less lonely. Ofcourse it has been lonely for a long time—always really—because we moved there while I had to be away during the war, then Gregory was overseas and the year and a half he was back he was never there in his heart, but away somewhere fighting old battles never settled at the Merton breakfast table. I keep hoping desperately that it will have been possible to break the chain of heartbreak and misfortune that seemed to come from Beatrice's family into the lives of her children, and that Cathy will be safe. It's hard to know. She is such a sunny steady child now, taking Gregory's new wife as one more good experience, looking forward to his having a baby, being extra sweet and gentle to me, that there seems good hope of it. What this all means for Gregory it is hard to tell. He has a good if rather odd job, teaching psychiatrists in the Veterans Administration Hospital, Palo Alto, California. I know practically nothing about his wife except that she is some fifteen years younger and said to be farm bred—a good background in this country, likely to provide patience and courage.

Philomena is all through her exams, and off for a few days holiday in Cambridge before getting back to work. She is managing a complicated schedule of classes, part time work and home responsibilities, with more invitations out than she has time to accept and is, I think, inspite of the troublesome wisdom teeth, in very good shape.

Thanks so much for the wedding photographs which everybody enjoyed. We have just celebrated Father's seventy seventh birthday, looking at coloured moving picture of Mother and the grandchildren.

Love from Cathy and myself,

Affectionately,

To Geoffrey Gorer, December 13, 1951, from New York.

Dear Geoffrey,

. . . I am so glad Cathy's letter pleased your fancy. She has simply leapt into puberty with her whole self. . . . Her conversation reads like a study outline on "Your daughter in her teens," and there isn't a point she has missed, biological or cultural. Half a year is condensed into a few hours, then that one is over. It fits with her old trick of getting four teeth at once, learning both types of toilet training within 24 hours, etc., but it is very odd to experience—definitely "the word made flesh." It's pleasant that she is pretty and endearing through it all, with a repentant kiss two minutes after a 180-degree mood swing. . . .

To Marie Eichelberger, September 2, 1953, from her second field trip to Manus. Here, Mead, almost fifty-two, reflects for the first time on becoming a grandmother figure.

Marie dear,

This is to welcome you back and hope that it has all been as good as you hope.[14]

Tomorrow is the 30th anniversary of my first wedding. I had never quite realised that I would have been past my silver wedding. I am literally a grandmother in this village. My contemporaries are shrunken little old crones, the ones who know about cutting umbilical cords and singing dirges over the dead. I think I can find out all about being a grandmother here—in its wider significance—in a way that I won't find out from Cathy's having children because all that will be too special and individual. But when I have to sit up fighting for the life of a child whose mother or father I held in my arms, or forgive a churlish woman whom I remember as the ugly girl whom everyone had decided was a thief because she took one banana—when I constantly see in the children the features of the parents—not always their parents—the children of yesterday, then I get some hint of the role of grandparents in the world as the people who can forgive everything without effort, who no longer need to urge and judge, reward and punish. I hope that somehow this will give me some hint on how to handle the great distance between myself and today's students. Things with the Schwartzes are going very well at present. They both have all the necessary gifts, they just have to learn so much.

I am keeping extraordinarily well. I am brown in all the places the sun gets to which means I look rather spotty. I've had one headache—from getting nothing to eat until 10 o'clock one morning because of following the early morning fishing, and no colds, and only 2 days of fever. I have only wakened up with eyes that were heavy and didn't want to open about three times since I've been here and that has been when I've been up almost all night. In many ways I can do so much more than I could 25 years ago. Unfortunately ofcourse our sights have been raised since then. It's still a problem how to get the work done I want to do, and at the same time really participate, as I have never done before, participate and care, about what happens in the lives of over 200 people, all of whom are within earshot. Food is healthful and adequate if somewhat monotonous. I live on cucumbers so I am afraid that is one delicacy I will not be begging for at your house. However I look forward to everything else your house has to offer me. And I do hope all has gone well and you have come back with full eyes.

Lovingly,

To Mary Catherine Bateson, thirteen years old, September 14, 1953, from her return field trip to Manus.

Cathy darling, It was lovely to get your letter from Florence and know that you really had enjoyed your summer and found Italy enough to plan a honeymoon there. I am so glad it has turned out well.

Today I got up with a sort of it's autumn, there are leaves to be burnt and soon the children will be going to school feeling. It is exactly the middle of my trip and I feel very pleased that it is the middle—there is a lot to do yet, and it will be lovely to be home again. So I put up a lot more pictures—old ones of me here 25 years ago that they hadn't seen yet, and new ones of you and Mary and Kevin and Rhoda and Daniel, to reassure me that you were all there waiting for me. And you should be back in New York about now, and in a week or so starting back to school.

Will you telephone Aunt Marie when you get this letter and write Bompa a note and give him my love and say I will try to write soon. But it is very hard to write letters because my neck is always aching from typing notes and I am at least a million notes behind, and no matter how hard or late I work, I am still behind. It's ten minutes to 12 now and I've been up since 6 and the thermometer registers 96. I sit and look at your pictures and try to imagine you here. And they say, "yes she has seen pictures of herself taken often ever since she was a little girl and so she knows what she looked like. But we have never seen any pictures of ourselves as children and when you bring them to us we are not like Maria Katerina, we can not recognize them. If you had hidden her pictures from her she wouldn't be able to recognize them either."

Goodnight my darling, be happy. I love you so much.

Your

Cathy, fifteen years old, was in France for the summer, in the care of Mead's friends the Hellers. Cathy exasperated her mother by neglecting to write. To Mary Catherine Bateson, July 22, 1955, from New York where Mead was staying with Eichelberger.

Cathy darling,

I am beginning to worry at not having a letter, even though I try to work out a schedule by which you would have waited a day or so after getting to Brittany to let your first impressions form. Please write at least once a week, to me until I leave on August 13, and then to me at the Hotel Hilton, Istanbul, Turkey, until you come.

And please write Geoffrey and give him your address so he can make plans.

It continues to be boiling here. A pigeon got into the apartment the other night—don't tell Martha, she knows they do and hates it—and left calling cards everywhere. The book is getting done, inspite of casualties. . . .[15]Aunt Fanny has been shipped off to Aunt Mabel. My new secretary is settling in.

Have a beautiful time, darling, and please write. Your second set of books were sent off last week—that completes the lot.

Lovingly,

To Mary Catherine Bateson, August 8, 1955, telegram.

PLEASE CABLE COMPLETE REPORT YOUR PRESENT STATE HEALTH AND HAPPINESS.

LOVE,

MUMMY

To her Aunt Fanny McMaster, September 27, 1955, from New York. In her seventies, McMaster could no longer care for her foster daughter with Down syndrome. With Mead's help, Sally had been situated in a school in Texas.

Dear Aunt Fanny,

I was simply delighted to get back and find that everything had been so happily settled, including Judy's[16] full participation and your beginning to fly. It's all most excellently contrived. And I don't think you should mind Sally having cried, she ought to cry, it shows she's a loving child, sad at parting with someone who has been all the world to her. It would have been something to worry about if she hadn't cried and clung to you. Loving one home is the way she'll learn to love another.

And I know how empty you must feel, after these long years of responsibility. I even found I was feeling a little empty myself when I found myself in London, the last conference over, my book done, no house yet for my mind to attach itself to, Cathy happily off with friends—a temporary sort of vacuum. The holiday was a good one—if a little long. I really am a poor tourist. But Cathy ate it up, and I found a great deal to enjoy very much. In fact Cathy ate it up so hard she practically [gave up] eating . . . [orig] solid food. But she did enjoy her summer.

As I understand it Beth arrives on Monday and will be greeted, and wined and dined appropriately. Priscilla and I will share the honors.

All my love,

Mary Catherine Bateson, almost seventeen years old, had gone with her mother to Israel. She wanted to stay and take language classes. This turned out to be the time of the Suez Crisis. To Geoffrey Gorer, November 5, 1956, explaining the situation.

Dear Geoffrey,

. . . Cathy is at present staying on in Israel. This is what she wants to do, and Gregory and I have given her permission to stay, subject to the Palgis' continuing judgment . . . [orig] that is, it is not a blanket permission for an indeterminate future. There is also the possibility that all Americans will be ordered home. But she seems to me old enough to make her own decisions on a matter like this.

I feel rather weak and dizzy from the unprecedented events of the last few days, better than we could have hoped for in Eastern Europe,[17] perhaps better in the Near East. It's strange how localized one can become. In talking Cathy's decision over with Gregory over the telephone—made possible by Betty[18] calling me the other night to ask for news of Cathy—he said, "I don't suppose this is the big blow-up," and I answered, "Well they are massing in Jordan," and never even thought of the atom bomb. Somehow the bomb is an over-arching concept between vague powers, while this whole thing is so local in feeling.

Would you mind telephoning Pat Llewellyn-Davies . . . and either Nora or Erasmus Barlow. . . . and tell them that we have heard from Cathy, she wants to stay, and we are letting her stay for the present. If it's too much trouble for some reason, don't, but I know I won't have [time] to write to them this afternoon, or perhaps not for several days. I am just unGodly busy. I have a wonderful new secretary, but one secretary can't manage my life now. . . .

I manage pretty well mostly, but I find it hard to go into Cathy's room.

Love,

To Mary Catherine Bateson, seventeen years old, in Israel, May 11, 1957, from New York.

Cathy darling,

Your letter with your summer plans arrived the day after I sent my cable, and as I haven't had any answer to the cable I assume that you figured that it would. It made me so happy. Perhaps your letters wouldn't be quite such a delight if they weren't so rare but I rather doubt it.

I do realize that your grades are MAGNIFICENT. I am having your school head's letter to me copied and sent to you.

Now about plans for the summer. . . . I could meet you anywhere you liked between the 20th and the 28th [of August], and then you could visit the various people you want to see in England, and we could fly home together—or not—as seemed best, around the 10th [of September]. I'll be either going from Copenhagen to wherever the Hellers are, or to England, and I'll be wanting to end up in Portugal, so we could make any plan for meeting you like. For instance Florence, at the Annalena[19]. . . [orig] or the lovely little Inn in Devon where Edith Cobb and I stayed in 1948—if you want it to be just us.

Or we could stay with Nora Barlow which would give us a lot of time to ourselves, as she and Alan are so undemanding, and we could spend hours walking around the garden talking. I'd think tentatively about Friday the 23rd we could meet somewhere where El Al touches, like Paris. You know how Uncle Geoffrey fusses about plans, and in this case there is also Margaret Lowenfeld[20] to cope with who wants me to do things in Denmark so if we could settle on a place and a date soon, at least to aim at, it would be easier. If you can decide for instance between Florence, Gellers, Devon, Boswells, and agree on a date, then while I am in England I can settle things with everyone. So if you can decide, cable me by May 30th if possible, and anyway before June 7.

I leave for England on Wednesday. Your letter made me so happy, I just walked around on air. It is a delight to be your mother.

Lovingly,

To her godson Daniel Metraux, eleven years old, July 16, 1960, from Palo Alto, California.

Dear Daniel,

Just to write July 16th gives me a little thrill because when I was a child, we lived in a town full of Italians and they had a big celebration on July 16, Our Lady of Mount Carmel's Day. They built arbors all along the roads, and a whole amusement park came to town, and there was a band and a parade and fire crackers. Then the other people in the town—its name was Hammonton NJ—got worried for fear their children wouldn't appreciate the 4th of July, so they tried to have a bigger parade and more fire works and so all the children in the town had what amounted to—two fourths of July.

I am out here giving a couple of lectures, making one TV show, (with some Stanford University students—Stanford is where Gregory teaches), and visiting a friend of mine[21] who has 4 sons and no daughters. So every room in this house is a boy's room. The one I am sleeping in has pennants from all over the United States, and the 48 star American flag and 49 star, and then there are pictures of teams, and piles of baseball gloves and track shoes, and 4 bicycles out in the patio, and copies of a magazine called *World Youth* piled up on things. The bulletin board has special funny kinds of thumb tacks—including [a] little skeleton from Mexico. In fact, you actually met these boys in Mexico last Christmas—there are four of them, Miles, Neil, Kent and Laird, and you played Risk together and drank Coca-Colas. Their mother has a microbus and they all drove to the meetings in it at Christmas. Next year, the oldest is going to a very good school in New Mexico, and the rest are going to South America. So they will have a chance at some fieldwork, but not quite as young as you did. I guess you were the youngest field worker who ever went into the field. They all send you their best!

I was delighted to know that you are enjoying camp so much and that everything was going so well. Now that Eric[22] is married, and Cathy back from Beirut,[23] life is settling down again and the next big event after I get back will be the weekend your mother and I go up to Conn. and your mother goes up to see you, while I stay and visit the Cobbs.

I hope I'll find some more issues of the *Waverly Times*, Country edition when I get back to Waverly Place.[24]

    Love,

To Daniel Metraux, fifteen years old, July 23, 1964, from New York.

Dear Daniel,

I waited to write to you until I had at least a little news. It has been a very quiet summer. Your mummy[25] and I have been working very hard, turning out work at a tremendous pace, but one article after another doesn't make much news.

I am enclosing a review of Father Flye's [26] book about the writer James Agee. Tonight a little village coffee shop called the Lion's Head is giving a party for him—a surprise party, to celebrate. You'll probably want to keep this. James Agee was 10 when he met Father Flye, and later they became such great friends, and now, after he died, Father Flye is still here—very cheerful too this summer—to write about him. He had kept his letters so faithfully.

Last week I had several interesting experiences. I went to lecture to a group of people in New England, who are the descendants of French Canadians who came to this country during the last 60 years. They call themselves Franco Americans and now they have a special skill to offer the United States—the ability to speak French.

Then I went down to Virginia to a Center named after Woodrow Wilson, in the town where he was born, where they mix up people with all sort of handicaps, lost arms and legs, paralysis, strokes, and those who are not quite as bright as other people. The not so bright ones—called mentally retarded—look after the can't walk so well or can't cut their meat ones. They are called students, not patients, and everyone there is learning to make a living.

Then I spoke to the Peace Corps groups who are going to Ethiopia. There were fifteen Ethiopians, black skinned, semitic-like features, beards, teaching them the language. And I spoke to a group of men and women from the foreign service about how people could change fast and still have a sense of who they really were. It was quite a two days, and I had to charter two small planes in order to make my connections in time.

I am so glad you are having such a good summer. I leave on Friday and your mother says maybe you will telephone Thursday night so I can say goodbye.

Affectionately,

Mead's daughter had married Barkev Kassarjian in 1960. They were going to spend a year in the Philippines. To Mary Catherine Bateson, September 25, 1966, from New York.

Cathy darling,

It was lovely to have a letter from Aunt Marie before I left Montserrat saying you had called her and were on the point of leaving, with Barkev finished and gone on ahead, and then to find your cable when I got back to the office. A letter that I wrote you from Montserrat to the Center, was forwarded by them to Isabel, which so alarmed her that she actually wrote me. And Aunt Marie had just had a letter with a $500 check in it, returned, marked "not known," which she had sent you at the place you were living. Might be a good idea to write both of them with your new address. . . .

You should also have had a bulletin letter with a certain amount of information of a general sort in it. Hal Conklin was sad that you never came to see him but I said you'd been stuck in Washington. I mentioned the question of teaching Philippinos (not spelled right) about savages, but he thinks they have been able to be proud of their primitive ancestors. It will be interesting to see what you find. But I am convinced that studying what is happening to English will be very worthwhile. One of the crucial points seems to be that as a major language (perhaps only when in contact with another) breaks into an oral dialect, people lose consciousness of what is happening. In Montserrat, if one can find out the dialect word, or if one can overhear a conversation, then one can get translations, but you can't move from English to the dialect word—which is usually English too but an older vintage pronounced differently. I did quite a little work on it. We also realised that Daniel's pell mell type of speech, with words tumbling over each other, is in fact, Montserrat English. Rhoda thinks it may help to make it conscious. I don't know.

Aunt Marie will doubtless continue to send you all the news; Aunt Jane has cancer of the throat and has been having very painful treatments; Uncle Larry is still in hospital and they plan to try another rhythm conversion next Wednesday; Uncle Geoffrey has been re-pronounced—by still another specialist—as fine. Daniel is safely in Beloit which he describes as "the nearest thing to Heaven."

It would be pleasant to correspond with you, not with a sense of urgency but about things that seem to matter from time to time. . . . Don't make any drastic decisions on the basis of my possible visit for I have a strong feeling it could fall through. . . .

Our new apartment is on the 16th floor, 82nd street side, charming and light, and adequate in space. I found an apartment over at West End and 72nd that I fell in love with because it was really like a house, odd, unsystematic, with a terrace, but—it was really too odd and too unsystematic, and the Beresford is really extraordinarily well run, solid, and so close to the Museum. Rhoda is happy about it all, and there will be money left to start Daniel on a piece of Vermont land near the Mitchells. All my love my darling, and I'm so glad the summer is over, so happily.

[End of letter]

To Mary Catherine Bateson, March 17, 1967, after a visit with her and her husband in the Philippines.

Cathy darling,

I realized as I wrote the date that this is St. Patrick's Day and that [it] is probably too late to call Aunt Mary—7:35. There is a blizzard blowing up, fine powdered snow, so fine and dry that when I had it in a dust pan I was uncertain where to dispose of it. A wastepaper basket seem[ed] more suitable than anything wet. Very odd.

It was such a lovely visit and I am so glad I came. Really all the fates conspired to make it happen, and it was right all the way. Aunt Fanny collapsed into a nursing home soon after I was there, but we had this last visit in her little house. Jess was well by the time I got back to Los Angeles. Jim stayed until almost the day I got back and had filled Rhoda's life with concerts.

I live, very contentedly on pictures in my head, of real things, and my eyes are still filled with clear pictures, you dressed for the Visayan ball, the sheen on Barkev's hair that no Cambridge barber ever attained, [Barrio] Magsaysay and the project, the reflections of green on marble in the new Amerphil (?)[sic] building, and all the long unhurried talks we had. Doxiades has written an article about man learning to live humanly in an inhuman—and inevitable environment—the modern city. I wonder what the difference is between learning to live—by better design, etc.—bringing life with one, which is not snuffed out. Rhoda was fascinated by the streets inside the project that therefore made people shut their windows. What is the project called? It will be easier to describe.

Aunt Marie was deeply touched and tearfully happy that you did want letters and had thought of her visiting you. She is at present suffering over my income tax and the Annual Meeting of IIS which coincide this year; I am moved to relieve her of the burden and then realise that if [she] put the same energy into something less pain-making she might not get as much rational satisfaction. . . .

I am now putting in shape all the articles I drafted while I was there and smiling with reminiscent delight when I come on your gammas where periods ought to be.[27]

We are also xeroxing a portion of Rhoda's project plan for you.

Give my love to Barkev and thank him for being such a skilled "mother keeper." Remember me to Julie and Innocencio,

Lovingly,

Your rationed mother.

To Aunt Eleanor (Mrs. E. B. Fogg), April 7, 1968. Eleanor Bartholomew Fogg was the first of Uncle Leland Fogg's three wives.

Dear Aunt Eleanor,

I was so sorry not so see you on this visit to California, but I expect to be out there again in early June and I hope to have more time then. This time I just had the one day in which to stop and see Uncle Leland and drive on down to La Jolla and see Aunt Fanny. My brother Dick and his wife, Jess, drove me down. We found Uncle Leland a little pale but otherwise quite himself, with a familiar twinkle in his eye over this last escapade. I told him that he had effectively demonstrated that he really wanted to live. But he does have rather few resources and the days are long.

Aunt Fanny is still astonishing, reading books I haven't even heard of yet, and making plans for greater activities as soon as she feels a little stronger.

It is twenty years since Cathy's father and I separated, and I cherish today the fact that we are now good friends and able to give her, and hopefully will be able to give her children, a sense of family behind them. He married again, twice in fact—and I am grateful that he has a wife who is both friendly to me, and good for him as he gets older. As these breaches heal, so I believe people's minds heal a little too. I believe it has been a comfort both to Aunt Beth and Uncle Leland that they are now friends again after the years of estrangement. I believe it might mean a great deal to Uncle Leland if he were to hear from you—or see you. I hope you don't mind my saying this. Your house and your beauty meant a great deal to me as a little girl, and events like this carry me back to the early days of your marriage and to the welcome that I felt in your home. Dear Aunt Eleanor, I hope all goes well with you.

Affectionately,

To Daniel Metraux, twenty years old, November 5, 1968, from New York. Daniel was spending a year as a student in Japan.

Dear Daniel, it's election day, your mother and I have voted but we are staying away from the TV until evening and just hoping—Humphrey has gained a great deal of momentum, pulled even with Nixon—about—but the question in everyone's mind is, is it enough? During the last two weeks some of us have been working hard to correct some of the TV mistakes that have been made—Humphrey doesn't like TV and also his way of using it was odd. Most speakers are better when they improvise than when they read—but he wasn't; the minute he started to improvise, he lost his audience. Sending in memos here and there into the confusion which surrounded the Humphrey headquarters was rather like sealing messages in bottles and sending them out to sea; people promised to call back but never did; things appeared to happen that we had suggested, but had it anything to do with our suggestions? The President of the United States was reputed to be speaking but no one knew when; call the White House; it's Sunday, nobody in the news room, in desperation ask for Mrs. Johnson's secretary, get her secretary's secretary, we don't know anything except the president's in New York. "Dr. Mead do you think the V.P. has a chance?"

Last night there was this tremendous telethon with HH and RN answering calls from all over the country. HH's gay and related to people. Nixon's a gloomy futuristic performance all alone in a pool of light. I went to a meeting of a committee with a lot of top doctors on it; they looked at the button on my coat and said, Come in. But Annie, our Barnard student, came up to report that the whole Museum newsroom was voting for Wallace.[28] Only on the Catholic point, I'm sure. But lots of people, on all sides, don't know he's a Catholic. It hasn't paid anyone to play it up; he wants the southern protestant vote, his opponents don't want to give him the Catholic vote, and so "religious bigotry" is absent from the campaign.

The book cases are all built and on Thanksgiving we will get our books, a pretty horrendous possibility. And for Christmas, Cathy and Barkev are going to Hawaii, and your mother and I are thinking of leaving town. Next year you and Judy[29] will be back.

You may have thought it odd of me to send you a Japanese book for your birthday, but I thought it just might not come your way. It was sent to me as Ruth Benedict's executor. I wonder if you saw Ken.[30]

We are enjoying your very full bulletins very much. I am tempted to file them in so many places that clearly the only thing to do is to file them with themselves.

Your mother is working on her paper for the Anthropological meetings in Seattle, and asking me questions about the notes I wrote 29 years ago and don't remember too well. Dr. Gold[31] and the birds are well. Tulia[32]—after weeks of vicissitude, now has a telephone. And in twelve hours from now we will know—maybe. *Redbook* now wants my columns to become more timely and political, and the next one to be written is on the presidency. It's odd not to know what we will be writing tomorrow.

You sound as if you were getting a tremendous lot out of Japan. I am so glad.

Lovingly,

To Aunt Fanny, her mother's sister, January 31, 1969.

Dear Aunt Fanny,

I am just finishing up my annual two week trip in Cincinnati and things have gone so well that I have an unhurried morning to tidy life up. I got your note. I will be coming out to see you on Friday, March 21, or early Saturday morning March 22. I haven't got the detailed itinerary yet, but I hope to have all Saturday with you and most of Sunday, I have to leave for Phoenix on Sunday afternoon. I am not making a long California trip this year, just [coming] to see you and give the Phoenix consultation and meeting which will pay for the trip.

This year's two weeks here in Cincinnati have been fine. I came with a cold which I shook off, I've written almost a whole small book, and seen a lot of interesting things. It's an ideal life, work until noon, absolutely undisturbed, then from then on life completely planned for and an even balance of listening and talking myself.

Cathy is fine and a little impatient to have a baby. . . . She is working hard on the book which is to come out of the conference she helped her father with last summer. Barkev is teaching at the Harvard School of Business. I have had a busy winter between student troubles and faculty troubles at the new Fordham college, and delays in the construction of my hall.[33] But life goes on at the same quick pace.

I am trying to develop a design for the future in which the elders and the young can play an equal part. I have decided we should think of the future as a child about to be born, set in the center of life, around which young and old can cluster without hierarchy, and the elders can bring wisdom and the experience of change and the young the new knowledge that only they have of the present-day world. How do you like that figure of speech. I find I am, at the moment at least, very pleased with it.

See you in March and you should get the exact time soon.

Lovingly,

To Daniel Metraux, still in Japan, February 8, 1969, from New York.

Dear Daniel,

I apologize for not having written sooner to thank you for your congratulatory letter. If you were around you would get worn out making congratulations because this year—as my last year—the Museum has apparently decided to retrieve all the years of relative neglect, and I spend my days with interviewers and photographers. Very tiring.

I am glad you and Judy decided to stay, for a number of reasons. In the first place I think you would have found later that your stay was too short for such a complicated culture. It is a familiar experience in field work to reach a point—quite early—in which you feel you understand everything and there are—in effect—no mysteries. In primitive field work it comes at about 2 months. It doesn't last and is—in field work—followed by a period of despair . . . [orig] a sense that it is all unknowable. In the second place, drastic revisions of plans that involve drastic revisions of other people's plans often are very unsatisfactory. It is true that it is a sign of strength to revise one's commitments, but it is also a sign of strength to be able to balance other people's too. Your mother is very much interested in Judy's new plans and happily thinking of books which may interest her. Third, sticking to your original plans will make next summer make more sense if you want to get a job, as summer is the time when there are student jobs available—while regular employees go on vacations.

I think you didn't quite understand what I meant by sending you Ruth Benedict's book in Japanese.[34] Obviously if one has read a book in a language one knows well, one doesn't reread it—except as a chore—in another language. But I thought you might have found it interesting to take a single passage—chosen from the English text—like, for example the discussion of *giri* or *on*, and looked at the Japanese, and perhaps got some new cross cultural insight. The Japanese translator must have had to struggle with the unfamiliar statements about his own culture's ideas.

In field work it is getting a first grasp of the language and general outlines of a culture that give one this habituated sense, and in a new country, I suspect it is travel, such as I have done with Uncle Geoffrey in places like Iran. If you go all over a country fast, you get a sense of "having seen it all," which means a kind of closure at one level.

I am going to be most interested in your comments on Japanese students. Your mother and I have just done two columns for *Redbook* on student power, and the MAN AND NATURE lectures which I am giving at the Museum this year are on the generation gap. I have two students who have worked with students in Japan, and what seems to be a point is that disgruntled students in Japan can demonstrate against the U.S., and a particular party in power, while in this country students can only, effectively, attack all of their own government as they have no foreign country to blame for the Vietnam situation. I think that must make a considerable difference.

In general, Nixon is keeping a sense that things are happening, going, without committing himself very far, and there is quite a little quick back tracking going on, the new sec. of the interior reversed himself on stopping oil drilling on the California coast—which he first refused to do. They are going to take another look at the AIBM [sic]—after Boston area town meetings, and after saying all foreign policy would be coordinated in the White House, and under the Security Council, there are reassurances about the role of the Department of State.

Uncle Geoffrey's man, William, and Uncle Geoffrey, both fainted—and hit their heads—probably not a coincidence but due to the flu.

Everyone enjoys your bulletin letters very much.

Last night your mother and I went to see the *Yellow Submarine* with the Beatles, and thoroughly enjoyed it. It is Saturday morning, your mother is doing accounts, I am writing to you, and Doctor Gold is supervising from the table near the window. Almost all the pictures are up, the books are all in, and it is beginning to look as if we lived here. So far I have escaped the real flu but had a mild gastric virus this week which might have turned into flu and been contagious. I prefer malaria as a disease—. . .

To Geoffrey Gorer, November 11, 1969, from New York, on her new grand-daughter, Sevanne Margaret Kassarjian.

Dear Geoffrey,

. . . The baby flourishes. She is called Vanni, and she is extraordinarily alert without being either hyperactive or really on the active side at all. She has been perceiving patterns since birth practically. I am increasingly impressed with how poor our theories are about what little babies can see and integrate. Both parents are ecstatic, and the baby sitter plan, a flat on the next floor with students, and 9-month-old baby, and the wife takes the baby half the day, works very well indeed. Cathy is teaching and doing research.

Life has been very hectic. Rhoda at last had to face a deadline over the book she has been doing with Tao for years and years and—as you know— she finds all deadlines harassing. Tulia went up to care for the baby for ten days which has not improved life at home.

I testified on psychotropic drugs for a Senate committee on monopoly with a recommendation for legalizing marijuana as a minor clause, and it has got picked up and become a cause celebre—the time was exactly ripe. I have learned from the uproar that we must never pass a law to legalize anything—I've known this for abortion ofcourse—but only repeal laws against things. Americans seem to think the word legalize means *sanctify*. It was worth having this point reiterated. So I have appended a paragraph to the Senate testimony saying that in the light of the discussion I wish to emphasize that we should repeal the laws against marijuana rather than use the phrase *legalize*. I put the age very low and now people are saying it is too low—implicitly accepting the rest. But even if it were to be done tomorrow, the damage that has been done to the young who have been made into heroin addicts or poisoned by these bootleg psychedelic mixtures. . . .

To Peggy Rosten Muir, her niece, December 9, 1969, from New York, in response to a letter asking for information on her mother, who had committed suicide in early December 1959.

Dear Peggy,

I am typing this letter because no amount of love makes my handwriting really legible beyond the shortest bread and butter letter.

Your letter sent my imagination wandering, and there will be many things which I can tell you or find for you, that will make your mother more vivid in the days before you were born, and when you were little. I'll try to put some of these together for you, but it won't be until February perhaps before I can do very much because most of our family pictures and letters are buried deep still, never excavated after we moved here.

You are quite right. Your mother was very brave, and beautiful. I spent last weekend with Liza who told me how suddenly she felt unbelievably saddened, but as she never realizes what day of the calendar it is, she couldn't think why and then she remembered. You know that we did bury her ashes among the daffodils across from your last house in Connecticut and I am sure the daffodils are still blooming there next spring. Your father was away and so it wasn't possible to ask his permission for you to go and his relationship to ritual has always been a little uncertain.

My dear, there will be "things for you," as in the Edna Millay poem, "I'll come again to Camelot." Do you know it?

[End of letter, no closing or signature]

To Geoffrey Gorer, January 14, 1972, at her sister Elizabeth's in Cambridge, Massachusetts, on her granddaughter.

Dear Geoffrey,

. . . This is now, however, a pleasant interlude turned towards the future. I am spending three weeks up here, staying with Elizabeth who has been weak and depressed after a bad attack of flu and needed cheering up, and being a grandmother on call, filling in as Cathy tried to get off for Vietnam.[35] She is burdened down with packing, getting their house ready to rent, doing an index for her book, finishing a research paper for her child language study, getting grant operations in, and there [are] all the shots. Vanni plays giving and receiving shots all day long. So I write at Elizabeth's and read manuscripts for people, and then go over whenever Cathy needs me. It's a very condensed bout of the usual grandmother behavior but it has given me a chance to see Vanni over several weeks which is really the only way to get to know so small a child. She's extremely precocious, and generally delicious. People delight in her wherever she goes so taking her to shops and things, although fraught with possibilities of trouble, still is fun. I can still lift her and walk as far as she can—which is all that [is] required. . . .

To Mary Catherine Bateson, living and working in Iran with her husband and small daughter, April 14, 1972.

Cathy darling,

Your letter to Aunt Marie sounds as if things were still very dismal. I am trying to make all sorts of assumptions—that you will be able to get servants as soon as you get the language, etc. but still I hate to think of you having such an impossible time. Have you considered trying to get an Armenian widow and daughter, for instance, who might form a nucleus of a household who wouldn't be frightened at night.

I am also wondering whether, if things stay as difficult, I shouldn't plan to spend all the time between Delos and Cardiff in Tehran, living somewhere near you—in a hotel—and taking the baby most of the day. I could do it; I would simply bring all my Mundugumor material along and write that monograph. It isn't bulky and I have copies so there isn't even the risk of losing the original. I will have to spend a week with Geoffrey somewhere, but that is the only commitment I would have. And he says he'd be glad to look at Iran, parts we didn't see before—that is ofcourse if the earthquake hasn't made a lot of things impossible in Iran. I talked to Lita about the grant question, and if necessary, I think she and I can trade off some grants. Nothing has been decided yet about Wenner Gren grants—they are just forming panels of consultants and I said I'd be one. You will have got the material I sent you by now—I am puzzled as to why we haven't heard from Feachin O'Doherty.[36]

Life here at the moment is so overcrowded that I can hardly breathe. Crises everywhere—the whole field of environmental activities is crumbling under the pressure of new responsibilities and inadequate funds, just like the rest of the economy which is primarily troubled by the population explosion not the usual causes of a depression. SIPI[37] is in a Crisis, Maurice Strong's office in the UN is in a crisis, UN nearly bankrupt, everyone trying to raise money from the same sources. I try to think it's a good idea you are out of it but then life for you there sounds bad.

Rhoda has been reading the draft of my mss this week in Vermont—this is the first time she has looked at it and how I am going to get the necessary revisions done now, with lecture trips every week, and a trip to a short Stockholm conference from May 5–12—I just don't know. I suppose the Lord will provide somehow. I succeeded in taking my auto. away from McCall's—now the Saturday Review Press—and giving it back to Morrow. Rhoda has decided to defend—to Morrow—my right to write this book in my own style, which is a magnificent solution—I think.[38] Ralph [Blum]'s novel—written as from a 17-year-old boy about his grandfather—for his future grandson is out, and I think it is good. I'll send you a copy. Daddy's book[39] is supposed to come out any moment but I haven't seen it. Be sure and let me know if there is anything more you want me to do about your book.[40] There isn't much time now to do anything.

Aunt Marie just read your St. Patrick's Day letter to Aunt Mary over the telephone. Everyone shares letters so it's good you can write personal notes to each. Vanni dances through my mind. I am so impressed with the need to be in constant touch with a small child to know what they are thinking—and it's equally so with Aunt Fanny. She is perfectly lucid but she brings in anything from a span of 95 years and I am supposed to follow.

I'm glad Barkev is finding things good. Kiss Vanni for me, and I do so hope things will look up.

Lovingly,

To Mary Catherine Bateson, still in Iran, August 25, 1972, from Cambridge, England.

Cathy darling,

This is Philomena's home emergency typewriter, the ribbon is old and it jumps. So I am afraid the general appearance of this letter will not fit the subject matter.

I had 18 days in Utrecht, devoted to religion, and with the presence of several people whose spiritual counsel I value. . . . But 18 days of religion—of the most beautiful Greek Orthodox music—that goes right into the marrow of one's bones, as the new discotheque music does—of discussions with Father Anthony,[41] of listening to Indians saying: "One isn't baptized into the church one is baptized into humanity," has left me without any yearning to spend more time on a life of meditation and prayer than I now spend. It confirms the choice I made originally of living in the world, and doing that which most needed to be done.

But in thinking this over, it seems to me that what you have been fussing about in saying your work didn't have any direction, was simply that you haven't yet made a vocational choice that is positive. In the end of the 1950s vocational choices lacked direction, the main point was good human relations and interesting work, as when you asked: "Would it [be] better to teach poetry to Physicists or physics to poets?" Both you and Barkev had multiple gifts and multiple possibilities. You took linguistics as interesting and turned to anthropology because you were not caught by the current trends in linguistics, etc. You used every opportunity that came your way skillfully, BUT you haven't really made up your mind where your own work fits into some scheme of commitment which is large enough to relate all the pieces to each other. I think that is partly what you meant when you said that religion seemed to be the only thing which would get you out of bed with a will. And until you have that worked out you are pretty well bound to be dissatisfied with your work, no matter how good each piece is individually.

I have never thought that I should have done something else, and my choice of anthropology was sufficiently broad so that each piece—kinship, language, art, child rearing, fitted into the whole. But in the 1920's a religious [vocation]—in the sense of finding out what one could do best in God's world—was more pressing than it was in the late 1950's when you were making your choices, and certainly far less pressing than it is now. Doing interesting work, receiving professional recognition, having vivid intellectual relationships with one's colleagues, are certainly not satisfying enough in this beleaguered world. (The fly swatter has just come in handy.)

There is something about these nights in air conditioned rooms—thank heaven this isn't—that has made for a great deal of dreaming. By the way Daddy was here, stayed with the Leaches, Nora has a cloud of astonishing fair hair and Lois has bleached her hair to match it.[42]

. . . I had a few days of extreme loose endedness when I got everything done for which I had saved time, Uncle Geoffrey decided his household was in too great disarray for a visit, and I had a whole week on my hands, with nowhere to go, as I couldn't even reach Philomena. But the angels took firm command. I got a room in Cambridge, Philomena came back and reserved a room at the same hotel I'd booked at Utrecht, I just caught Erasmus [Barlow], and he will drive Nora [Barlow] and me down to their Norfolk cottage—not the one which you visited—tomorrow. Claudia will get back before I go. I have now found a way to check bags so I don't have to lift them down from the high racks which made me dread English trains, and I'll be going to Wales a day early to see Mary Fisher Langmuir Essex.[43] I find I can sleep almost without limit so this will be a placid week.

Father Anthony talks on television, on TV and writes, with his picture on his books, which is strangely right. Never in history have the same people read and listened to prophet[s], and it makes a new combination. Rhoda insisted that the picture of me to go on the jacket must be recognizable to the

TV audience (we are using one of Ken's with the miniature Manus village in the foreground), and I think she was right, reading the words of someone one has seen and known make such a difference.

Vanni dances and dances through my dreams. This week I will try to find a bookstore for a first installment of books.

Love to all of you,

To Mary Catherine Bateson, June 10, 1976, from Habitat Vancouver, Canada.

Cathy darling,

I can now make fairly definite statements about summer plans. The Athens meeting of Ekistics was cancelled and replaced by a French meeting at the end of September, so I will not be coming to Europe this summer. Present plans are that Rhoda and I will spend a month in Vermont from about the 25th of July to the 25th of August, and that Rhoda will go to Europe at the beginning of September with Eric Metraux. I will be free to be anywhere in the U.S. that seems a good place for you and me and Vanni to meet the last week in August, and the first two weeks in September, with only a couple of engagements—to limit my mobility at all—Aug. 31–Sept. 1 in the NY-Washington area.

You and Vanni can come to NY and stay in the apartment—Tulia should be back from Haiti—and seeing Marie—or if you plan to spend some time in Boston, I can go up and stay with Liza, if you, for example, were staying with Mary, or Jerry and Polly [Polyxene Cobb]. You could come to Vermont, where there is a motel where a lot of children stay—and a swimming pool— you and Marie and Vanni could stay there and I could come down from Rhoda's school house. This was a suggestion of RM's—a possibility for having MEE and Vanni near each other where Vanni would have other children to play with. I think something could be found in the way of play groups— Ken Heyman's children will be on Fire Island, Maddy[44] has a house an hour up the Hudson, etc. but it will take planning.

Several people here from Japan with warm reports of your participation. Habitat itself, I think, has gone pretty well—a big anti-nuclear development that no one anticipated and a good deal of sense about future plans. But we are just at the stage—three days from the end, where everyone is critical and apprehensive. I developed a toothache at the beginning which has given me a bad neuralgia at time[s] and RM has found Tsung Yi Ling (president of WFMH) trying. Percy Lee (you remember her wedding when you were a teenager) is here assisting me, rather randomly and the telephone never stops ringing. Carter now seems the certain Democratic choice and he has come out flat-footed against nuclear spread—and he was trained as a nuclear engineer. I think we are in a very dangerous turning point on proliferation. There seems little doubt that the greed of the nuclear industry—not enough sale for reactors in the developed world—may destroy us all. The initiative failed in California, but it did get some 35% which is remarkable. (Needed 2/3).

It will be lovely to see you and have time to talk.

Lovingly,

To Daniel Metraux, August 28, 1976, from New York.

Dear Daniel,

I am back in New York, expecting Cathy and Vanni within a couple of hours. I came down from Vermont on Monday and I just spoke to your mother on the telephone. The surveyor has now come and papers are all being drawn up to buy Mr. Stone's land. . . .

I don't know whether anyone has told you that Aunt Marie had an exploratory operation which was very hard on her, even tho it turned out that there was nothing wrong. But she now is almost back to full strength, coming in to work everyday. And she suggests—and I am sure she is right—that as you finish a draft of a chapter of your thesis, you send it—that is a carbon or a xerox—back to the Museum for safe keeping. There is a famous story about a manuscript on The Theory of Probability, with 3 copies, all of which got lost, and slides that were destroyed in the old Tokyo earthquake and in the San Francisco earthquake and fire. So send them back, as you finish a draft, and we will put them away for you. Send them to Aunt Marie for it will give her pleasure to preside over their safe keeping.

This coming week, God willing, Mrs. [Edith] Cobb's manuscript[45] goes to the press after some few hundred more hours spent on it. It is certainly expensive to associate with mavericks. It will come out in June, with a jacket design by her oldest granddaughter. Whether she will understand it has happened isn't certain. But she gave all of us a great deal of pleasure, intellectual and otherwise, for many years.

It has been fun having your articles come out in the *Transcript*, sort of double take, like watching someone on TV who is also in the room. The Akins have been very neighborly, bringing us flowers and vegetables. It was a funny month, only 3 sunny days and one swim, but the new TV is splendid so we could watch the convention. Reagan's timing is superb. I think that Carter will win, certainly, but not by a great land slide. I am on his committee of advisors, the committee of environmentalists for Carter and the committee of Scientists for Carter, but I don't plan to get into the scrum too much unless things look bad.

Your mother spends a great deal of time planning the new annex to the house, new wing I think should be its name. She found a small school house that she thought perhaps could be moved but it was too expensive to do so, and would have involved cutting it into quarters, and cutting down a lot of trees to move it in.

We all enjoy the pictures of Katie so much.

Love to Judy,

To Mary Catherine Bateson, in Iran, November 15, 1976, from New York.

Cathy darling, I've been the delinquent letter writer. I first had news of you all from Geoffrey, then your letter to Geoffrey and then your long letter to me, and then—life got terribly busy and also very confused as to timing and planning. I kept hoping that I would be able to say more definite things about something.

I had to go to France twice because the Ekistics exec. board was cancelled at the last minute after I had made plans with everyone else, the Margaret Lowenfeld Trust, UNESCO, Clemens, Uncle Geoffrey. Aunt Marie has been overworked with Philomena's departure to Angola, her sister-in-law['s] second cataract operation and the illness of Mrs. Gilbert's husband, so that Mrs. Gilbert hasn't been able to help her much. She seems to have recovered from the operation perfectly well but her customary anxiety is sharpened. I was to have gone to a Conference in Japan but they changed the date and it conflicted with the IIS Board meeting and I didn't dare change that date. When we are questioned by anyone in authority she has a terrible fit, and I have to find some form of transition which I can afford, that she can tolerate.

Then there have been endless bustles and tribulations and what not over my birthday;[46] with everybody having secret agendas. We have now—I hope—reduced it to: a Museum drive for money for anthropology, moving my hall and reinstalling it—they have closed it entirely only five years after it was opened—which will be launched with the help of *Redbook* and William Morrow, a small conference at Tarrytown to replace a grandiose scheme of Bob Schwartz for a birthday ball and 200 people, an article in the *Saturday Review* by Jean Houston, and a profile in the *Washington Post*. I HOPE. But there are also lurking secret plans, one from Sociology at Columbia which I just heard about by accident.

. . . Rhoda has gone through a whole series of plans about the Vermont house and is still in the midst of them and in the midst of plans for a session on the Sepik River at the AAA next week which Rhoda and Bill Mitchell planned, and Daddy is very reluctantly coming. He is being invited to the Tarrytown conference which Schwartz thinks he can make into a Mini Delos, but that is right after Christmas and I doubt he will come. Eric Bateson[47] has been here, and is thoroughly bored at Bloomington and will, I think, shift to TC.[48] He misses the intellectual fare he has been accustomed to—as all you children have. The book on my letters from the field is now in Rhoda's hands and proceeding apace.

In the end I had to put a lot of time and effort into Carter's campaign, as the margin was so narrow that every TV appearance might be crucial, so in addition to inventing a senior-junior political partnership for the senior citizen campaign, I did four days of campaigning, mostly in Western Pennsylvania, which we won by a narrow margin. Now there is hope of more sense in the world. New Directions, the new citizens organization on American foreign policy—partial counterpart of Nader and Common Cause, is just launched, with Russel Peterson who was head of Environmental Quality and is a fine person, as head. I am chairman of the advisory council, with several vice presidents, Norman Cousins, Father Hesbergh, John Gardner (head of

Common Cause), etc. I will be devoting my next chunk of good works time to that—and it will tie together UN things and domestic things and the AAAS role in foreign affairs which I have helped develop.

. . . The world is a mixture of older people falling to pieces and young people anxious to do things and not finding any way to do them, no grants, no field funds, no jobs of the sort they had been led to believe. And this additionally to the tremendous new nuclear dangers from the breeder. Tulia, fortunately, is in wonderful shape, cheerful and plump, Daniel is getting his thesis written, Liza has found a new drug which has got her out of the extreme pain she has had all summer, Philomena, though apprehensive and foolishly over equipped, is excited about going to Angola. These are the good bits of news. I wonder how much the income tax law which takes away all the benefits of living abroad is going to matter to you. I saw Mike Wallace the other night; he has just done a TV interview with the Shah and in a week or so will air a story of how miserable Americans are in Teheran, and that without knowing the new tax law has passed. I am trying to think what initiative could be taken by someone, perhaps the Dutch, to point out how disastrous this is to our foreign relations and to all the best people abroad, but I haven't figured out anything to do about it yet. *World Enough* is still buried without a trace, so now we all have one buried book each. Have you heard from Lindisfarne about the proposed paperback of your book?

Is there anything you would like me to do in any way? I have a very energetic student who is working [on] parent-child perinatal relationships, especially fathers, and I have her picking out things in the literature which might relate to the article you are doing for Margaret.[49] Or have you done it? If you have, I can quit, as I know you don't intend to go much further with early parent-child relationships at this moment. But there is good stuff coming out. So let me know whether you have finished and sent off that article or not or need new articles. We can xerox them and send them off to you.

I was meditating rather gloomily on how I appeal to depressed people, and despairing people, and late risers (?)[sic] and then adding that you have always represented pure delight. Somebody brought me a lovely picture from Lindisfarne of you with a dress embroidered like a chain of flowers around your neck. I am using one of Polly's pictures of the 3 of us for my Christmas card. The USIA film won a big prize.

I guess that's about it for now.

Lovingly,

[P.S.] Phyllis Palgi is having bad heart trouble. She is trying to get her thesis finished before open heart surgery.

To Mary Catherine Bateson, February 18, 1977, from New York.

Cathy darling,

I had the oddest dream the night before last. I was in a strange country, as a visitor, and I think Geoffrey was there, and I was going about, meeting stranger[s], packing and preparing to leave and suddenly I said: But I am in Iran and I haven't seen Cathy, and then there was a desperate attempt to get in touch with you when I didn't know your telephone number, or so I thought then. I do know it ofcourse.

Life goes on at its hectic pace, with more and more casualties among people we know: Edith Cobb's book should soon be in print and she is barely alive; Luther's wife just died, at almost 90; Reo's wife is in hospital. Mrs. Gilbert—Aunt Marie's devoted secretary—'s husband is very ill and hard to nurse. Mathilde[50] was just here and had managed to get the flu, as had June [Goodfield], with whom I have [sic: had] dinner last night, with her Portuguese friend, who is delightful. Daniel is back but we haven't seen Judy and the baby because they both came down with the flu at Judy's mother's.

The way summer looks now is that I should be able to get to Iran by something like the 15th to 20th of May, if that still makes sense. I will be going to Nairobi in early May, by way of Athens and then on my way back stop in Iran, come back to the USA for June, and probably leave again for July. I have to be in Bali the 18th to the 24th at least—for a conference which will pay my way. None of this is hard and fast yet, but it seems to be shaping up. I have been asked to go to Chatauqua Lake—where my grand-mother used to take Bompa when he was a small boy, in the summer, for intellectual refreshment. It's tailored for the elderly culture cravers, and I hope I can take Aunt Marie with me. She hardly ever gets a holiday, but at least I've got her a bookkeeper and an accountant, and relieved a lot of her anxieties.

We are just finishing up the manuscript for the book of my Letters from the Field, 1925–1975, with all the usual last minute rush and panic. I have to leave for Denver for my last responsible AAAS meeting tomorrow and Rhoda is, God willing, going to get off a day later. I did get her off to London and Paris, to take a look at Alfred's archives, see other Metraux, Clemens, Geoffrey, and the Lowenfeld Trust.

Last night in the middle of the night, the things I must do before April first looked impossible, but tonight, after a good steak meant to restore us, life doesn't look so difficult.

Marie has been overwhelmingly happy over Vanni's school work. It was a real brain wave to send it. She gloats at watching "a kind grow."[51] I've been reading a book called *Mister God this is Anna*, a dialogue between a child and a young mathematician—the matter reminds me of you but the intensity is Vanni's. Very confusing. Does Vanni like getting post cards?

Love to Barkev and Vanni,

[P.S.] I received a very beautiful book on Iran from Her Imperial Majesty and I have sent her *World Enough*.

Epilogue

"I care so much more for the state of a relationship, for the essence, than for the accidents, and time and space and social organization are all accidents in comparison to the degree of understanding that exists with someone I love."

—*Mead to Geoffrey Gorer, undated letter, circa 1952*

Margaret Mead had a genius for nourishing and cherishing relationships. During her life, she created or attempted to create wide and varying networks among the people she knew. As many of the letters in this book show, she loved to connect people who shared similar interests. She attempted this across the divides within anthropology and across the divides among disciplines, as well as in personal friendships outside her professional circles.

When relationships hit snags, her letters show that she worked to overcome the problems and to keep the connections alive. She valued her close relationships as living things. As she wrote Geoffrey Gorer in the letter above: "Love between two people is like a child, it should be cherished as unashamedly, by both."

When field trips prevented the constant personal meetings that nourish friendships, she produced general bulletin letters meant for circulation and filled them and her personal letters with vivid, concrete details of her life and her current thoughts to keep relationships of all kinds from stagnating through distance. Pictures of her friends' children accompanied her or were sent to her on field trips, and she often had a wall of pictures wherever she stayed, of both children and friends.

She could be said to have had a vision of life as a great interconnected network and of herself as doing the job of a transformer, the piece of electrical equipment which shunts energy and information from one connection to another, in the process transforming the communication itself not into electrical impulses, but into something that might transform social thought, or later, the world.

Mead worried about her reputation because she wanted her work to make a difference in the world. She tried to protect her reputation in various ways, such as by writing refutations of bad reviews or in private and public arguments with colleagues. She had seen the way the work of Franz Boas and Ruth Benedict had been attacked as they got older and after their deaths. At a low point on a revisit to Manus, she wrote to Rhoda Metraux, October 24, 1965, wondering if her work would be attacked as theirs had been. "I think the paranoid ideas about Ruth increased with her frailty," she reflected, "but was it rather with the age and sense of failing strength of those who had them, or do the two things so work together that it's silly to even raise the question? Yet, one can after all, exert the most influence on one's own side of the scale. . . . But it's these projections that are so worrying. Will they go on after I am dead, as Linton and Reo still felt about Ruth, after she was dead. At least while one is alive, one has possibly a chance to modify them, to try to prevent the paranoid from injuring one and never forgiving themselves. . . . Linton did do Ruth real injuries, like his denunciation of her to the FBI. . . . And I've seen, at least for a long time, that I mustn't be vulnerable, mustn't suffer or get ill, or let myself be hurt. . . ."

Mead's work was indeed reevaluated after her death. Her most prominent critic was New Zealand–born anthropologist Derek Freeman, who took issue with her early Samoan field work.[1] As he had earlier in his dealings with Tom Harrisson in Borneo, Freeman attacked Mead's Samoan research with messianic zeal.[2]

Freeman and Mead studied different parts of Samoa at different times and held substantially different views on the interaction of biology and culture. He had done field work in Western (British) Samoa in the 1940s, and Mead had done her field work in American Samoa in the 1920s. He worked primarily among the men, not with predominantly female adolescents, as

Mead had done. Mead felt the differences in their findings were due to the differences in time, place, and population studied, and thought both were valid pieces of a larger picture that would someday become clear. Freeman's idea of science, however, mandated that there could be only one version of truth, and therefore her work had to be wrong. Their ideological differences played out on a large stage in the 1980s and 1990s as Freeman published his challenges to Mead's work, and by extension, to the whole of Boasian anthropology. The arguments among anthropologists about the validity of Mead's Samoan work after her death have been as much a debate over what constitutes science and truth as over her work itself.

The reevaluation of Mead's Samoan work corrected some errors and omissions but largely led to her vindication. This was the conclusion that Louise Lamphere, then president of the American Anthropological Association, drew in her letter to the *New York Times* in response to their obituary for Freeman in 2001. She disagreed that Mead's reputation was permanently damaged, stating that "most serious scholarship casts grave doubt on his [Freeman's] data and theory."[3]

Anthropologists, like all social scientists, are focused on work in the present tense. Mead's field work lives on in the history of anthropology, where it remains a testament to what was happening in the discipline at a certain time and within a certain context, much of which she influenced. She led the way in many areas during her career. She became the first anthropologist to study gender and child-rearing cross-culturally, and was a pioneer in visual and applied anthropology, as well as in experimenting with psychological testing and concepts and promoting their use within anthropology. Ideas from her writings, such as the role of culture in shaping adolescence and the relativity of gender itself, were first links in chains of scholarship that are still ongoing in many disciplines beyond anthropology.

During her lifetime, Mead became an iconic figure. This book reveals, through Mead's own words, how rich her life was in all its joys, frustrations, love, and hopes. These letters reveal the intense, complex woman behind the icon.

Notes

INTRODUCTION

1. See, for instance, her letter of November 7, 1934, to Gregory Bateson, included in Chapter 2.

2. This quotation appears in her letter of November 14, 1967, to Reverend Richards Beekman, in Chapter 4.

3. Margaret Mead, *Blackberry Winter: My Earlier Years* (William Morrow, 1972), p. 80.

4. She discusses this in her letter to Gregory Bateson, January 10, 1935, in Chapter 2.

5. She's referring to the day the ship from Pago Pago came with mail.

6. See Gerald Sullivan, "A Four-Fold Humanity: Margaret Mead and Psychological Types," *Journal of the History of the Behavioral Sciences* 40 (2) (Spring 2004): 183–206. See also Lois W. Banner, *Intertwined Lives: Margaret Mead, Ruth Benedict, and Their Circle* (Alfred A. Knopf, 2003), especially Chapter 11.

7. Rhoda Metraux, "Foreword," copyright 1966, in Margaret Mead, *An Anthropologist at Work: Writings of Ruth Benedict* (Equinox Books/Avon, 1973), p. iv.

8. Unless otherwise noted in the text, letters included in this volume are from the Margaret Mead Papers and South Pacific Ethnographic Archives at the Library of Congress. Mead's godson Daniel Metraux and others supplied some additional letters for the authors as they worked on this project. Photocopies of all of the letters supplied by Metraux—and four of the originals—are now held by the Library of Congress.

CHAPTER 1

1. Her father offered to send her to the University of Pennsylvania for the second semester, then to either Vassar or Wellesley after that.

2. Mead's brother; Dorothy was a friend from home in Pennsylvania.

3. Aunts Beth and Fanny were her mother's sisters; Aunt Mabel was the wife of her mother's brother Lockwood.

4. Her roommate Ruth Greenfield's home, where she expected to spend Christmas.

5. *The Young Visiters* [sic] was a best-seller of 1919, written by a nine-year-old, and published with uncorrected spelling and punctuation. People found it delightful and avant-garde. Alice West was another friend from home in Pennsylvania. Bessie was a young hired girl who worked for Mead's family.

6. Margaret did not generally use a nickname. She sometimes shortened her name to "Mar" while writing to her father, and he called her that in his letters. She used "Mar" in letters to him in college and later in life when feeling unusually nostalgic; in her other letters, she is Margaret.

7. A book of poems by one of Mead's favorite authors, Rabindranath Tagore.

8. This usually referred to her sisters, Elizabeth and Priscilla, younger than she. Her brother Richard, also younger but closer to Mead in age, was usually referred to by name, as he is here.

9. The DePauw celebration later in the month on May 22, 1920, of which her pageant was a part.

10. Delta Upsilon, her father's fraternity, allowed alumni to stay at the chapter house.

11. By her second semester, Mead had had considerable success and made friends. But over the summer, she would decide to attend Barnard College in New York City.

12. Here she is talking about May 1, the traditional May Day.

13. Katharine Rothenberger was Mead's closest friend at DePauw; the others mentioned here were fellow students Hilda Varney, Mary Mutschler, Mead's then-roommate Mildred Benton, and Christine Booth, a young Englishwoman who later became a missionary. The Dean was the dean of women, Katherine Sprague Alvord, and Miss Sophia Steese was the head of Mead's dorm, Mansfield Hall, as well as director of physical education for women.

14. Léonie Adams was one of Mead's apartment-mates at Barnard who became a lifelong friend. She graduated a year ahead of Mead and went on to become a prominent poet. Adams was poet laureate (then called consultant in poetry to the Library of Congress) of the United States in 1948–49.

15. Louise Rosenblatt, another of Mead's close Barnard friends, became an influential literature professor, best known for advancing the reader response theory of literary criticism.

16. Mead's Uncle Lockwood Fogg and his wife, Mabel, were lending Mead and Cressman their cottage for part of their honeymoon.

17. In *Blackberry Winter*, p. 34, Mead says they called a brick factory in Trenton, New Jersey, "the plant." She also could be referring here to his long-time side business of making blocks of compressed coal dust, trying to find a use for coal waste products, which she also speaks of there.

18. *Anent* is a largely archaic preposition, meaning "concerning" or "about."

19. Herskovits was a little ahead of Mead in graduate school in anthropology at Columbia. He later became well known for his work on Africa and on African survivals in the Caribbean and the United States.

20. Her sisters Elizabeth and Priscilla had just started attending the School for Organic Education in Fairhope, Alabama. Grandmother Mead went with them and stayed in the town as a link to the family.

21. Mead was living with the chief pharmacist's mate, Edward R. Holt, and his family at the naval dispensary in Ta'u village. After his premature death a few years later, Mead remained close to his widow, Ruth, and their children, Arthur and Moana.

22. Although Mead wrote "are" here, the context of the paragraph shows that the word should be "aren't."

23. The Lanes were a medical doctor and his wife; the others were naval personnel at the dispensary where she had been living on Ta'u.

24. Mead, who had one ovary removed as a teenager, had been told by a doctor in France in 1926 that she would not be able to have children due to a tipped uterus. She married Reo Fortune believing that she could not have children. She subsequently was able to get pregnant, however, and suffered numerous miscarriages before finally carrying a baby to term in 1939, while married to her third husband.

25. Marie Eichelberger, whom Margaret met at Barnard in the early 1920s and who became her devoted lifelong friend, took care of many of the practical details of Mead's life at various times.

26. Beginning in her Barnard years, Mead had trouble with pains in her arms, which were diagnosed as neuritis. Her Grandmother Mead and sister Elizabeth suffered similarly. At times, Mead thought her condition might be a manifestation of neurosis, but she was also treated for it as a physical problem, including a near fatal surgery on her sinuses in 1929. The exact origin of the problem seems never to have been determined.

27. Reo Fortune.

28. Mead's younger brother, Richard, married his first wife, Helene, in February 1932. They married quietly, not telling his family until afterward.

29. Mead is referring here to a field bulletin. Beginning with her first field trip in 1925–26, Mead typed general accounts of her experiences in the field on multiple carbons and mailed them to family and friends at home. These bulletins were then circulated to other people, sometimes retyped. Many of them were later published in *Letters from the Field* (Harpercollins, 1977).

30. Clark Wissler, head of the Anthropology Department at the American Museum of Natural History in New York City and Mead's boss.

31. Mead was trying to sell an anonymous account of her shipboard adventures as a short story to *Harper's Magazine*.

32. This is Nora Barlow, granddaughter of Charles Darwin and a friend of the Bateson family (misspelled "Norah," in the original). Nora Barlow and Gregory's mother, Beatrice Bateson, had come to visit Mead and Bateson in Bali.

33. This is shorthand for "in our culture."

34. H. G. Wells, *Joan and Peter: A Story of an Education* (The Macmillan Company, 1918).

35. Mead is not referring to race as customarily understood here but to classifications from a theory of culture and personality she had been developing, a theory referred to as "the squares." In this context, "cross-racial" refers to people of contrasting temperamental types.

36. Arthur Steig, Bill's brother.

37. Karen Horney, a neo-Freudian psychoanalyst known for her emphasis on cultural factors in personality.

38. Bateson (influenced by Ruth Benedict) was using "ethos" in this period to refer to "the expression of *a culturally standardized system of organization of the instincts and emotions of the individuals.*" [See *Naven* (1936).] Mead is using the term to describe the emotional tone and world views of the families involved.

39. *Schismogenesis* was a term Bateson had coined to describe the development of gaps or schisms in human communication.

40. She published such a book, but not for many years: *Male and Female: A Study of the Sexes in a Changing World* (William Morrow, 1949).

41. Burns, Philp & Co., Ltd., was an Australian company through which they shipped supplies and on whose boats Mead, Fortune, and Bateson traveled the Pacific to get to their field sites.

42. Ruth Benedict.

43. Dr. Spock would become the foremost child expert of his generation, a household name to millions of American mothers.

44. Mead worked with this study, which was run by progressive educator Dr. Caroline Zachry.

45. Clinical psychologist Theodora Mead Abel, called Tao (Tay-ō), was a friend of Mead's from Columbia. They had studied psychology together.

46. Myrtle McGraw, also called Psyche, was a psychologist who specialized in very early childhood development. She was a friend of Tao's. The odd sentence structure is in the original.

47. Mead had enlisted her mother to do an index for a book and paid her for it. Her mother did this kind of work on more than one book. Mead also enlisted her godmother, Isabel Lord, for the same kind of work. Leland Fogg was one of Emily's brothers.

48. Helene's daughter by an earlier marriage.

49. Mead's daughter, Mary Catherine Bateson.

50. Courtesy of Daniel Metraux.

51. Geoffrey Gorer, British social scientist and long-time friend.

52. Metraux's infant son, Daniel Metraux.

53. Mead had had to leave Ruth Benedict's bedside in her last illness to give a lecture in Washington, D.C. She was reliving the anxiety of that time, the fear that Benedict would die before she returned.

54. Anthropologist Edward Sapir.

55. This refers to a teaching license. Elizabeth was an artist and art teacher.

56. Lawrence K. Frank.

57. From Mead's theory of the squares; Kilipak had been one of the boys who worked for her and her then husband Reo Fortune on her first Manus field trip.

58. To colleague Bella Weitzner at the Museum, Mead wrote, November 14, 1953, "Actually it is a virulent flu and I have taken this occasion to get my young people [Ted and Lenora Schwartz] to write proper memos about the state of their notes, by doing one myself also . . . on the supposition that notes might be the only thing that came back from this trip."

59. World Health Organization.

60. A shelter without walls.

61. Janet Rich Stephens, her brother Richard's step-daughter by his second wife, Jess.

62. Her mother's sister, Elizabeth Fogg Upton Vawter.

63. Elizabeth's two children, Jeremy (called "Jemmy") and Lucinda ("Lucy").

64. Her Uncle Lockwood Fogg's wife, son, and daughter-in-law.

65. Her brother's second wife, Jessica Wilbraham Rich, was his first wife's sister.

66. Anthropologist and friend Rhoda Metraux.

67. Aunt Beth's second husband, William A. Vawter II.

CHAPTER 2

1. Mead was in love with anthropologist Edward Sapir, but she had decided to break off the relationship on the way to Samoa.

2. Her college apartment-mate, who became a nationally known poet.

3. This refers to the poem "A Decade" by Amy Lowell, which speaks of relationships as zesty, sparkling wine or nutritious daily bread.

4. One who jeopardizes or interferes with the success of an endeavor.

5. Ethel Goldsmith, a friend from Barnard College.

6. Reo Fortune's name is pronounced "Ray-ō" and Mead sometimes, especially early in their relationship, referred to him as "Ray."

7. Sociologist William Fielding Ogburn, one of her college mentors.

8. Birth control—it's unclear what she actually used.

9. The word is unclear in the letter.

10. Pliny E. Goddard was her immediate boss at the American Museum of Natural History and a friend of Franz Boas.

11. He had a large private collection of textile samples and costume designs that included some from prehistoric times. Part of the collection was color photographs of both ancient and modern textiles, which he took himself.

12. A schoolboy interpreter originally from Pere village, lent to them in Rabaul.

13. This is an Anglicization of the Pidgin English expression *"samting nating,"* which means something trivial or commonplace.

14. This is a reference to mental sets or attitudes.

15. "Problem" here refers to their research topic.

16. Elizabeth Brown, who had studied anthropology under Malinowski in England, where Bateson met her; later Mrs. Elizabeth MacKenzie.

17. Englishwoman Betty Stephenson Cobbold, known as Steve, who came to the Sepik with Bateson, got sick, and left for Australia.

18. His father was English biologist William Bateson, a pioneer in genetics.

19. This word is unclear in the letter.

20. Max Bickerton, originally Reo's friend, who became Mead's friend as well.

21. E. D. Robinson, known as Robbie or Sepik Robbie, the government district officer at that time.

22. This word is unclear in the letter.

23. A note at the bottom of the page in Mead's handwriting says "Pidgin for revenge." According to linguist Lise Dobrin, "The Pidgin source for this concept, *bekim*, is a more general term referring to reciprocation or compensation, including its negative forms such as revenge" [personal communication, August 10, 2005]. Mead sometimes used the term to talk about balance and reciprocity in her personal relationships, specifically to explain behavior that seemed to balance out someone else's behavior.

24. A fairly mosquito-proof room, 9 feet by 10 feet by 10 feet, made of wood and copper wire.

25. Government rest house, according to Mead in *Letters from the Field* (Harper & Row, 1977), p.65.

26. Ruth Benedict's husband, Stanley Benedict, and a lover named Thomas Mount.

27. Léonie Adams had an affair with literary critic Edmund Wilson in the late 1920s that ended badly for her.

28. Anthropologists Gladys Reichard, Pliny Earle Goddard, and A. R. Radcliffe-Brown. A mixed type was a bisexual.

29. She is referring to the language of the squares.

30. This word is unclear in the letter.

31. Australian friends Timothy and Caroline Tennant Kelly.

32. A friend from Barnard College days. See Chapter 4.

33. A. P. Elkin, professor in the Department of Anthropology at the University of Sydney, Australia.

34. In a letter to Bateson, marked "Australian coast, early December" [no other date, probably 1933], Fortune talked to him about hitting Mead and justified it by saying she had used Samoan invective at him.

35. Jean Houston, ms., "The Mind of Margaret Mead," clean copy, p. 441, Q18, Margaret Mead Papers, Library of Congress.

36. Her secretary at the American Museum of Natural History.

37. This is a reference to the language of the squares theory.

38. College friends Eleanor Pelham Kortheuer, Louise Rosenblatt, and Hannah Kahn, known as David.

39. Benedict's lover, Natalie Raymond.

40. This is a variant spelling of Mira, the name they gave to a woman Fortune was involved with in Australia in 1933, after the Sepik trip.

41. Mead applied for a Guggenheim Fellowship but did not receive one.

42. Reo was at that time in England, as was Bateson.

43. Bateson's friend, Noel Porter, who was skeptical about their relationship.

44. There was a job opening in anthropology at the University of Cairo in Egypt.

45. Jeannette Mirsky, who was editing the manuscript.

46. In Biblical terms, she saw herself as a Martha-figure, busy at many things, rather than a Mary-figure, meditative and not concerned with the practical world.

47. Isabel Gordon Carter, who had been in graduate school at Columbia with Mead, and her husband, Hugh Carter.

48. Mead uses *E*, *W*, *N*, and *S* here as abbreviations for the points of the compass in her squares theory. Critical to her theory was the idea that people on the same axis (*E* and *W*, *N* and *S*) had a certain type of relationship to each other and that people on the other axis appeared differently to them than did those on the same axis. Mead, further, felt that southerners were most accepting and understanding of the theory of the squares. She was more likely to attempt to explain it to people she deduced were southerners than to others, who might not understand.

49. William Troy, Léonie's husband.

50. British Museum.

51. Ian Hogbin, an Australian anthropologist.

52. This word is unclear in the letter.

53. Anthropologists Bronislaw Malinowski, Raymond Firth, Ian Hogbin, and Lucy Mair.

54. British anthropologist A. R. Radcliffe-Brown, then at Chicago; American anthropologists Ruth Benedict at Columbia and W. Lloyd Warner, then at Harvard.

55. Australian government anthropologist in New Guinea E.P.W. Chinnery and Judge Monte Phillips, both friends of Mead and Fortune.

56. Betty Stephenson Cobbold.

57. British Museum.

58. *Eidos* was a concept Bateson developed in his study of the Iatmul, the culture he had lived in while on the Sepik at the same time as Fortune and Mead. While *ethos* referred to the emotional tone of a culture, eidos referred to the cognitive aspect of culture.

59. Mead is talking about the paired relationships of people diagonally opposite from each other on the squares diagram. While each member of the pair was antithetical to the other, they also shared qualities.

60. Mead and Bateson arranged for him to give a couple of talks in the United States as an excuse for him to come over in the spring of 1935.

61. David Vorhaus, her lawyer and friend.

62. American anthropologist Jules Henry and his wife, Zunia. Fey was a word associated with the squares meaning a creative or other-worldly person, the opposite of a Turk.

63. Sociologist William Fielding Ogburn.

64. Sociologist John Dollard.

65. Mead edited a book, *Cooperation and Competition Among Primitive Peoples* (McGraw-Hill, 1937), and insisted the royalties be divided equally among the participants.

66. This letter is from the Lawrence K. Frank Papers, Modern Manuscript Collection, History of Medicine Division, National Library of Medicine, Bethesda, MD; MS C280b.

67. This word is unclear in the letter.

68. This is a reference to the squares, which Bateson actually had been involved in developing.

69. This word is unclear in the letter.

70. Working on a book on Balinese drama with Walter Spies while Mead and Bateson were in Bali.

71. Eileen Pope, Fortune's second wife.

72. A woman Bateson had an affair with before meeting Mead in New Guinea.

73. Then Bett Brown, who had an affair with Bateson in the early 1930s.

74. E. P. W. Chinnery, government anthropologist in New Guinea.

75. A moment on the New Guinea field trip that brought them together.

76. Courtesy of Daniel Metraux.

77. English anthropologist Raymond Firth.

78. This is the Scottish woman raised in England Luther married after Margaret— Dorothy Cecilia Loch.

79. Anthropologist whom Mead introduced to Manus in the 1950s.

80. Mead's long-time friend, Marie Eichelberger, now retired from her career in social work.

81. Mead's daughter.

82. These words are unclear in the letter.

83. Anthropologist Harry Shapiro, her boss at the American Museum of Natural History.

84. Anthropologist Bill Mitchell, then a graduate student.

85. Luther Cressman, *Prehistory of the Far West: Homes of Vanished People* (University of Utah Press, 1977).

86. Martha West, daughter of good friends Sara and Allen Ullman.

87. In a letter dated February 16, 1978, Cressman sent Mead two photos from a trip he took to France in late 1925, where he visited their mutual friend Louise Rosenblatt, who was studying at the University of Grenoble that year. He recalled that the photos were taken in Grenoble, around Christmas.

CHAPTER 3

1. This again refers to the Amy Lowell poem, which talks of a relationship in terms of wine and bread.

2. Ariel refers to a poem of the same name that Edward Sapir wrote about Mead, casting her as the spirit of Shakespeare's *The Tempest*.

3. She's referring to the day the ship from Pago Pago came with mail.

4. Eda Lou Walton was a friend deeply involved in the poetry scene of the 1920s.

5. She created these materials to use in the psychological testing of Samoan girls, as part of her field methodology on this trip.

6. George H. Sherwood, head of the American Museum of Natural History.

7. William Morrow, Mead's publisher.

8. This refers to the character of Hans Castorp in *The Magic Mountain* by Thomas Mann. The book was published in German in 1924; in Britain (in English) in 1927; and the United States in 1929.

9. Ruth's husband, Stanley Benedict.

10. With her husband, Stanley, and with Mead.

11. This was the company that shipped their supplies.

12. Mead and Benedict shared an apartment during the summer of 1928.

13. Frances L. Phillips had become editor-in-chief of William Morrow and Company the previous year. The Blue Ribbon was a cheaper edition of Mead's book *Coming of Age in Samoa*, priced for a mass readership.

14. Barter was Fortune's younger brother, whom they were helping with his education.

15. This is a term derived from biology, meaning a cell with four chromosomes in each set. The squares were based on a division of four and were originally conceptualized as having particular physical as well as temperamental characteristics. The fourfold divisions were conceived of as having distinctive yet dialectically related adaptive, temperamental, and ethnological qualities. A tetraploid in this context embodies the whole of the squares within himself or herself; therefore, he or she has all of the strengths as well as weaknesses involved in balancing ways of interacting with others and with the larger world.

16. Anthropologist Ruth Bunzel, whose major work was on the Zuni people in the Southwest.

17. Eleanor Phillips, Louise Rosenblatt, and Marie Eichelberger were all friends of Mead's from Barnard who had gotten to know Benedict.

18. Her Barnard friends, Léonie Adams and Leah Josephson Hanna.

19. Bernard Mishkin, an anthropology student from Columbia attempting his first field work in the Pacific.

20. These were her friends, Jane Belo, who had been in Bali with them, and Louise Bogan, the well-known American poet.

21. This phrase, originally used by a Serrano Indian informant to Benedict about the state of Serrano culture in California, is used here in reference to Mead and Benedict's relationship.

22. Courtesy of Daniel Metraux.

23. Mead is referring to the response her daughter, Mary Catherine Bateson, made to the phrasing of the cable, probably, "SAFE LOVE."

24. Jim Mysbergh, who worked with Bateson during World War II and spent much time living at Mead's and Metraux's residences in subsequent years.

25. Both Lisbeth and EH refer to psychologist Elisabeth Hellersberg, who was guiding Bateson through his analysis. Martha was Mead's friend Martha Wolfenstein.

26. Her friend, artist Allen Ullman.

27. Courtesy of Daniel Metraux.

28. Anthropologist Paul Radin; Florence was anthropologist Edward Sapir's deceased first wife.

29. Tulia Sampeur was Metraux's Haitian housekeeper, who cared for her son as a child.

30. Although the letter says January 8, 1948, the postmark on the envelope and the next page place it in 1949.

31. Courtesy of Daniel Metraux.

32. Metraux previously had a very brief affair with Bateson, and Mead is emphasizing that it is not a point of conflict.

33. Jane Belo.

34. Courtesy of Daniel Metraux.

35. Her niece, Peggy Rosten, daughter of her sister Priscilla and her husband, Leo Rosten.

36. These words are unclear in the letter.

37. Courtesy of Daniel Metraux.

38. Courtesy of Daniel Metraux.

39. Metraux's friend, Murray Keller, was looking for a place for her to live when they got back from Haiti.

40. Courtesy of Daniel Metraux.

41. Courtesy of Daniel Metraux.

42. Judge Monte Phillips of New Guinea. This took place during Mead's first field trip with Fortune; while Fortune returned to do some further work on the Dobu people, Mead stayed in New Guinea, where she interviewed Mrs. Pheobe Parkinson, who was half Samoan.

43. Caroline Tennant Kelly.

44. Edith Cobb.

45. This word is unclear in the letter.

46. Courtesy of Daniel Metraux.

47. Gladys Reichard, whom Mead had known since graduate school, was a professor of anthropology at Barnard. At the time, Mead hoped Metraux could get the job after Reichard retired, but this never happened.

48. Kenneth Emory of the Bishop Museum in Honolulu.

49. Psychologist Ruth Valentine, who had been Ruth Benedict's lover. Ruth Tolman was the wife of physicist Edward Tolman and Ruth Valentine's intimate friend.

50. Courtesy of Daniel Metraux.

51. John Kilipak was one of Mead and Fortune's Manus cook boys on their first field trip there, now grown up.

52. This is an example of Mead's relentlessly positive attitude, as she had endured numerous frustrations on the trip.

53. Psychologist Theodora Abel.

54. After her divorce, Mead briefly contemplated Gorer as a potential husband. The fact that he was primarily homosexual was not a bar to marriage in her thinking.

55. Leila Lee was an office worker for the Research in Contemporary Cultures project, who became Mead's secretary.

56. Courtesy of Daniel Metraux.

57. This letter was handwritten.

58. Charles Wagley was professor of anthropology at Columbia University and an executive officer in the department in that period. He had been a student of Boas and Benedict. Mead became adjunct professor of anthropology at Columbia University in 1954.

59. Harry Shapiro was chairman of the Anthropology Department at the American Museum of Natural History in that period and Mead's boss.

60. In the Rhoda Bubendey Metraux Papers at the Library of Congress.

61. Metraux's brother.

62. This is a reference to Metraux's study of the Cultural Structure of Imagery, which drew on field materials from Montserrat and Papua New Guinea.

63. Pidgin English.

64. Courtesy of Daniel Metraux.

65. Anthropologist Yehudi Cohen.

66. Ted Schwartz, his second wife, anthropologist Lola Romanucci-Ross, and their children.

67. Anthropologist Kenneth Emory of the Bishop Museum and his wife Marguerite.

68. Colleagues who worked at the Bishop Museum in Hawaii.

69. Bateson's third wife.

70. This word is unclear in the letter.

71. Courtesy of Daniel Metraux.

72. Recommended by faith healer Carmen de Barraza.

73. Mead's son-in-law and her granddaughter.

74. This place is unclear in the card, but the information comes from Mary Catherine Bateson.

CHAPTER 4

1. Ruth Greenfield, Mead's first-semester roommate.

2. Misdated 1920.

3. Viola Corrigan and Eleanor Pelham Kortheuer, called Pelham by the group.

4. Luther Cressman was a part-time minister at St. Luke's, and Mead worked with the children there.

5. Léonie Adams, Leah Josephson Hanna, and Viola Corrigan are the friends named only by their first name here.

6. Mead knew Nearing, the writer, social radical, and peace activist, through her family and corresponded with him as a child.

7. Eleanor Pelham Kortheuer, Luther Cressman, Eleanor Phillips.

8. Eichelberger suffered from tuberculosis. She was Marie Bloomfield's college "little sister" and it was through her that Mead became friends with Marie Bloomfield.

9. Mead's friend from home Sarah Slotter, whom Mead had known since her Buckingham Friends School days.

10. Mead was writing a master's thesis which utilized data on an Italian community her mother had studied for years in New Jersey.

11. Misdated 1922.

12. *Blackberry Winter* draft, "Barnard," 6, I205, Margaret Mead Papers, Library of Congress.

13. *Malaga* refers to a formal traveling party of people of high rank to another village.

14. The letter says from Pago Pago, Samoa, but in the letter Mead indicates she is at Ta'u.

15. Lee Newton.

16. She was being glib here, because she did learn Samoan invective and used it later in her life.

17. Eda Lou Walton, *Dawn Boy: Blackfoot and Navajo Songs* (E.P. Dutton & Co., 1926).

18. She is referring to the novel *Thunder on the Left* by Chistopher Morley (Doubleday Page and Company, 1925).

19. Phillips came to the United States in 1934 and stayed with Mead in New York, meeting Ruth Benedict, Jeannette Mirsky Barsky, and Bill Whitman. At the time this letter was written, Eddie Barsky was involved in the Spanish Civil War.

20. Anthropologist Douglas Oliver and his wife were in New Guinea in the 1930s doing field work; Mead had been helpful to Eleanor Oliver, who was not trained as an anthropologist, in suggesting points to look for and things she could do. She did this in other cases as well, such as with Lenora Schwartz in Manus.

21. Kickernicks, a type of underwear.

22. Ash Can Cats, her Barnard College friends; since Samoa Mead had sent her family and friends letters they could pass around to each other, called bulletin letters—many of these have been published in *Letters from the Field*.

23. Caroline Tennant Kelly in Australia.

24. Marie had a long career as a social worker in various states and with different agencies.

25. Well-known lyric poet Louise Bogan, whom Mead, Benedict, and Adams had gotten to know.

26. Both of Margaret's sisters had Jewish husbands.

27. Her Barnard friend Ethel Goldsmith was killed in a flood in Cincinnati in 1937.

28. Léonie's husband, literary critic William Troy.

29. This is Leah Josephson Hanna and not Lee Newton because Mead asks about her sister Denah and her husband.

30. She is probably referring to the *Hawaiian Clipper,* part of the Pan Am fleet of flying boats, which crashed into the Pacific on July 28, 1938.

31. There are only two wishes.

32. He had been accepted back into the Australian military as a colonel and was posted to England.

33. *Blackberry Winter* draft, "College, Barnard," 3, 1205, Margaret Mead Papers, Library of Congress.

34. This would become *Male and Female.*

35. F. O. Matthiessen, an English professor from Harvard, who also participated in the seminar.

36. The Hellers had twin boys, Michael and Ivan.

37. U Khin Zaw, Director of the Burma Broadcasting Service, Rangoon, Burma.

38. Colin McPhee, who studied Balinese music; his former wife, Jane Belo Tannenbaum; Mead's daughter, Cathy; Margo Rintz, who had lived with Mead and her daughter as an adolescent when her family was having problems in the late 1940s. Rintz married anthropologist Burt Tollerton.

39. Philomena Guillebaud, one of the English children Mead and Bateson took in during World War II, now grown up.

40. Anthropologist Ted Schwartz.

41. This refers to her 1951 Australian lecture tour, sponsored by the New Education Fellowship.

42. Barkev Kassarjian, Catherine Bateson's husband and Mead's son-in-law, was attending Harvard Business School.

43. The Pugwash Conference on Science and World Affairs, first held in Pugwash, Nova Scotia, in 1957, and annually thereafter, among scientists concerned with the dangers of nuclear weapons and the potential for global warfare.

44. Lyndon Baines Johnson.

45. This letter was written before the Gulf of Tonkin events and the resolution that put the United States at war with Vietnam.

46. Frank was from Ireland, and Mead was writing on St. Patrick's Day.

47. Liza was Mead's sister, Elizabeth Steig.

48. House cooling was a Mead family tradition started by Mead's mother, since they moved so often, as a way to say goodbye to a familiar place and move on to the new.

49. She's referring to the house Rhoda Metraux owned and that the two of them shared at 193 Waverly Place, in Greenwich Village.

50. This is Mead's friend, the Duchess of Argyll, formerly married to Clemens Heller.

51. Mead's daughter and her husband, Barkev Kassarjian.

52. This is Sara and Allen's daughter, still living in the United States.

53. Jane's husband, Frank Tannenbaum.

54. Mary and Larry's son and his wife.

55. This letter courtesy of Martha Ullman West. It is not included in the Library of Congress collection.

56. Mead's only grandchild, Sevanne Margaret Kassarjian, called "Vanni" [Von-nē].

57. Bateson's third wife and new daughter, Nora.

58. Sara and Allen Ullman's daughter Martha and her husband, Frank West.

59. Her father, Allen Ullman, was an artist.

60. Mead is probably referring to the book *What the Woman Lived: Selected Letters of Louise Bogan, 1920–1970*, ed. Ruth Limmer (Harcourt Brace Jovanovich, 1973).

61. Leah Josephson Hanna, of their college circle of friends.

CHAPTER 5

1. *Blackberry Winter* draft, "War and Postwar," 8–9, I204, Margaret Mead Papers, Library of Congress.

2. See, for example, this Web site that examines the Mead-Freeman controversy: www.ssc.uwo.ca/sociology/mead; see also *The Journal of Youth and Adolescence,* Fall 2000.

3. A. R. Radcliffe-Brown, renowned English anthropologist, then head of the Department of Anthropology at the University of Sydney, Australia.

4. "His own culture" here means the particular culture he is studying.

5. Dr. Herbert E. Gregory, director of the Bishop Museum, Honolulu, Hawaii.

6. Published as "Social Organization of Manu'a," *Bernice P. Bishop Museum Bulletin,* 76, Honolulu, Hawaii, 1930. Reissued 1969.

7. New Zealand anthropologist Peter Buck, associated with the Bishop Museum at this time.

8. New Zealander Elsdon Best, who died in 1931, was an ethnologist at the Dominion Museum in New Zealand for twenty years and wrote extensively about Maori life. Mead was interested because she had written a guide for the American Museum of Natural History, *The Maori and their Arts*, published in 1928.

9. Kroeber's wife.

10. Here Mead is talking about her monograph *Kinship in the Admiralty Islands,* which came out in 1934.

11. Mead used this spelling early on while in the field and later standardized it as Mundugumor. The people are today called the Biwat.

12. She is referring to the Arapesh group they had studied previously.

13. In Mead's theory of the squares, the Turk, who also occupied the western position, was the strong manager type.

14. Caroline Tennant Kelly.

15. Radcliffe-Brown.

16. This refers to the Hanover Seminar.

17. This would become his book *Naven* (Cambridge University Press, 1936).

18. Raymond Firth, then head of the Department of Anthropology at the University of Sydney where Carrie was a student.

19. Judge Phillips from New Guinea.

20. Kelly's husband and her mother, whom they called the Duchess.

21. Founder of the Technocracy movement of the 1930s.

22. Culture and Personality.

23. She is referring to psychologist Gardner Murphy, but misspelled his name.

24. Mead found the phrasing "Personality and Culture" redundant.

25. His wife and daughter.

26. This became the book *Cooperation and Competition among Primitive Peoples* (McGraw Hill Book Company, 1937), which Mead edited.

27. Mead is probably referring to Jacob Levy Moreno, founder of psychodrama, here.

28. General Education Board of the Rockefeller Foundation.

29. John Dollard.

30. Mead consistently spelled his name this way through their whole friendship.

31. *Middletown in Transition* traced changes in that community resulting from the Depression. Later, Mead did something similar with her 1953 restudy of the Manus of New Guinea.

32. *Cooperation and Competition among Primitive Peoples* (McGraw-Hill Book Company, 1937), edited by Mead.

33. Dollard's influential book *Caste and Class in a Southern Town* (Yale University Press, 1937).

34. He later changed his name to Erik Erikson and became a well-known psychoanalyst.

35. John Dollard.

36. Katharane Mershon.

37. The book is *A Black Civilization* (Harper and Bros., 1937).

38. A. R. Radcliffe-Brown, John Dollard, and W. Lloyd Warner.

39. Mead is referring to her article "Public Opinion Mechanisms among Primitive Peoples," *Public Opinion Quarterly* 1, 3 (July 1937): 5–16.

40. The word *mother's* has been typed over the word *father's* here.

41. In the original she used the word "without," but the context indicates the word should be "with."

42. Psychoanalyst Erich Fromm, who had become a friend.

43. The additional letter was not with this one in the file.

44. William Steig's brother.

45. Gorer went to Philadelphia and Mississippi to visit Dollard while he was doing research for his book *Caste and Class in a Southern Town*.

46. Misdated 1938.

47. Misdated 1938.

48. She is referring to the interdisciplinary conference organized by Larry Frank in Hanover, New Hampshire, in 1934. This conference played a critical role in the development of the "culture and personality" approach in the social sciences.

49. Milton Erickson.

50. Claire Holt, who worked with Mead, Bateson, Belo, and others in Bali.

51. Shorthand for "in our culture."

52. She was executive secretary of the Committee on Food Habits, part of the National Research Council.

53. Later the Institute for Intercultural Studies (IIS).

54. Australian anthropologist Phyllis Kaberry.

55. Rhoda Metraux.

56. Edith Cobb.

57. Catherine, born late in her mother's life, represents a blessing to her parents and mercy from the gods.

58. This term refers to Bateson. It is an equivalent term for "Mr." used for white men in Bali.

59. This term refers to Gregory's mother, Beatrice Bateson. *Njonjah* is a term equivalent to "Mrs." used for white women.

60. This refers to family friend Nora Barlow, who accompanied Beatrice on a trip to Bali in late 1936. Mr. and Mrs. McPhee (Colin McPhee and his then wife, Jane Belo), Mrs. Mershon (Katharane), and Miss Holt (Claire) were all Westerners living in Bali in the 1930s who worked with Mead and Bateson while they were there.

61. She is referring to her responsibilities at the American Museum of Natural History.

62. She and Fortune did fieldwork there for three months in late 1932, not in 1931.

63. Social anthropologist Sigfried Nadel.

64. He had mentioned in his letter that he was reading a lot, catching up on neglected reading.

65. Robert K. Lamb.

66. He had mentioned in a postscript to his letter that rereading it had made him realize how interdependent he was.

67. Riesman's wife.

68. Virginia was Wilton's wife. Their son, Harris, was born the next month.

69. Alonzo Moron, president of the Hampton Institute (now University). Mead served on the board of this historically black institution.

70. Freeman had used this term in his letter when expressing his surprise at Mead's reaction during the seminar to his having neglected to give her his manuscript the previous day.

71. The Kellys' adopted son.

72. Actually the Oceanic Institute; Mead got the name wrong.

73. Anthropologists Clifford and Hildred Geertz, then married.

74. Biological scientist C. H. Waddington, a founder of human ecology.

75. This phrase is unclear in the letter.

76. This word is unclear in the letter.

CHAPTER 6

1. I Madé Kaler had been Mead and Bateson's Balinese secretary/assistant when they were in the field there. He had met Beatrice Bateson when she and her friend Nora Barlow had visited Bali at Christmas in 1936. *Tuan* or *toean* is a Balinese term for white man; he is asking about Gregory Bateson's whereabouts.

2. Nora Barlow's niece.

3. Philomena Guillebaud.

4. Nanny Helen Burroughs's daughter.

5. *And Keep Your Powder Dry: An Anthropologist Looks at America* (William Morrow, 1942).

6. He was with the OSS.

7. The British edition of *And Keep Your Powder Dry*, retitled *The American Character* (1944). Iso and Hedwig were refugee musicians whom Bateson's mother helped.

8. Buzz bombs.

9. Staughton Lynd, Robert and Helen Lynd's son.

10. Jane Belo.

11. Clemens Heller, who was one of the organizers of the Salzburg meetings; he and his wife became good friends of Mead's.

12. Anthropologist Clyde Kluckhohn and his wife, Florence.

13. It was called Sunte House.

14. Eichelberger had just returned from taking Mead's daughter, Cathy, on a trip to Europe—primarily to Italy with a stop in Great Britain.

15. *New Lives for Old* (William Morrow, 1956).

16. Her niece and Mead's cousin, Judith Upton Hoyt.

17. She is referring to the Hungarian Revolution.

18. Betty was Bateson's second wife.

19. The Annalena was the pensione where Emily Mead stayed with Elizabeth in Florence and then later where Cathy stayed with Marie Eichelberger on their Italian trip.

20. A colleague in child development.

21. Patricia Grinager.

22. Daniel's half-brother from their father's first marriage.

23. She and her husband had gone to Beirut on their honeymoon.

24. This is a newsletter that Daniel wrote.

25. Mead used this English pronunciation and spelling of Mommy with her own daughter and when talking to Daniel about his mother, Rhoda Metraux.

26. He was the priest at St. Luke's Episcopal Church, which was Mead's church and Daniel's school.

27. Cathy had a phonetic typewriter for doing linguistic translations, which her mother had used while visiting.

28. George Wallace, Governor of Alabama.

29. Daniel's future wife.

30. Photographer and friend Ken Heyman.

31. This was one of a series of cats in the household, all named Gold.

32. Their Haitian housekeeper.

33. The Hall of Pacific Peoples at the American Museum of Natural History, for which Mead was responsible. In 1971, after years of preparation, the hall had a grand reopening as the Margaret Mead Hall of Pacific Peoples.

34. She is referring to Benedict's *The Chrysanthemum and the Sword* (Houghton Mifflin, 1946).

35. She means Iran.

36. Mead and Catherine were planning the Wenner-Gren conference on ritual, and O'Doherty had been invited to participate.

37. Scientists' Institute for Public Information.

38. Mead's autobiography, *Blackberry Winter.*

39. *Steps To an Ecology of Mind* (Ballantine Books, 1972).

40. *Our Own Metaphor* (Knopf, 1972).

41. Father Anthony Bloom, the metropolitan of western Europe for the Greek Orthodox Church.

42. Nora Bateson, Bateson's daughter with his third wife, Lois. The child was named after Nora Barlow.

43. A colleague in child development.

44. Madeline Lee, one of Priscilla and Leo Rosten's daughters.

45. *The Ecology of Imagination in Childhood* (Columbia University Press, 1977).

46. There were many celebrations in honor of Mead's seventy-fifth birthday.

47. Bateson's stepson, Eric Vatikiotis-Bateson.

48. Columbia University Teachers College.

49. Margaret Bullowa was Catherine's research colleague at MIT.

50. Mead's friend, the Duchess of Argyle, former wife of Clemens Heller.

51. She misspelled "mind" or is using the German word for a child.

EPILOGUE

1. While the Freeman controversy has ensured that Mead's first field work, which brought her to prominence in the 1920s, remains best associated with her legacy, most of Mead's field work throughout her career was conducted in Papua New Guinea. For a recent scholarly retrospective on the life, work, and legacy of Mead (as well as Ruth Benedict), see Dolores Janiewski and Lois Banner, eds., *Reading Benedict/Reading Mead: Feminism, Race, and Imperial Visions* (Baltimore: Johns Hopkins University Press, 2004).

Additionally, as part of the 2001 centennial commemoration of Mead's birth, the Library of Congress mounted an exhibition on Mead's work that drew on the Library's Mead Collection. An online version of that exhibit, *Human Nature and the Power of Culture,* is available at: www.loc.gov/exhibits/mead.

2. Freeman's obsessiveness was well known during his lifetime but only recently has fuller documentation of his psychological issues been published. See Hiram Caton, "The Exalted Self: Derek Freeman's Quest for the Perfect Identity," in *Identity: An International Journal of History and Research* 5 (4) 2005: 359–384; the related article by Caton titled "Conversation in Sarawak: Derek Freeman's Awakening to a New Anthropology," available online at www.anthroglobe.ca; and an analysis of Caton's findings in Peter Monaghan, "Archival Analysis: An Australian Historian Puts Margaret Mead's Biggest Detractor on the Psychoanalytic Sofa," *The Chronicle of Higher Education* 52 (19) (January 13, 2006): A14. For further information on Tom Harrisson, see Judith M. Heimann, *The Most Offending Soul Alive: Tom Harrisson and His Remarkable Life* (Honolulu: University of Hawaii Press, 1998).

3. Louise Lamphere, "Margaret Mead's Critic" (letter to the editor), *New York Times,* August 12, 2001, section 4, page 12. John Shaw, "Derek Freeman, Who Challenged Margaret Mead on Samoa, Dies at 84" (obituary), *New York Times,* August 5, 2001, section 1, page 32.

Selected Works of Margaret Mead

Books are listed chronologically by year of publication.

Coming of Age in Samoa: A Psychological Study of Primitive Youth for Western Civilization. New York: William Morrow, 1928.

Growing Up in New Guinea: A Comparative Study of Primitive Education. New York: William Morrow, 1930.

The Changing Culture of an Indian Tribe. New York: Columbia University Press, 1932.

Sex and Temperament in Three Primitive Societies. New York: William Morrow, 1935.

Cooperation and Competition among Primitive Peoples (ed.). New York: McGraw-Hill, 1937.

And Keep Your Powder Dry: An Anthropologist Looks at America. New York: William Morrow, 1942.

Balinese Character: A Photographic Analysis (with Gregory Bateson). New York: New York Academy of Sciences, 1942.

Male and Female: A Study of the Sexes in a Changing World. New York: William Morrow, 1949.

Growth and Culture: A Photographic Study of Balinese Childhood (with Frances Cooke Macgregor and photographs by Gregory Bateson). New York: Putnam, 1951.

The School in American Culture. Cambridge, MA: Harvard University Press, 1951.

Soviet Attitudes Toward Authority: An Interdisciplinary Approach to Problems of Soviet Character. New York: The RAND Corporation/McGraw-Hill, 1951.

Primitive Heritage: An Anthropological Anthology (ed. with Nicolas Calas). New York: Random House, 1953.

The Study of Culture at a Distance (ed. with Rhoda Metraux). Chicago: University of Chicago Press, 1953.

Childhood in Contemporary Cultures (ed. with Martha Wolfenstein). Chicago: University of Chicago Press, 1955.

New Lives for Old: Cultural Transformation—Manus 1928–1953. New York: William Morrow, 1956.

An Anthropologist at Work: Writings of Ruth Benedict. Boston: Houghton Mifflin, 1959.

Continuities in Cultural Evolution. New Haven, CT: Yale University Press, 1964.

American Women: The Report of the President's Commission on the Status of Women and Other Publications of the Commission (ed. with Frances B. Kaplan). New York: Scribner, 1965.

Family (with Ken Heyman). New York: Macmillan, 1965.

Culture and Commitment: A Study of the Generation Gap. Garden City, NJ: Natural History Press/Doubleday, 1970.

A Rap on Race (with James Baldwin). Philadelphia: Lippincott, 1971.

Blackberry Winter: My Earlier Years. New York: William Morrow, 1972.

Ruth Benedict. New York: Columbia University Press, 1974.

Letters from the Field, 1925–1975. New York: Harper & Row, 1977.

Index

Letters are indicated by **boldface** page numbers following the names of recipients.